Computational and Visualization Techniques for Structural Bioinformatics Using Chimera

CHAPMAN & HALL/CRC
Mathematical and Computational Biology Series

Aims and scope:
This series aims to capture new developments and summarize what is known over the entire spectrum of mathematical and computational biology and medicine. It seeks to encourage the integration of mathematical, statistical, and computational methods into biology by publishing a broad range of textbooks, reference works, and handbooks. The titles included in the series are meant to appeal to students, researchers, and professionals in the mathematical, statistical and computational sciences, fundamental biology and bioengineering, as well as interdisciplinary researchers involved in the field. The inclusion of concrete examples and applications, and programming techniques and examples, is highly encouraged.

Series Editors

Proposals for the series should be submitted to one of the series editors above or directly to:
CRC Press, Taylor & Francis Group
3 Park Square, Milton Park
Abingdon, Oxfordshire OX14 4RN
UK

Published Titles

An Introduction to Systems Biology: Design Principles of Biological Circuits
Uri Alon

Glycome Informatics: Methods and Applications
Kiyoko F. Aoki-Kinoshita

Computational Systems Biology of Cancer
Emmanuel Barillot, Laurence Calzone, Philippe Hupé, Jean-Philippe Vert, and Andrei Zinovyev

Python for Bioinformatics
Sebastian Bassi

Quantitative Biology: From Molecular to Cellular Systems
Sebastian Bassi

Methods in Medical Informatics: Fundamentals of Healthcare Programming in Perl, Python, and Ruby
Jules J. Berman

Computational Biology: A Statistical Mechanics Perspective
Ralf Blossey

Game-Theoretical Models in Biology
Mark Broom and Jan Rychtář

Computational and Visualization Techniques for Structural Bioinformatics Using Chimera
Forbes J. Burkowski

Structural Bioinformatics: An Algorithmic Approach
Forbes J. Burkowski

Spatial Ecology
Stephen Cantrell, Chris Cosner, and Shigui Ruan

Cell Mechanics: From Single Scale-Based Models to Multiscale Modeling
Arnaud Chauvière, Luigi Preziosi, and Claude Verdier

Bayesian Phylogenetics: Methods, Algorithms, and Applications
Ming-Hui Chen, Lynn Kuo, and Paul O. Lewis

Statistical Methods for QTL Mapping
Zehua Chen

Normal Mode Analysis: Theory and Applications to Biological and Chemical Systems
Qiang Cui and Ivet Bahar

Kinetic Modelling in Systems Biology
Oleg Demin and Igor Goryanin

Data Analysis Tools for DNA Microarrays
Sorin Draghici

Statistics and Data Analysis for Microarrays Using R and Bioconductor, Second Edition
Sorin Drăghici

Computational Neuroscience: A Comprehensive Approach
Jianfeng Feng

Biological Sequence Analysis Using the SeqAn C++ Library
Andreas Gogol-Döring and Knut Reinert

Gene Expression Studies Using Affymetrix Microarrays
Hinrich Göhlmann and Willem Talloen

Handbook of Hidden Markov Models in Bioinformatics
Martin Gollery

Meta-analysis and Combining Information in Genetics and Genomics
Rudy Guerra and Darlene R. Goldstein

Differential Equations and Mathematical Biology, Second Edition
D.S. Jones, M.J. Plank, and B.D. Sleeman

Knowledge Discovery in Proteomics
Igor Jurisica and Dennis Wigle

Introduction to Proteins: Structure, Function, and Motion
Amit Kessel and Nir Ben-Tal

Biological Computation
Ehud Lamm and Ron Unger

Optimal Control Applied to Biological Models
Suzanne Lenhart and John T. Workman

Clustering in Bioinformatics and Drug Discovery
John D. MacCuish and Norah E. MacCuish

Published Titles (continued)

Spatiotemporal Patterns in Ecology and Epidemiology: Theory, Models, and Simulation
Horst Malchow, Sergei V. Petrovskii, and Ezio Venturino

Stochastic Dynamics for Systems Biology
Christian Mazza and Michel Benaïm

Engineering Genetic Circuits
Chris J. Myers

Pattern Discovery in Bioinformatics: Theory & Algorithms
Laxmi Parida

Exactly Solvable Models of Biological Invasion
Sergei V. Petrovskii and Bai-Lian Li

Computational Hydrodynamics of Capsules and Biological Cells
C. Pozrikidis

Modeling and Simulation of Capsules and Biological Cells
C. Pozrikidis

Cancer Modelling and Simulation
Luigi Preziosi

Introduction to Bio-Ontologies
Peter N. Robinson and Sebastian Bauer

Dynamics of Biological Systems
Michael Small

Genome Annotation
Jung Soh, Paul M.K. Gordon, and Christoph W. Sensen

Niche Modeling: Predictions from Statistical Distributions
David Stockwell

Algorithms in Bioinformatics: A Practical Introduction
Wing-Kin Sung

Introduction to Bioinformatics
Anna Tramontano

The Ten Most Wanted Solutions in Protein Bioinformatics
Anna Tramontano

Combinatorial Pattern Matching Algorithms in Computational Biology Using Perl and R
Gabriel Valiente

Managing Your Biological Data with Python
Allegra Via, Kristian Rother, and Anna Tramontano

Cancer Systems Biology
Edwin Wang

Stochastic Modelling for Systems Biology, Second Edition
Darren J. Wilkinson

Bioinformatics: A Practical Approach
Shui Qing Ye

Introduction to Computational Proteomics
Golan Yona

Chapman & Hall/CRC Mathematical and Computational Biology Series

Computational and Visualization Techniques for Structural Bioinformatics Using Chimera

Forbes J. Burkowski

University of Waterloo

Ontario, Canada

CRC Press
Taylor & Francis Group
Boca Raton London New York

CRC Press is an imprint of the
Taylor & Francis Group, an **informa** business
A CHAPMAN & HALL BOOK

MIX
Paper from responsible sources
FSC® C014174

CRC Press
Taylor & Francis Group
6000 Broken Sound Parkway NW, Suite 300
Boca Raton, FL 33487-2742

Printed on acid-free paper
Version Date: 20140519

International Standard Book Number-13: 978-1-4398-3661-3 (Pack - Book and Ebook)

Library of Congress Cataloging-in-Publication Data

Burkowski, Forbes J., author.
Computational and visualization techniques for structural bioinformatics using chimera / Forbes J. Burkowski.
pages cm. -- (Chapman & Hall/CRC mathematical and computational biology series)
Summary: "Goals of this book while attending a workshop or conference on Structural Bioinformatics you may overhear tidbits of conversations that are interspersed with phrases such as "phosphofructokinase regulation", "singular value decomposition", or "class instantiation". The usage of such terminology, arising from biochemistry, mathematics, and computer science respectively would not be surprising in this setting because these three areas of investigation have become the core of expertise required for the study of structural bioinformatics:"-- Provided by publisher.
Includes bibliographical references and index.
ISBN 978-1-4398-3661-3 (acid-free paper) 1. Structural bioinformatics--Simulation methods. 2. Molecular structure--Mathematical models. 3. Proteins--Analysis--Data processing. I. Title.

QH324.2.B864 2014
572'.330285--dc23 2014006988

Visit the Taylor & Francis Web site at
http://www.taylorandfrancis.com

and the CRC Press Web site at
http://www.crcpress.com

To my wife
Karen

Contents

Preface, xvii

Acknowledgments, xxiii

Author, xxv

CHAPTER 1 ■ Introduction: Macromolecules and Chimera 1

1.1 MOTIVATION 1

1.2 WHY CHIMERA? 2

1.3 GETTING STARTED WITH CHIMERA 4

1.4 OVERVIEW OF PROTEIN STRUCTURE 4

1.4.1 Amino Acids and Primary Sequence 5

1.4.2 Secondary Structure 10

1.4.2.1 Alpha Helices 10

1.4.2.2 Beta Strands 12

1.4.2.3 Loops 16

1.4.3 Tertiary Structure 17

1.4.3.1 What Is Tertiary Structure? 17

1.4.3.2 Tertiary Structure of Myoglobin 18

1.4.3.3 Tertiary Structure beyond the Binding Pocket 21

1.4.4 Quaternary Structure 27

1.4.5 Protein Functionality 29

1.4.6 Protein Domains 31

1.5 EXERCISES 33

REFERENCES 35

CHAPTER 2 ■ Accessing and Displaying Molecular Data with Chimera 39

2.1 MOTIVATION 39

2.2 PYTHON SKILLS 40

2.3 PYTHON SCRIPTING 41

2.3.1 Script 2.1: Using Raw Input to Get a PDB ID 44

2.4 CHIMERA OBJECT HIERARCHY 44

2.5 ATTRIBUTES FOR MOLECULE OBJECTS 48

2.6 ATTRIBUTES FOR SEQUENCE OBJECTS 49

2.6.1 Referencing a Sequence Object 49

2.6.2 Identification of a Sequence 50

2.6.3 Sequence Entries 50

2.6.4 Getting Residue Objects from a Sequence 51

2.6.5 Residue Maps for Sequences 52

2.7 ATTRIBUTES FOR RESIDUE OBJECTS 53

2.7.1 Referencing a Residue Object 53

2.7.2 Identification and Characterization of Residues 54

2.7.3 Location of a Residue 55

2.7.4 Deriving Substructure 56

2.7.5 Other Useful Attributes 57

2.8 ATTRIBUTES FOR ATOM OBJECTS 58

2.8.1 Referencing Atom Objects 58

2.8.2 Identification of Atoms and Atom Properties 59

2.8.3 Working with Atom Coordinates 60

2.8.4 Atom Attributes Related to Chemical Structure 60

2.8.5 Atoms and Bonds 61

2.8.6 Atom Attributes Related to the Display 62

2.9 ATTRIBUTES FOR BOND OBJECTS 62

2.9.1 Referencing Bond Objects 62

2.9.2 Bond Attributes 63

2.10 WORKING WITH A BATCH OF FILES 64
2.10.1 Dealing with Files 64
2.10.2 Script 2.2: Getting a Batch of Files from the Protein Data Bank 66
2.10.3 Script 2.3: Getting a Batch of PDB Files by Using Chimera 67
2.11 HILITER CLASS 67
2.12 HBONDDICTS CLASS 69
2.12.1 Using the HBondDicts Class 69
2.13 EXERCISES 71
REFERENCES 86
WEBSITES FOR PYTHON 86

CHAPTER 3 ■ Algorithms Dealing with Distance 87

3.1 MOTIVATION 87
3.2 CALCULATING INTERATOMIC DISTANCES 87
3.3 APPLICATIONS 90
3.4 RAPID DETERMINATION OF ATOM MEMBERSHIP IN SHELLS 90
3.4.1 Implementation Strategy 91
3.5 CONTACT MAPS FOR PROTEINS 95
3.5.1 Script for Generating Contact and Distance Maps 97
3.6 INERTIAL AXES 99
3.6.1 Motivation 99
3.6.2 Mathematical Analysis 100
3.6.3 Script for Computing the Inertial Axis 104
3.7 DEHYDRONS 108
3.7.1 Motivation 108
3.7.2 Script for Identifying Dehydrons 110
3.8 EXERCISES 112
REFERENCES 132

CHAPTER 4 ■ Algorithms Dealing with Angles 135

4.1 MOTIVATION 135
4.2 BOND ANGLE 135
4.2.1 Calculating Bond Angles in a Script 136
4.3 DIHEDRAL ANGLES 137
4.3.1 Defining Dihedral Angles 137
4.3.2 Mathematics for the Computation of a Dihedral Angle 139
4.3.3 Computation of a Normal 139
4.3.4 Calculating the Phi Dihedral Angle 142
4.3.5 Sign of the Dihedral Angle 142
4.3.6 Calculating the Psi Dihedral Angle 144
4.3.7 Calculating Dihedral Angles in a Script 144
4.4 RAMACHANDRAN PLOTS 145
4.4.1 Scripts for Generating Ramachandran Plots 149
4.4.2 Script 4.1: Generating a Ramachandran Plot for a Protein 150
4.4.3 Background Densities for Ramachandran Plots 151
4.4.4 3D Ramachandran Plots 152
4.5 LEAST SQUARES PLANE 154
4.5.1 Motivation 154
4.5.2 Least Squares Plane: Mathematical Analysis 154
4.5.3 A Script for Computing the Least Squares Plane 159
4.6 EXERCISES 165
REFERENCES 179

CHAPTER 5 ■ Structure Overlap and Alignment 181

5.1 MOTIVATION 181
5.2 INTRODUCTION 184
5.2.1 Specifying the Problem 184
5.3 TECHNIQUES FOR STRUCTURAL COMPARISON 185
5.4 SCORING SIMILARITIES AND OPTIMIZING SCORES 186

5.5 SUPERPOSITION ALGORITHMS 186

5.5.1 Overview 186

5.5.2 Characterizing the Superposition Algorithm 188

5.5.3 Formal Problem Description 189

5.5.4 Computations to Achieve Maximal Overlap 190

5.5.5 Summary 197

5.5.6 Measuring Overlap 198

5.5.6.1 Calculation of the Root Mean Square Deviation 198

5.6 A SIMPLE SCRIPT TO DO STRUCTURAL SUPERIMPOSITION 199

5.7 PROTEIN SEQUENCE ALIGNMENT 202

5.7.1 Partial matches 203

5.7.2 Gaps 205

5.7.3 Summary 206

5.7.4 Computation 206

5.7.4.1 Subproblem Specification 207

5.7.4.2 Scoring Alignments 207

5.7.4.3 Suitability of the Subproblem 208

5.7.4.4 A Global Alignment Example 211

5.8 VARIATIONS IN THE GLOBAL ALIGNMENT ALGORITHM 212

5.9 PERCENT IDENTITY COMPARISON 213

5.10 LOCAL ALIGNMENT 214

5.11 A SCRIPT TO DO SEQUENCE ALIGNMENT AND THEN STRUCTURAL OVERLAP 216

5.12 DEALING WITH WEAKER SEQUENCE SIMILARITY 218

5.12.1 Structural Alignment by Overlapping Pairs of Secondary Structure Elements 220

5.12.2 A Script for Structural Alignment by Overlapping Secondary Structure Elements 224

5.13 EXERCISES 226

REFERENCES 242

CHAPTER 6 ■ Potential Energy Functions 245

6.1 MOTIVATION 245

6.1.1 Empirical Observations 246

6.1.2 Mathematical Modeling 246

6.1.2.1 Energy Terms for Bonded Atoms 247

6.1.2.2 Energy Terms for Nonbonded Atoms 249

6.1.2.3 Total Potential Energy 250

6.1.3 Computational Issues 251

6.2 SOME SIMPLE SCRIPTS TO DO ENERGY CALCULATIONS 254

6.2.1 Linear Repulsive Energy Terms 254

6.2.2 An Energy Function Based on the Lennard–Jones Formula 255

6.2.3 Computing Energy Using Amber Parameters in Chimera 256

6.3 EXERCISES 258

6.3.1 Comparing Lennard–Jones and Piecewise Linear Energy Functions 258

REFERENCES 260

CHAPTER 7 ■ Rotamers and Side-Chain Conformation 263

7.1 MOTIVATION 263

7.2 SIDE-CHAIN PACKING: COMPUTATIONAL ISSUES 264

7.3 ROTAMERICITY 265

7.4 ACCESSING ROTAMERS USING A PYTHON SCRIPT 267

7.5 ROTAMERS AND DIHEDRAL ANGLES 271

7.6 A SINGLE SIDE CHAIN AND ENERGY CONSIDERATIONS 271

7.7 SIDE-CHAIN PACKING AND DEAD-END ELIMINATION 275

7.7.1 Goldstein's DEE: Basic Strategy 276

7.7.2 Goldstein's DEE: A More Efficient Strategy 278

7.7.3 Side-Chain Packing 278

7.8 EXERCISES 278
7.8.1 Dead-End Elimination 278
REFERENCES 279

CHAPTER 8 ■ Residue Networks 281
8.1 MOTIVATION 281
8.2 THREE-DIMENSIONAL (3D) VISUALIZATION OF RESIDUE NETWORKS 282
8.3 ALLOSTERY AND CONTACT REARRANGEMENT NETWORKS 283
8.4 EXERCISES 285
8.4.1 Spheres Representing Residues 285
8.4.2 Spheres and Spindles for Residue Networks 286
8.4.3 Spheres and Spindles for Residue Networks: Graphical User Interface (GUI) Implementation 289
8.4.4 Graphs for Residue Networks: GUI Implementation 290
8.4.5 Contact Rearrangement Networks: Evaluating $RF(i, j)$ 290
8.4.6 Displaying the Residue Network for Rearranged Contacts 292
REFERENCES 295

APPENDIX A: SIMPLE DIALOGS, 297

APPENDIX B: SCENOGRAPHICS, 317

APPENDIX C: THE GRAPH CLASS, 345

APPENDIX D: 2D AND 3D PLOTS, 365

APPENDIX E: DYNAMIC PROGRAMMING, 395

INDEX, 409

Preface

GOALS OF THIS BOOK

While attending a workshop or conference on structural bioinformatics, you may overhear tidbits of conversations that are interspersed with phrases such as *phosphofructokinase regulation*, *singular value decomposition*, or *class instantiation*. The usage of such terminology, the aforementioned terms arising from biochemistry, mathematics, and computer science, respectively, would not be surprising in this setting because these three areas of investigation have become the core of expertise required for the study of structural bioinformatics:

- Biochemistry provides realistic knowledge about protein structure and functionality.
- Mathematics helps us to build models that represent protein structure and functionality.
- Computation allows us to apply the mathematical models to detect patterns and make predictions about biological systems. As well, computation is necessary for the creation of visual displays that may be used to demonstrate various aspects of protein structure and function.

This book addresses all three areas with an emphasis on computational techniques:

- Topics in the book deal primarily with protein structure, and there are many exercises that are grounded in biological problems at the molecular level.

- The book encourages mathematical analysis because it provides a firm foundation for subsequent computations.
- Computational techniques are covered by providing and analyzing several Python scripts that execute within the Chimera environment. This supplemental material (over 140 scripts) is available from the book's website (http://structuralbioinformatics.com), and it provides solutions for the various exercises at the end of the chapters. Due attention has been given to the modularity of the scripts with the goal of providing the reader with a toolkit of Python classes that can be directly used in the reader's own applications or used as starting scripts that provide base functionality via class inheritance.

What are the typical computations that would be done with a script? A script working with Chimera can easily use data downloaded from the Protein Data Bank (PDB). For example, if your algorithm requires protein atomic coordinates, a script can access a PDB file relying on Chimera to parse the content and make these coordinates available as input to the algorithm. If the algorithm is modifying or generating atom coordinates, then Chimera can be used to provide a visualization of the altered molecule. Moreover, Python scripts can be designed to add various graphical constructs that help to describe various aspects of protein structure and function. Visual representation of an abstraction (e.g., a plane that has some relevance to a biological construct) helps us to discover configurations that are made clearer with the appropriate visualization. In addition to providing visual output, changing the display can be indispensable during debugging and for final validation of the script that implements the algorithm.

SCRIPTING WITH CHIMERA

This book will help any reader who wishes to use the scripting capabilities of Chimera. There are over 60 exercises that involve the development of Python scripts. Most of these scripts are available as supplemental material (downloadable from http://structuralbioinformatics.com). There are several approaches to the use of these scripts:

- To gain experience with Python scripting in the Chimera environment, the reader may choose to work through an exercise and then compare this answer with the provided script. This *learning by doing* helps to build confidence, leading to the solution of more challenging problems. Some of the exercises duplicate the calculations done in research papers. Doing such an exercise is not meant to verify or double-check the result but rather to gain experience in the solving of such problems so that the reader can tackle similar problems in the future.

- In many cases, the exercises demonstrate the various applications of linear algebra. For example, eigenvectors and eigenvalues are not just theoretical mathematical abstractions, but they can be used to build visual objects that are then displayed in the Chimera window. Last but not least, scripting is fun to do. There is a compelling sense of accomplishment when the execution of a script runs a graphical user interface (GUI) or produces an immediate visual feedback in the Chimera window.
- The reader may use the script as a starting point for the development of a similar application. There are many situations, especially during GUI development, when a working script can show the reader how some code snippet handles a problem that is difficult because it involves some exasperating implementation issue that would otherwise be resolved only after extensive searches on the Internet.
- The reader may use classes from the `StructBio` toolkit in the development of an application. In addition, the ability to display some structural entity related to the algorithm can help in the debugging of a complex program that involves several difficult and complicated steps. I have spent several months in the design and development of these classes with the hope that they would be useful for the future scripting efforts of both myself and the readers of this book. These classes provide several tools that may be used for computations such as structure overlap, data plotting, *scenographics*, and display of residue networks.

INTENDED AUDIENCE

This book is mainly intended for people who wish to build Python scripts to extend the capabilities of Chimera. It can also be used by students, typically third- or fourth-year undergraduates, who have completed introductory courses in bioinformatics and wish to go further in the study of structural biology. Prerequisites include introductory linear algebra, elementary calculus, bioinformatics ("biology 101," sequence analysis, etc.), computer programming, and an introduction to algorithms. We assume that the reader can do elementary Python programming. An introduction to Python is not included in this book because there are several excellent sources for such information (see the references at the end of Chapter 2).

BOOK CONTENT

Chapter 1 presents some introductory material on protein structure, and subsequent chapters cover various selected algorithms and topics that are typically seen in structural bioinformatics. The algorithms are mainly used as a source of exercises that demonstrate scripting techniques in Chimera. As such, the chosen algorithms are not meant to provide a comprehensive coverage of the algorithms used in structural bioinformatics.

Chapter 1 introduces protein structure. This is essentially a review of the material that may have been covered in the earlier biochemistry courses of a bioinformatics program. The concepts of primary, secondary, tertiary, and quaternary structures are introduced and explained. The material serves to underscore the most salient aspects of protein structure while providing several figures that illustrate the visualization capabilities of Chimera. Exercises cover the use of Chimera menu invocations to produce various types of displays related to protein structure.

Chapter 2 introduces Python scripting for the Chimera environment and rapidly progresses to a description of the Chimera object hierarchy. Various objects and their attributes are covered so that the programmer can start writing scripts dealing with molecular structure including hydrogen bonds. Some elementary file I/O is discussed so that scripts can deal with multiple PDB files. A final exercise shows how polygonal surfaces can be added to a display (an introduction to scenographics).

Chapter 3 covers scripts that rely on the calculation of distances (most often, interatomic distances). Exercises deal with distance shells, contact maps, inertial axes, and dehydrons. The `Solids` and `Ellipsoid` classes are introduced to provide more scenographics. Generation of contact maps introduces the reader to scripts that can display plots.

Chapter 4 deals with scripts that rely on the calculation of angles (usually dihedral angles). The chapter provides more material for plotting, using Ramachandran plots as an example. Lagrange optimization (introduced in Chapter 3 for the generation of inertial axes) is applied again for the creation and display of a least-squares plane. More examples are presented for GUI implementations using Tkinter classes.

Chapter 5 covers the supporting theory and implementation of scripts to do structure overlap. This will duplicate the current functionality of Chimera, but the idea is to learn the mathematics behind that functionality. This has intrinsic value and, more significantly, it gives the reader the skills to generate an overlap class that can be used in larger applications. For example, if you need to do several hundred structure alignments, then menu invocations will not be effective; you need a script that can repeatedly call a function to do overlap.

Chapter 6 is a short discussion on energy functions. Its main importance is to provide a simple energy calculation for a script in Chapter 7.

Chapter 7 introduces rotamers and the Chimera functions that work with them. Goldstein's dead-end elimination is covered as a prelude to side chain packing. Scripts provide examples of more extensive GUI development, including facilities to do strip charts. Scripts to do plotting of three-dimensional (3D) surfaces are introduced.

Chapter 8 covers residue networks and provides examples for the use of the `Graph` classes in the `StructBio` package. The final exercise describes a script that provides a visual display of a network for residue contact rearrangements.

Appendix A discusses the implementation of GUIs that can be used for Python applications that require a more sophisticated interaction with the user. Because the Chimera window will still be the main window of concern, these GUIs are typically modeless dialogs that set up data input for the application. The appendix gives a quick introduction to Tkinter along with several sample scripts set up as a *widget buffet*. These can be extracted, modified, and placed into the reader's applications.

Appendix B introduces the scenographics toolkit. This includes several classes that can be used to place various lines, surfaces, and solids into the Chimera display. An extremely versatile set of classes under the heading "Parametric Surfaces" includes classes for Frenet frame surfaces, extrusion surfaces, surfaces of revolution, and ruled surfaces. The `LabelGroups` class can be used to place labels into the display at specified 3D positions.

Appendix C introduces the `GraphBase` class, which can be used to implement abstract graphs via the `Vertex` and `Arc` classes. Methods are included for the computation of shortest paths and for the extraction of various subgraphs. A set of derived classes (`Graph`, `Node`, and `Edge`) are provided for the construction of 3D graphs that will be displayed in the Chimera window. Instantiation of these classes will also generate the underlying `Vertex` and `Arc` objects so that various graph algorithms can be applied to the constructed graph.

Appendix D covers various scripts to do plotting including bar charts, density plots, scatter plots, plots of parametric curves, and 3D surface plots. Plots can be shown in their own windows or within a modeless dialog designed by the programmer.

Appendix E provides a review of dynamic programming to act as background material for the sequence alignment algorithms presented in Chapter 5.

Acknowledgments

Almost all the molecular graphics were done using UCSF Chimera. The primary reference for this molecular modeling system is as follows:

E. F. Pettersen, T. D. Goddard, C. C. Huang, G. S. Couch, D. M. Greenblatt, E. C. Meng, and T. E. Ferrin. UCSF Chimera—A visualization system for exploratory research and analysis. *Journal of Computational Chemistry*, **25** (2004), 1605–1612.

This book makes use of many PDB files. Please note the following reference that is to serve for all future use of PDB files:

H. M. Berman, J. Westbrook, Z. Feng, G. Gilliland, T. N. Bhat, H. Weissig, I. N. Shindyalov, and P. E. Bourne. The Protein Data Bank. *Nucleic Acids Research*, **28** (2000), 235–242.

I thank Ariel Fernández and Ridgway Scott for their supportive comments on the dehydron material in Chapter 3.

Special thanks to Eric Pettersen for his careful reading of the chapters and many helpful suggestions.

Also, thank you to Sunil Nair, Sarah Gelson, Joette Lynch, and Joselyn Banks-Kyle at CRC Press for their patience and helpful assistance. A final note of thanks goes to Dennis Troutman (project manager, diacriTech) for excellent copy editing and help during the final production process.

Author

Forbes J. Burkowski earned his PhD in applied mathematics at the University of Waterloo, Ontario, in 1970. He taught computer science at the University of Manitoba (1970–1982), and during this time he also pursued studies in organic chemistry and biology. In 1982, Dr. Burkowski returned to the University of Waterloo as an associate professor in the computer science department.

Over the past 7 years, he has served as the director of bioinformatics and has participated in the design of the undergraduate program in bioinformatics. He has taught the fourth-year structural bioinformatics course since 2003. Dr. Burkowski's research interests include side chain packing, drug design, and mathematical modeling of flexibility in protein structure.

CHAPTER 1

Introduction

Macromolecules and Chimera

1.1 MOTIVATION

Within a cell, life processes involve an overwhelming set of complicated interactions, most of them involving macromolecular structures such as DNA, RNA, and protein.* This final category of macromolecules is arguably the most important because it has the ability to contribute the most diverse functionality due to the structural complexity of its individual members. In fact, proteins act as enzymes that facilitate chemical reactions, regulators of cell processes, signal transducers enabling signals to be imported and exported into and out of a cell, transporters of small molecules, antibodies, structural elements that bind cells together, and motor elements to facilitate movement.

As such, proteins are the versatile building blocks and active molecules that form the components of all living systems. Proteins always act in concert with other molecules using a wide range of chemical reactions and mechanical reconfigurations. With over a billion years of evolution, these complex interactions have become sophisticated enough to accomplish all the coordinated events that constitute the life processes of a living cell. These evolutionary processes have derived structures, such as proteins, that are much more than simply static macromolecular

* In this book, we assume the reader is already familiar with the *Molecular Biology 101* topics such as cell structure, the Central Dogma of Molecular Biology, the genetic code, and the roles played by macromolecules such as DNA and RNA.

assemblies: their composition must ensure proper folding to a stable functioning conformation with an eventual allowance for proteolysis. Proper functioning will mean that the protein interacts with other molecules in a very selective manner demonstrating a high level of *molecular recognition*.

We study protein structure and its dynamic changes so that we can better understand the complicated functionality of macromolecules. These studies involve tools for the visualization of macromolecular structure and the subsequent mathematical modeling of structures using geometric analysis, linear algebra formulations, and statistics.

In this book, our *workbench* used for the study of macromolecules (in particular, proteins) will be Chimera: a software application that has been developed by the Resource for Biocomputing, Visualization, and Informatics (RBVI)* at the University of California, San Francisco (UCSF).†

1.2 WHY CHIMERA?

There are several molecular modeling systems with very sophisticated visualization capabilities. The decision to go with UCSF Chimera was based on several issues many of which relate to ease of use and suitability for a student environment. The following points list some of the more important reasons for choosing Chimera:

1. Visualization capabilities

   As a standalone application for the visualization of macromolecules, Chimera has several advantages:

   a. Changing perspective: Easy mouse movements give rotation and zoom.

   b. Sophisticated graphics: Atoms can be rendered in wire-frame, stick, ball and stick, or sphere representations. Ribbons for secondary structure can appear as flat, edged, or rounded. Molecular surfaces can appear as solid, mesh, or dot.

---

* http://www.cgl.ucsf.edu/chimera/ The RBVI is supported by grant NIGMS 9P41GM103311.

† http://www.ucsf.edu/.

c. Versatile selection abilities: The Select menu offers a variety of selection mechanisms. Selection by chain, chemistry, residue, structure, sequence, and so forth.

d. Visual effects: Users can give a scene a heightened spatial presence by means of perspective and fog. Surface effects include transparency, and coloring for electrostatic and hydrophobic values.

e. Labeling atoms and residues: A molecular scene can be augmented with the provision of various atom and residue labels.

f. Other sophisticated features: Chimera provides visualization of density maps, supramolecular assemblies, sequence alignments, docking results, trajectories, and conformational ensembles.

2. Multiplatform

   Chimera is available for Microsoft Windows, Linux, Linux 64-bit, Apple Mac OS X, SGI IRIX, and HP Tru64 Unix.

3. Maturity

   The first public release of Chimera occurred in March 2000. Since then there have been several production releases that represent a sequence of steady improvements for the software.

4. Scalability

   Chimera can be used for the visualization of large molecules.*

5. Pricing

   Chimera is free to academic and nonprofit users.

6. Computational tools

   More important for our purposes, Chimera is an extensible application that can be given more capabilities if the user develops Python scripts. Scripting capability allows the user to design, develop, and execute a variety of algorithms that can accept macromolecular data (e.g., atom coordinates), perform any needed calculations, and subsequently print results to screen or file. A script can also change the molecular scene being displayed.

---

* Not all visualizers survive the *1FFK test*; try fetching the PDB file 1FFK and render it using various display techniques.

1.3 GETTING STARTED WITH CHIMERA

To get a copy of Chimera, you can visit the UCSF Chimera homepage at http://www.cgl.ucsf.edu/chimera/ and click on the Download link in the left column. On the download page choose the current production release that is appropriate for your platform and follow the installation instructions, which are also available via a link on the download page. Documentation includes the Chimera User's Guide and the Chimera Programming Guide.

There are three ways that the user can interact with Chimera:

1. Menu invocations
2. Command line
3. Python scripting

The User's Guide contains a tutorial with a comprehensive description of both the menu and command line interactions. If the reader does not have any experience in the use of Chimera menus, then it is advisable to do some elementary exercises to gain familiarity with the basics. This book will concentrate on Python scripting and, as noted earlier, the scripts can be used to modify or create molecular visualizations. In some cases it is simply easier to use menu invocations to modify the molecular scene produced by the script. Exercises at the end of this chapter can be used to get some practice in the use of the menus. The command line interface is also covered in the tutorial of the User's Guide. Our work with Python scripting will make occasional use of command line interactions, but for the moment, they can be ignored.

The remaining sections of this chapter will provide a review of macromolecular structure in preparation for the next chapter, which will introduce the Chimera object model and Python scripting. While reading these sections, bear in mind that all the molecular visualizations have been prepared using Chimera menu invocations. If you are new to Chimera, it would be a worthwhile exercise to try and replicate some of these figures using Chimera and a protein of your own choice.

1.4 OVERVIEW OF PROTEIN STRUCTURE

We review the main aspects of protein structure by considering the various levels of description: amino acid primary sequence, secondary structure, tertiary structure, and quaternary structure. To make these topics a bit more interesting for the many readers that already understand these concepts, we include a bit of history.*

1.4.1 Amino Acids and Primary Sequence

During the 1800s, chemists knew about amino acids and realized that each consisted of a side chain (also called an R group) that is attached to an alpha carbon that is also bonded to amine and carboxyl functional groups, as shown in Figure 1.1.

The notion that proteins were linear chains of amino acids was proposed in 1902 by Emil Fischer and Franz Hofmeister. Both understood that amino acid residues were joined in a condensation reaction (Figure 1.2).

By 1906, approximately 15 amino acids were known to be constituents of proteins, but there was considerable debate about the maximum length that a protein could achieve. It was considerably later that proteins were understood to be very long chains of amino acids. In fact, lengths vary from fewer than 20 amino acids to more than 5000 amino acids. It is interesting to note that, despite the earlier discovery of the peptide chain, many scientists, Fischer included, were initially quite skeptical that proteins could have a very lengthy macromolecular structure. Even as late as 1952, Fredrick Sanger regarded the *peptide theory* as very likely to be

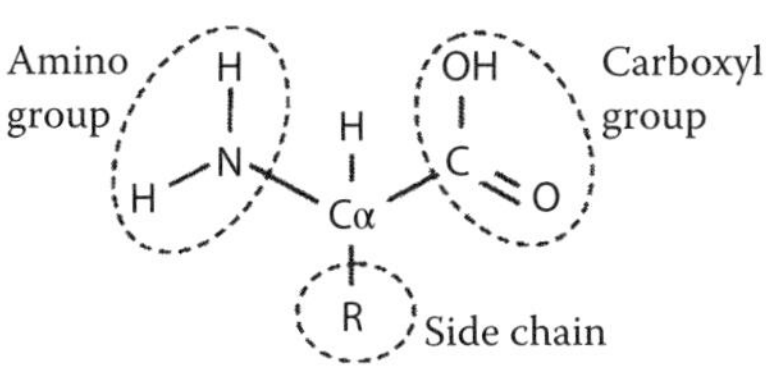

FIGURE 1.1 A typical amino acid.

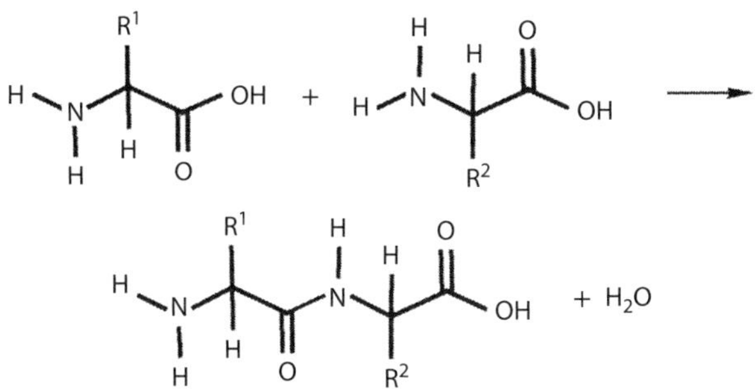

FIGURE 1.2 Condensation reaction.

---

* For more historical information, see [D105] for structural molecular biology and [TR01] for proteins.

true but nonetheless only a hypothesis that had yet to be proven [FR79]. The identities of all 20 amino acids were not completely known until 1940.

We now understand that proteins are composed of chains of such amino acids and this chain can involve several hundred amino acids. The first level of structural description of a protein is called the *primary sequence* and is simply a list of these amino acids mentioned in their order of appearance as one scans the protein starting with the amino end of the protein and proceeding to the carboxyl end.

In terms of chemical functionality, each of the 20 amino acids has its own *personality*. In an effort to simplify descriptions of amino acid behavior, it is convenient to place amino acids into various categories. There are several ways to do this. As an example, Figures 1.3 and 1.4 place residues into four categories: nonpolar hydrophobic, polar uncharged, polar charged positive, and polar charged negative.

All of the residues are shown in zwitterion form (an ionic molecule with separate positive and negative charges having a net charge of zero).

The given categorization should be regarded as somewhat fuzzy. For example, histidine has a relatively neutral pKa and so fairly small changes in cellular pH will change its charge. So, it could be placed in the polar uncharged group, but in Figure 1.4 it is put into the polar charged group because it can achieve a positive charge under physiological conditions. Similarly, glycine often appears in the hydrophobic category, but its side chain is essentially nonexistent, and so some biochemists will consider it to be in a category all by itself. It should be noted that hydrophobicity is of particular interest to researchers studying structural bioinformatics because it has been shown that the hydrophobic properties of amino acids contribute to the three-dimensional conformations of proteins.

Some amino acids have *special features* that also contribute to structural conformation. Proline has a peculiar structure due to the cyclic binding of its three-carbon side chain to the nitrogen atom that is normally part of the amine group in the other amino acids. This binding places limits on the flexibility of the backbone and so proline will often cause a bend in structures such as helices and beta strands (to be considered in Section 1.4.2). It is more likely to appear at the edges of such secondary structures or in loops that go between them. Another amino acid with important structural contributions is cysteine. It is possible for two cysteine residues to form a disulfide bridge by means of an oxidative reaction:

$$\mathrm{R\text{-}SH + SH\text{-}R' \rightarrow R\text{-}S\text{-}S\text{-}R' + 2H^+ + 2e^-}$$

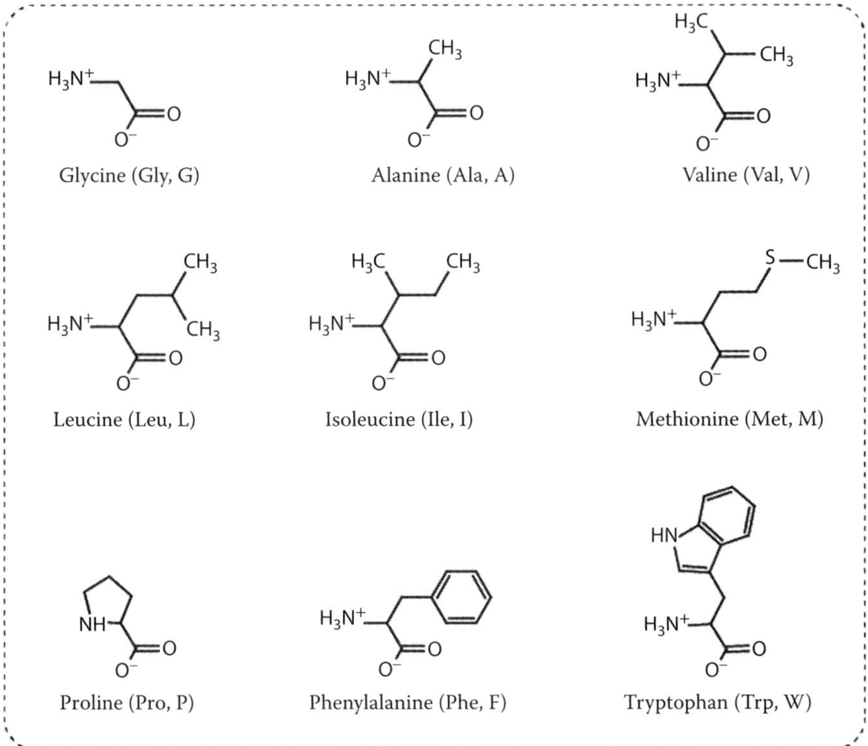

Nonpolar (hydrophobic) residues

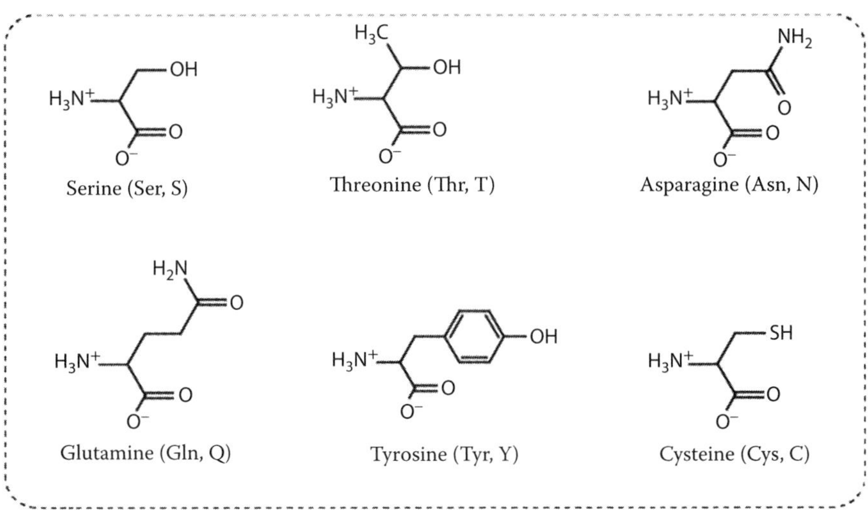

Polar uncharged residues

FIGURE 1.3 Amino acids: Hydrophobic and uncharged residues.

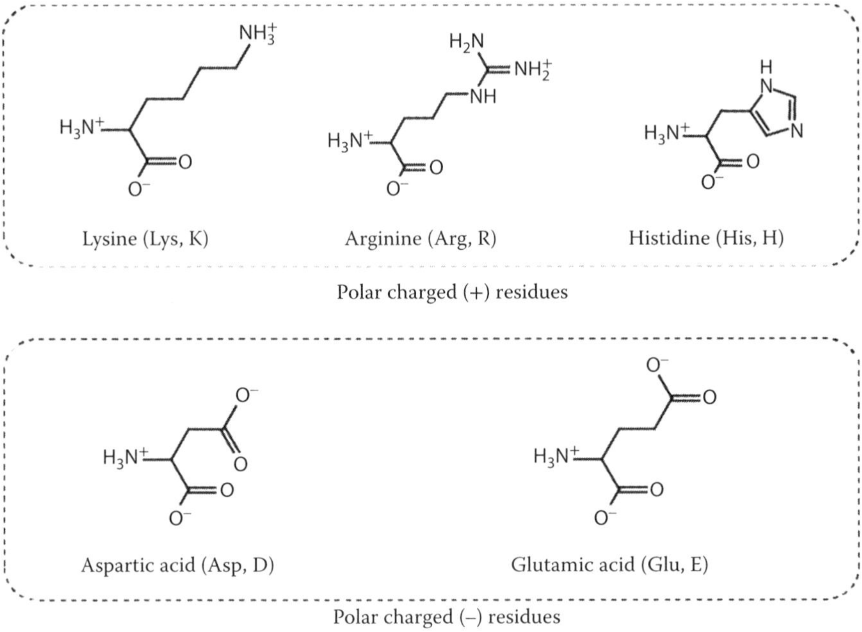

FIGURE 1.4 Amino acids: Charged residues.

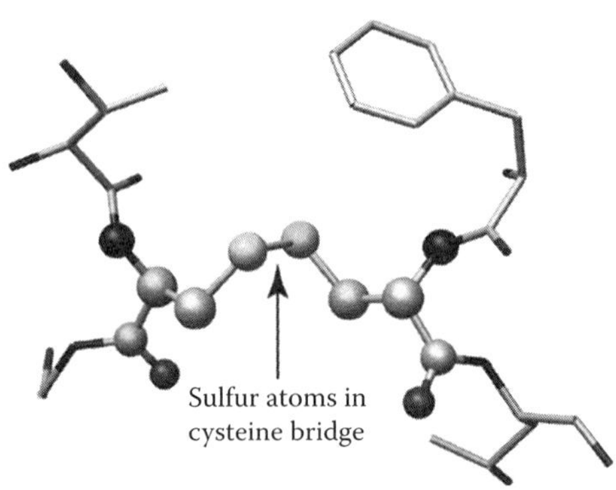

FIGURE 1.5 Cysteine bridge found in bovine pancreatic trypsin inhibitor.

This allows two cysteine residues in different parts of the protein to form a bridge that helps to provide some extra stability for the molecule. Figure 1.5 shows a cysteine bridge that has been extracted from the bovine pancreatic trypsin inhibitor with Protein Data Bank (PDB) code 6PTI.

In this figure, the cysteine residues including the two central sulfur atoms have been shown using a *ball and stick* visualization style while the four residues adjacent to the cysteine residues are simply shown in *stick* form.

It has been noted that small proteins tend to have more cys–cys bridges than larger proteins. An interesting example: snake venoms often have an unusually high preponderance of cys–cys bridges to enhance stability. Venom from the Malaysian krait contains the protein 1VYC that has three such bridges even though it is only 65 residues in length (Figure 1.6).

In this figure, we see that the protein uses cysteine bridges to provide more positional stability to various loop substructures that would tend to be more floppy without the bridges.

The alpha carbon of all amino acids, except for glycine, constitutes a chiral center. In most biological systems this chirality is described as being in the L-configuration (Figure 1.7). Amino acids with a D-configuration are not found in the metabolic pathways of eukaryotic cells but have been observed in the structure and metabolism of some bacteria.*

In Figure 1.7, we see that bonds attached to the alpha carbon between the NH_2 and COOH groups are coming up from the page, whereas the other two bonds are going down into the page as they go from the alpha carbon to the hydrogen and beta carbon of valine. The D-configuration of this amino acid would be the mirror image of what is seen in Figure 1.7.

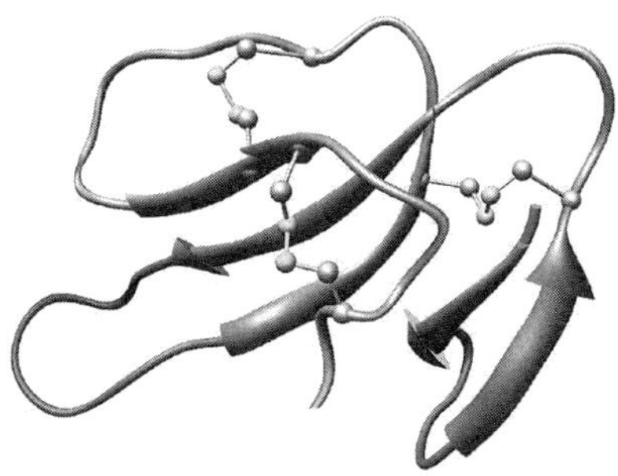

FIGURE 1.6 Cysteine bridges in bucain (a krait venom).

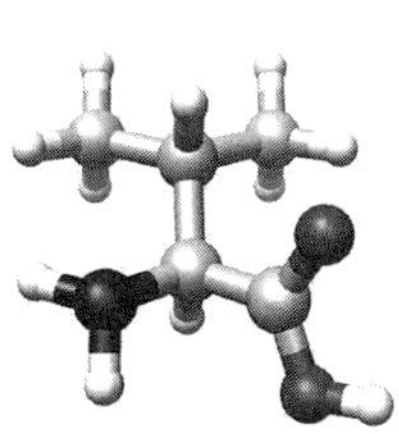

FIGURE 1.7 L-configuration for an amino acid (valine).

* A list of D-amino acid residues in the PDB can be found at: http://chemistry.st-andrews.ac.uk/staff/jbom/group/d-res.html.

As a final note to end this section, we mention that almost all life forms are restricted to the 20 amino acids listed earlier. However, recently it has been discovered that two other amino acids can appear, although rarely. They are selenocysteine and pyrrolysine. These unusual amino acids should not be considered as normal cysteine and lysine residues that have been chemically modified due to some posttranslational mechanism. The microbes that use these amino acids actually have specific transfer RNAs for them and they are inserted into the growing peptide chain just like any other amino acids [MG00].

1.4.2 Secondary Structure

The next level of structural description is referred to as *secondary structure*. The idea behind secondary structure is to simply characterize residues of the primary sequence, putting them into categories that reflect the most obvious description of the shape of the peptide chain. There is no attempt to establish the physical positions of residues in three-dimensional space as is done in the specification of tertiary structure. Generally there are three categories of secondary structure: alpha helices, beta strands, and loops.

The first two types of structure are mainly facilitated by the formation of hydrogen bonding between atoms in the backbone. The nitrogen atom in the backbone acts as a hydrogen donor and the carbonyl oxygen acts as a hydrogen acceptor.

1.4.2.1 Alpha Helices

Alpha helices are formed when the backbone of the amino acid chain winds into a helical conformation with hydrogen bonds used to stabilize the structure. Figure 1.8 shows the formation of one such hydrogen bond. In this figure, the alpha carbon atoms have been numbered so that one can see how the acceptor and donor are related to one another. Specifically, the carbonyl oxygen of residue 1 acts as the hydrogen acceptor for the nitrogen donor in residue 5. The alpha helix is distinguished by this pattern: residue n interacting with residue $n + 4$ through a hydrogen bond. This arrangement gives the alpha helix 3.6 residues per turn. So, if one views the helix from a perspective that is parallel to its axis, one sees an angular increment of approximately 100° between successive alpha carbon atoms.

It is also possible to have helical conformations that are more elongated. The 3_{10} helix has a hydrogen bond between the carbonyl oxygen of residue n and the nitrogen atom in residue $n + 3$. This gives three residues per turn and

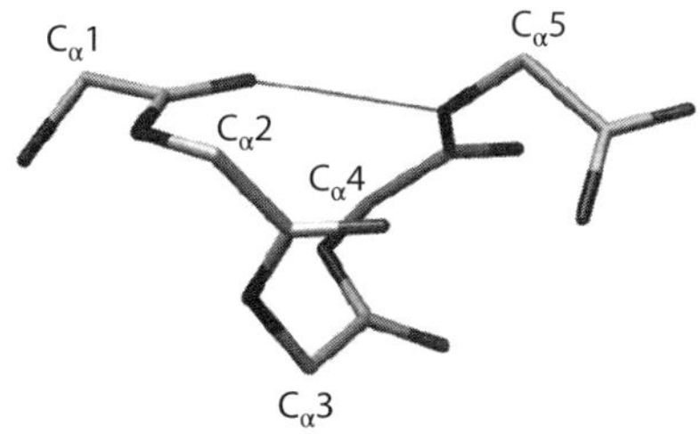

FIGURE 1.8 Formation of a hydrogen bond in an alpha helix.

is a type of secondary structure that occurs (rarely) in very short segments or at the ends of an alpha helix. The very rare pi helix has a hydrogen bond between residues n and $n + 5$ with 4.4 residues per turn. In this case, the helix is wide enough to have a hole along the axis of the structure and this is less energetically favorable when compared to the more compact alpha helix.

Looking at Figure 1.8, we can pretend to grasp the helix with the right hand so that fingers are pointing in a direction that is the same as that specified by an increase in residue number. If the thumb is pointing toward the carboxyl end of the protein, then we have a right-handed helix, otherwise it is designated to be a left-handed helix. The latter is also a rare event because it forces L-amino acid side chains into an excessively close proximity with carbonyl groups.

Figures 1.9 through 1.11 show an alpha helix that has been extracted from the protein 1QYS. The first figure shows only the backbone atoms with longer roughly horizontal lines representing the hydrogen bonds. The side chains have been removed because they would certainly obscure a view of the helical nature of the structure. Even with this simplification, it is still difficult to assess the helix structure and so Figure 1.10 presents a ribbon model of the structure. Hydrogen bonding as defined by the Chimera visualization software has been left intact. In many other figures to come, we will see ribbon diagrams that use a plain *linguini* type of ribbon to illustrate a helix. Finally, to provide another visual representation, Figure 1.11 shows a ribbon structure with the side chains sprouting off the helix.

It is interesting to note that Linus Pauling determined the structure of the protein helix in 1948. He is quoted as saying, "I took a sheet of paper and sketched the atoms with the bonds between them and then folded the paper to bend one bond at the right angle, what I thought it should be relative to the other, and kept doing this, making a helix, until I could form hydrogen bonds between one turn of the helix and the next turn of the helix, and it only took a few hours doing that to discover the alpha-helix." Pauling realized that the

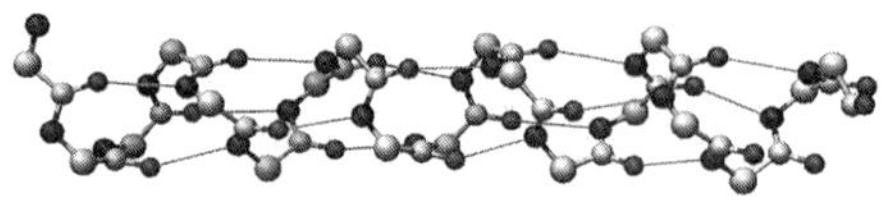

FIGURE 1.9 Hydrogen bonds in an alpha helix.

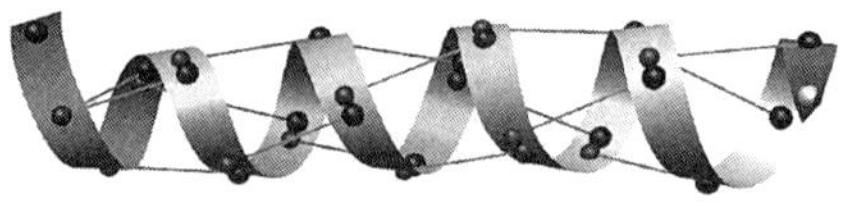

FIGURE 1.10 Ribbon diagram of a helix with hydrogen bonds.

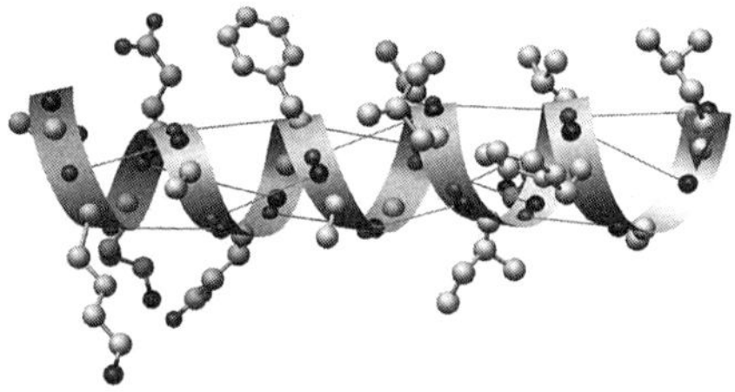

FIGURE 1.11 Ribbon diagram with side chains.

side chains would point outward from the protein helix.* In 1953, Pauling and Corey [PC53] published a paper that proposed a somewhat similar structure for DNA based on a triple helix structure. The reader may consult a report written by Stephen Lawson (http://lpi.oregonstate.edu/ss03/triplehelix.html) to get a balanced perspective on this paper "which would turn out to be one of the most famous mistakes in 20th-century science." In 1954, Pauling was given the Nobel Prize for Chemistry to recognize his work on molecular structure. Pauling was also known for his significant efforts on behalf of world peace and other issues related to humanitarian goals and social justice.

1.4.2.2 Beta Strands

Like alpha helices, beta strands are secondary structures formed by hydrogen bonding. In this case, however, the consecutive amino acids interact with other consecutive amino acids that may be quite distant in the primary sequence. The hydrogen bonds provide a sequence of linkages

* As noted by Anna Tramontano (Private communication): ... the reason why Pauling could predict the alpha helix first was because he was not trapped in thinking that the number of residues per turn had to be an integer number. This shows how common assumptions could mislead a lot of people! Also, he was the one who demonstrated the planarity of the peptide bond which is essential to come up with the model.

between two chain segments, which may have one of two orientations with respect to one another as follows:

- Parallel beta strands: The consecutive hydrogen bonds are such that the indices of amino acids on either end of the bonds are increasing in the same direction.
- Antiparallel beta strands: The consecutive hydrogen bonds are such that the indices of amino acids on either end of the bonds are increasing in opposite directions.

Figures 1.12 through 1.18 provide more details about beta strands.

In Figure 1.12, we see the backbone atoms of three beta strands. Side chains have been removed for clarity but the carbonyl oxygen atoms have been inserted to get the hydrogen bonds. The amino acids for each strand have their indices increasing as we go from left to right. This is indicated by the amide group on the far left and the carboxyl group on the extreme right. Notice how the hydrogen bonds are formed: the carbonyl oxygen atoms being hydrogen bond acceptors while the nitrogen atoms of the adjacent strand act as hydrogen donors.

Figure 1.13 presents a more simplified diagram of the same three strands each with a rounded ribbon arrow indicating the direction of increasing indices in the primary sequence.

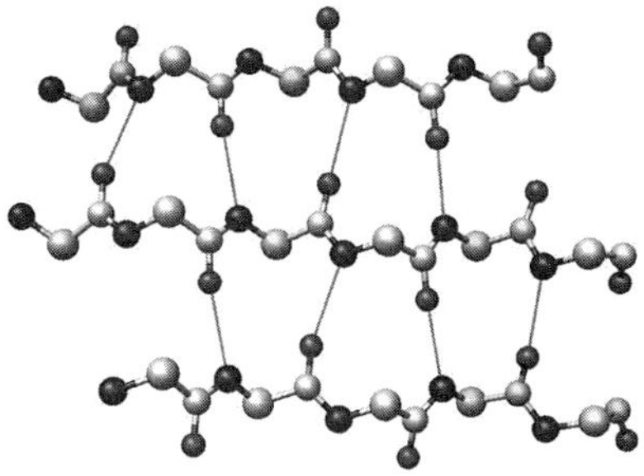

FIGURE 1.12 Hydrogen bonds in parallel beta strands.

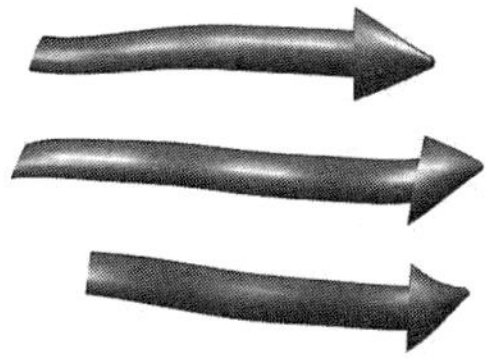

FIGURE 1.13 Ribbon diagram for parallel beta strands of Figure 1.12.

Figure 1.14 shows the same backbone atoms as those seen in Figure 1.12 but with oxygen atoms and hydrogen bonds removed. The view has been rotated so that the strands are almost behind one another and collectively they define a beta *sheet*. The alpha carbon atoms have been colored black so that they are more obvious in the figure. The importance of this view is that it shows how the strands tend to rise and fall in unison. In this way, the strands form a *pleated* sheet (sometimes described as a corrugated sheet). A significant consequence of pleating is that the side chains cannot originate from an alpha carbon that is in a valley of the pleat if they are to avoid steric clashes. In other words, they will seem to sprout from the alpha carbons in such a way as to avoid the atoms on either side of the alpha carbon. Consequently, the side chains originate from the peaks of the corrugations and so they appear to line up in rows when the sheet is observed on edge. This line up effect is shown in Figure 1.15.

Figure 1.15 shows the same three ribbons but with side chains present and drawn as *stick* structures. The scene has been rotated so as to make it more apparent that the three strands form a *sheet*. As just described, the side chains appear to sprout in rows that are determined by the pleats of the sheet and because of the nature of the pleats, these rows of residues alternate above and below the sheet. The ribbons are drawn in such a way as to eliminate the pleated structure of the sheet but the flatness of

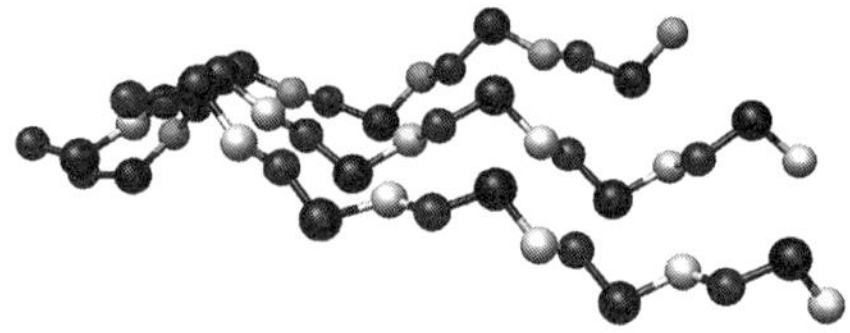

FIGURE 1.14 Side view of the previous ribbon diagram showing pleated conformation.

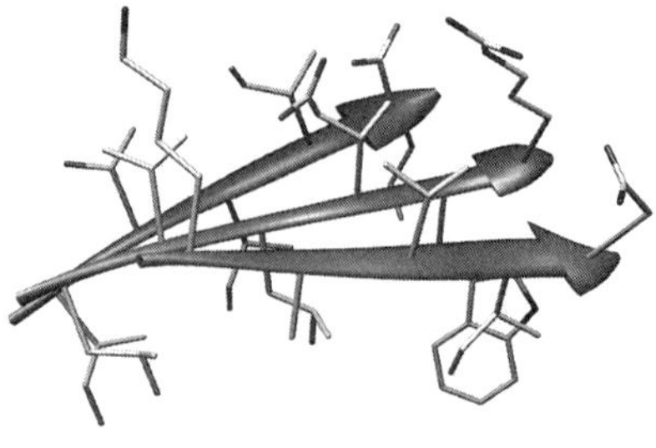

FIGURE 1.15 Ribbon diagram with side chains for parallel beta strands of Figure 1.12.

the ribbons brings out another significant observation: the sheet is not perfectly flat; it has a slight twist. This twist is not a peculiarity of this protein*; rather, it is typical of many beta sheets.

Let us now consider a set of beta strands that have an antiparallel conformation. All of the figures for this discussion have been prepared by viewing an antiparallel beta sheet within the protein 1QYS. As before, the beta strands make up a sheet but now they run in directions that alternate as illustrated in Figure 1.16. The hydrogen bonds that stabilize this sheet are shown in Figure 1.17. The reader should compare the appearance of these hydrogen bonds with that of Figure 1.12. In general, the successive hydrogen bonds that join two beta strands in an antiparallel configuration tend to be more uniform in their direction. They do not exhibit the slightly slanted appearance of hydrogen bonds in a parallel beta strand sheet as seen in Figure 1.12. The antiparallel construction of the sheet illustrated in Figure 1.17 has a pleated structure that is very similar to that seen in beta strands that make up parallel sheets. If the structure in Figure 1.17 is viewed from a line of sight that is almost along the surface of the sheet, the

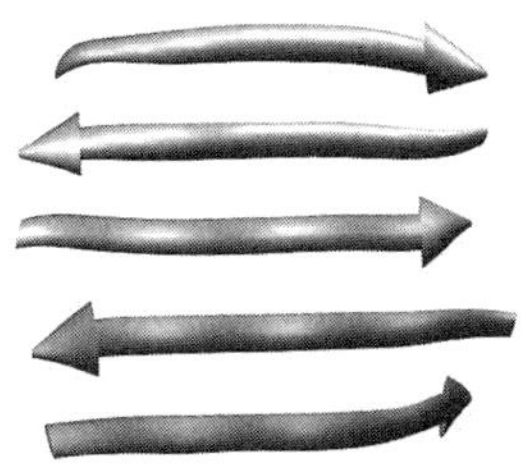

FIGURE 1.16 Ribbon diagram for antiparallel beta strands.

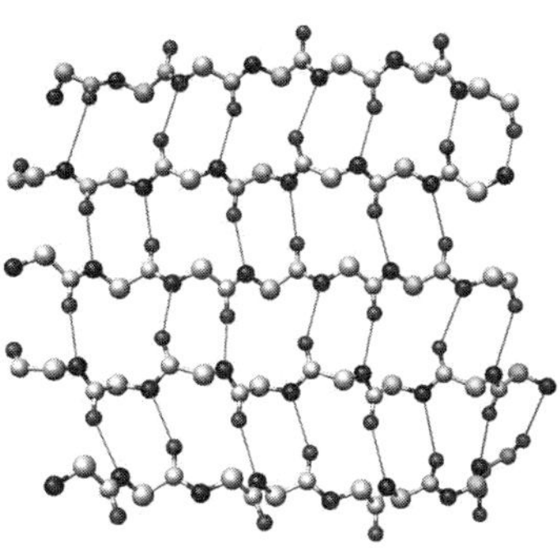

FIGURE 1.17 Hydrogen bonds in antiparallel beta strands.

* Figures 1.12 through 1.15 were done with UCSF Chimera, selecting a beta sheet from protein 1AY7.

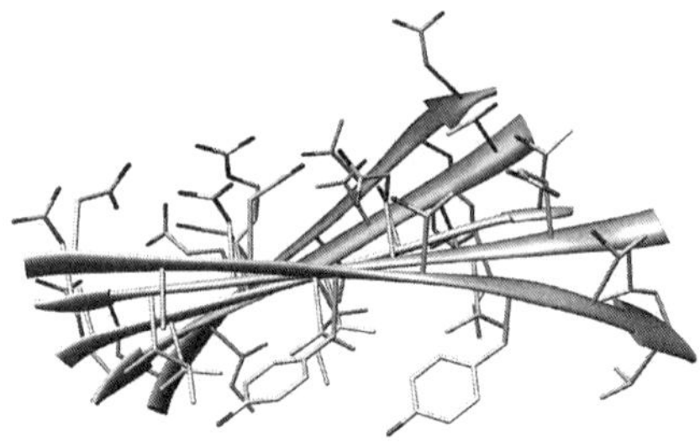

FIGURE 1.18 Ribbon diagram with side chains for antiparallel beta strands of Figure 1.17.

observer would see an arrangement of strands that is very similar to that of Figure 1.14. Because of this pleated formation we can observe the same patterns as before when looking at the orientations of the side chains. They seem to sprout off the sheet in rows and the successive rows alternate in their directions above and below the sheet. This is seen in Figure 1.18.

Sheet twist can also be observed for antiparallel beta strands. In fact, researchers have collected empirical evidence that such deformations are typically larger for antiparallel sheets than for parallel sheets ([EM04] and [HC02]). Figure 1.18 shows an illustration of this twist for the antiparallel case.

It should be stated that the ribbon diagram is very suitable when illustrating a twisted sheet. A diagram that shows only the atoms (without ribbons) is very complicated and the sheet structure becomes hidden in a forest of atoms that hide this pattern.

It is possible for a beta sheet to have a mix of both parallel and antiparallel strand pairs. An example of this can be seen in the salivary nitrophorin protein 1YJH.

1.4.2.3 Loops

The final category of secondary structure that we describe is the loop. Loops are chains of amino acids that have no particular hydrogen bonding patterns with other parts of the protein. There are proteins that consist of nothing more than a loop stabilized by disulfide bridges. A good example of this is flavoridin (PDB code 1FVL) which is a blood coagulation inhibitor from the habu snake. Another example is (PDB code 1ANS) a neurotoxin from the sea anemone, Anemonia sulcata. This protein has three disulfide bridges and while it does not contain any helices or beta sheets it does have seven beta turns.

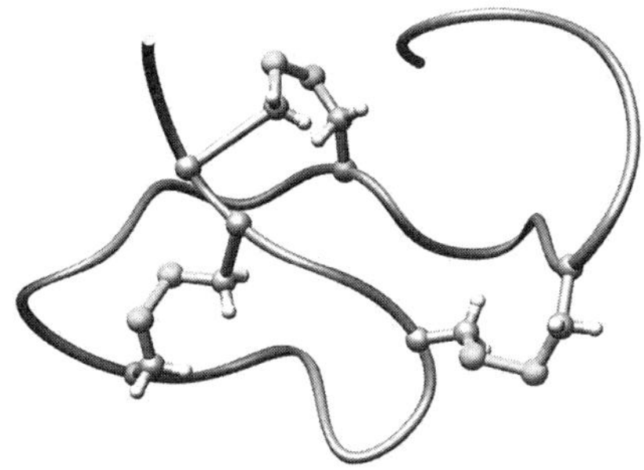

FIGURE 1.19 Loop structure and stabilizing disulfide bridges of the neurotoxin 1ANS.

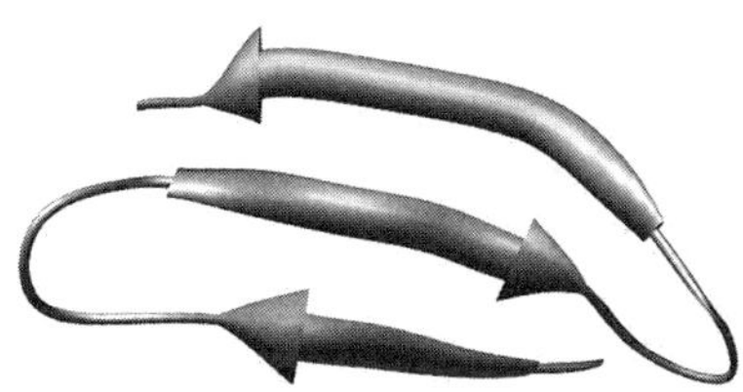

FIGURE 1.20 Loops connecting adjacent beta strands in an antiparallel sheet.

In Figure 1.19, the loop is portrayed as a ribbon diagram, whereas the disulfide bridges have been drawn in ball and stick form.

Proteins that are *all loop* are fairly rare. Most proteins will have beta sheets and helices forming a hydrophobic core and these secondary structures will be interconnected by loop segments. An example of this is given in Figure 1.20, where we see short loops connecting the adjacent strands of an antiparallel beta sheet within the *Escherichia coli* ribonuclease 1V74. This is the simplest way to connect such strands; it is possible to find other more complicated interconnections for adjacent strands involving helices and other beta strands.

For a sheet made of parallel beta strands, the adjacent strands may be connected with simple longer loops. However, in most cases, one will find that helices and other beta strands are used to connect adjacent parallel strands.

1.4.3 Tertiary Structure

1.4.3.1 What Is Tertiary Structure?

The tertiary structure of a protein specifies the positional relationships of all secondary structures. For example, the tertiary structure will show how the beta strands and helices combine to form the hydrophobic core of a globular protein. Interconnecting loops also play an important role and their placement is critical in various applications. Tertiary structure is essentially established once all the three-dimensional coordinates of the atoms are determined.

In Section 1.4.4, we will see that tertiary structures can combine to form larger quaternary structures. However, many proteins are single peptide chains and can be described by specifying the tertiary structure of that single chain. If we want to describe a protein's functionality from a biological perspective, then starting with the tertiary structure of the protein is a minimal requirement.

Because proteins interact with other molecules, they will have binding sites that must be organized in a way that facilitates some interaction while avoiding any other reactions that might produce harmful side effects. This is the most significant *raison d'être* of tertiary structure but there are other issues as well. To illustrate this, Section 1.4.3.2 provides a quick sketch of the myoglobin binding site and after that we cover some other topics related to tertiary structure.

1.4.3.2 Tertiary Structure of Myoglobin

The three-dimensional structure of myoglobin provides an interesting example of a tertiary structure. It has a biological functionality that has been extensively studied and it also has the distinction of being the first protein structure to be determined by x-ray analysis. This was reported by Kendrew et al. in 1958 [KB58]. The structure of the myoglobin molecule was not deposited into the PDB until later because the PDB was not established until 1971. Myoglobin appeared there in 1973 with identifying code 1MBN.

Figure 1.21 provides a view of the entire molecule. It has 153 residues that form eight helices interspersed with short loops. These helices are combined into a special arrangement called a *globin fold*. It consists of a hydrophobic pocket that contains a structure called a prosthetic heme group (Figure 1.22). The heme group for this figure was extracted from the PDB file for oxymyoglobin (PDB code 1MBO), which has a molecule of oxygen hovering over the iron atom at the center of the heme group. The function of myoglobin is to carry oxygen in muscle tissue. More specifically, it stores oxygen and carries it to the mitochondria for oxidative phosphorylation. For x-ray analysis, both proteins (1MBN and 1MBO) were extracted from the muscle of sperm whale (Physeter catodon), a mammal that has this protein in great abundance because of the necessity to stay under water for long periods of time. As any structural biologist can tell you, the myoglobin proteins and hemoglobin proteins share similar structures and functionality. Hemoglobin genes have been in the biosphere for at least 1.8 billion

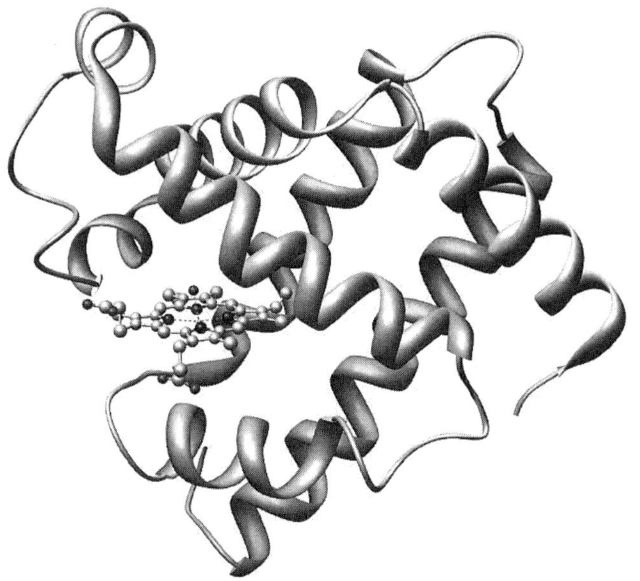

FIGURE 1.21 Tertiary structure of myoglobin (Protein Data Bank code = 1MBN).

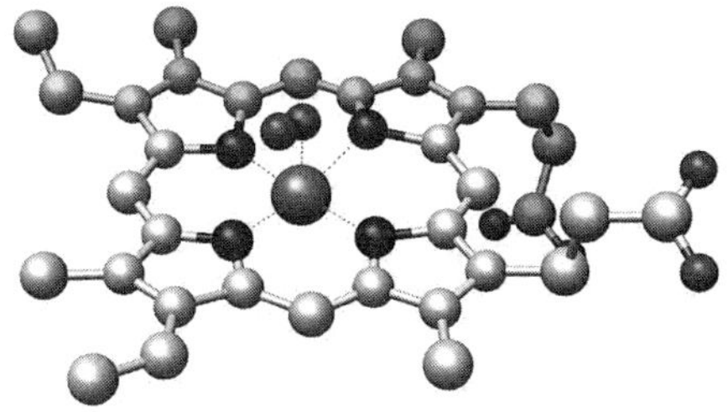

FIGURE 1.22 Heme group from 1MBO.

years and may have originated as proteins that were meant to sequester oxygen thus protecting the cell from harmful effects of too much oxygen produced by the photosynthetic processes [SG99].

At the center of the heme structure is an iron atom that is coordinated by the nitrogen atoms of the four pyrrole rings. In normal myoglobin the iron atom is ionized to the Fe^{2+} (ferrous) state. The heme group is not symmetric and contains two highly polar propionate side chains that appear on the right side of Figure 1.22. A close inspection of Figure 1.21 will reveal that these carboxylic acid side chains (ionized at physiological pH) avoid the interior of the hydrophobic pocket. This is also shown with more clarity in Figure 1.23. In this presentation, a surface has been put over the protein and it has been shaded so that portions of the surface near hydrophobic

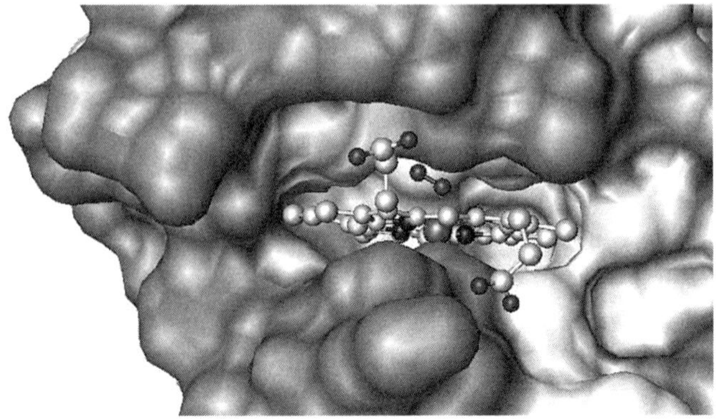

FIGURE 1.23 Heme group in its binding pocket.

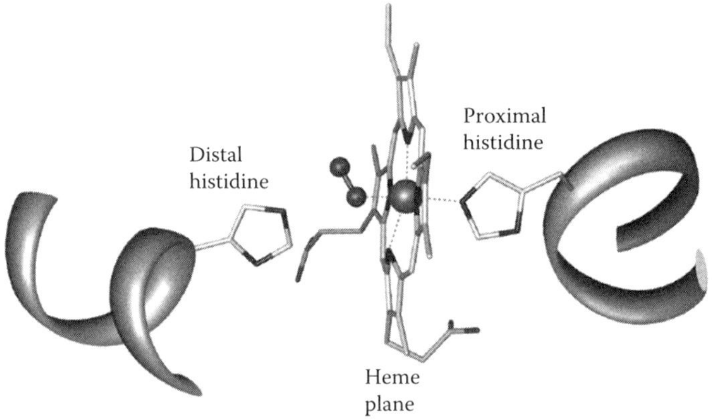

FIGURE 1.24 Heme group and histidine residues.

residues (as specified in Figure 1.3) are white while any surface over a nonhydrophobic residue is shown in gray tones. If you look carefully at Figure 1.23 you can see a gray bump immediately under the iron atom. This is the NE2 nitrogen atom of a histidine residue (H93). There is another gray bump just above the oxygen molecule due to another histidine residue (H64). These will be discussed further when we look at Figure 1.24.

Figure 1.25 is provided so that you can get a better appreciation for the construction of the heme binding pocket. This diagram shifts the perspective of the viewer so that the individual is looking at the heme group from a position that is below the plane of the heme. Furthermore, a portion of the surface has been cut away to reveal the residues that create the surface. The heme group (above the surface but in the pocket) is drawn in a ball and stick fashion while all structures below the surface are drawn in stick form. Hydrophobic residues appear in white and all other residues

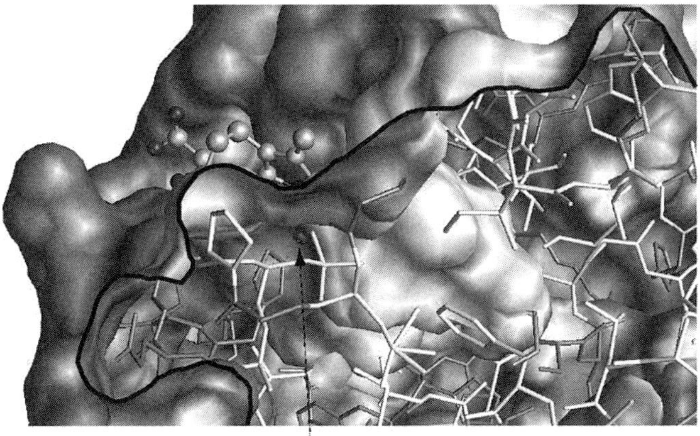

FIGURE 1.25 Heme group in pocket—cutaway view.

are have darker shades of gray. The dashed arrow indicates the position of the iron atom. If you look carefully, you can see H93 just below and to the left of the iron atom. The main point of this diagram is to see the several hydrophobic residues that surround the underside of the heme binding pocket. It should be noted that these nonpolar residues protect the Fe^{2+} from oxidation to the Fe^{3+} (hematin) state, which will not bind O_2.

Figure 1.25 shows the placement of the iron atom relative to the two histidine residues mentioned earlier. Coordination of the heme iron atom is done by the NE2 atom of H93 (called the *proximal* histidine) and the NE2 of H64 (called the *distal* histidine). The distal histidine crowds in on the binding space and causes the oxygen molecule to bind at an angle. This makes the bond to the iron atom weaker and so O_2 is more easily removed when needed.

1.4.3.3 Tertiary Structure beyond the Binding Pocket

> Perhaps the most remarkable features of the molecule are its complexity and its lack of symmetry. The arrangement seems to be almost totally lacking in the kind of regularities which one instinctively anticipates, and it is more complicated than has been predicted by any theory of protein structure.
>
> J. C. Kendrew et al. describing myoglobin, 1958

> Biologists must constantly keep in mind that what they see was not designed, but rather evolved.
>
> Francis Crick, 1988

In Section 1.4.3.2, we have seen that myoglobin tertiary structure provides a binding site that not only holds the heme group with its nonpolar residues but also has a particular geometry that manages the storage of O_2 in a very precise fashion. It is fairly clear that all the residues near the binding site participate in the support of these activities. However, there are many other residues that are considerably further away from the binding site. It is reasonable to ask about the participation of these residues in the primary functionality just described. Are they really needed? To add more significance to the question, it should be noted that the protein cytochrome B_{562} from *E. coli* is only 106 residues in length and it also has a binding site with a heme group. To be fair, it should be noted that the B_{562} binding site lacks the critically positioned distal histidine seen in myoglobin and so the functionality is somewhat simpler. However, it is possible to find other proteins that seem to be quite huge in comparison to their rather small binding sites.

It is perhaps this type of additional complexity that prompted Kendrew and his associates to express the statement quoted at the start of this section. This quote was taken from the *Nature* paper [KB58] and it gives us some perspective on the mindset of the early explorers venturing into the unknown territory of protein structure. The anticipated regularities of protein structure seem to be reminiscent of the earlier expectations of Copernicus and his assumption that planetary orbits would be circular. In line with the admonition made by Francis Crick [CR88] in the preceding quote, our current thinking provides a somewhat different perspective on the structure of the myoglobin molecule: It is the result of an evolutionary process driven by functional requirements with less attention paid to elegant design.

At the macroscopic level, we see many examples of this function-driven evolution with problematic design principles. For example, breathing and food ingestion share a common tract, much as reproductive systems and urine elimination share the urethra in male mammals. There are drawbacks to this tissue sharing: breathing will be accompanied by choking if food is taken into the trachea and in the case of shared reproductive/urinary systems, we can consider the unfortunate psychological associations due to such an arrangement. Even though tissue sharing provides a kind of time-sharing efficiency, this type of *tissue parsimony* does not always occur in the evolutionary process: everyone is born with a vestigial appendix that is not needed and men have nipples.

At the molecular level, we may also see redundant substructures in a protein simply because evolutionary processes have maintained them as legacy contributions inherited from an earlier protein (see [KD99] in which the authors entertain the idea that a particular folding topology may reflect a *vestigial structural remnant* of an ancestral SH2-containing protein).

Even though there is the possibility that various helices or strands might simply constitute vestigial *excess baggage*, there are other more significant reasons to have extensive and very complicated structure in a protein.

The functionality of tertiary structure relates to the following issues:

1. Formation and management of the binding site

   The residues (usually hydrophobic) create a pocket that promotes the docking of the protein with a ligand molecule. As just discussed in the myoglobin example, the positions of particular atoms in the binding site may be critical for the optimal functioning of the reaction. There is also the necessity of forming a binding site that does not combine with other ligands in a way that produces a deleterious effect. Protein evolution can do only so much in this regard and may not protect a cell against toxins that are relatively new to the organism. For example, hemoglobin will avidly take on carbon monoxide in lieu of oxygen molecules with disastrous results.

2. Flexibility

   There are situations in which a protein interacts with a ligand or another protein and this interaction is accompanied by a change in protein conformation that facilitates the reaction with yet another molecule. Allosteric proteins demonstrate this functionality and the overall change in the protein's conformation may involve residues that are quite distant from the binding site that initiates the change.

3. Control of a correct folding pathway

   In some sense, the folding of a protein involves many of the forces that contribute to the formation of a binding site. However, folding is functionality that must be accomplished *before* the binding site can be established. As such, the tertiary structure must have a level of organization that will ensure the correct fold. It has been experimentally observed that various peptide sequences will simply not fold. For those that do fold, it *may* be necessary to have a primary sequence with key residues that foster particular intermediate transitory folds that act as transitional conformations. Presumably, these are necessary to guide the folding process so that the final conformation is indeed the lowest energy conformation. The reader may consult [Sk05] for an opinion that stresses the importance of folding pathways and [Do03] for a discussion about transition state conformations that act as a quality control mechanism in protein folding. A discussion of ultrafast and downhill protein folding may be found in [Dy07].* Models for folding and misfolding can be found in [Do06]. Another aspect of the correct folding of proteins deals with cotranslational folding. This involves the assumption that proteins start to take on significant aspects of their tertiary structure during translation when they are still attached to the ribosome. Such an assumption also has implications about the folding pathway ([FB97] and [Ko01]). It should be stressed that there is still considerable controversy about the various phases of the folding process (e.g., [LB98]). Much of this should be settled as more data becomes available.

4. Stability of the protein

   After the completion of folding, a complex interaction of the residues must ensure that the final structure will be maintained without the possibility of a different conformation being formed. In the ideal situation, a protein folds to a tertiary structure that is characterized as the lowest energy conformation and it stays in this conformation except for any flexible changes that are necessary for its functionality. As described earlier, disulfide bridging enhances the preservation of conformation. In less than ideal situations, it is possible that the protein undergoes a refolding that produces a different conformation. The endogenous cellular prion protein PrP^{C} that can undergo a conformational change to become the scrapie isoform PrP^{Sc} gives a good example of such metastable behavior. The PrP^{C} conformation contains 43% alpha helix and no beta sheets, whereas the PrP^{Sc} form contains 20% alpha helix and 34% beta sheet. Unfortunately, PrP^{Sc} has a tendency to form aggregates in cerebral tissue and existing proteolytic mechanisms do not clear these aggregates. With unchecked accumulation, neurological damage occurs, ultimately leading to death.

These points emphasize that tertiary structure is necessarily complex because it must do a lot more than simply provide a foundation for the binding site. The additional responsibilities of tertiary structure include the support of delicate mechanisms that foster the proper behavior of the protein over its entire *life cycle*. This includes its creation by means of a guided folding pathway, the stable maintenance of its conformation and binding site, and its final susceptibility to a degradation process that will lead to the reuse of its amino acids.

---

* *Current Opinion in Structural Biology*, **17**(Issue 1) has several papers devoted to protein folding issues.

Meeting all of these requirements is a complex optimization problem with multiple objectives. Furthermore, it is usually quite difficult to explain the functional significance of any one amino acid in terms of its contribution to one or more of the four issues just described. The proximal and distal histidine residues in myoglobin, various other residues that act as secondary structure *breakers*, and particular residues that form hairpin structures [IA05] are exceptions to this statement, but in the general case, a particular residue serves along with several other amino acids to define the various functional aspects of the tertiary structure.

Despite the possible lack of regularity in a tertiary structure, the fold appearance of the structure may be replicated in another protein even though its primary sequence is reasonably different. In Figure 1.26 we can see an obvious similarity between myoglobin and a single chain within hemoglobin, which is also an oxygen carrying protein. This structural similarity is present even though the primary sequences have only 29% of their primary sequences matching identically. As noted in [OM97]: "At 30% sequence identity, proteins will almost certainly have the same overall fold." This is especially significant for the globin family where similar folds have been seen for even lower sequence identities (less than 15%). The alignment (Figure 1.27) was done using the TCoffee server [PO03]. This example underscores the observation that tertiary structure is more conserved than sequence in an evolutionary descent.

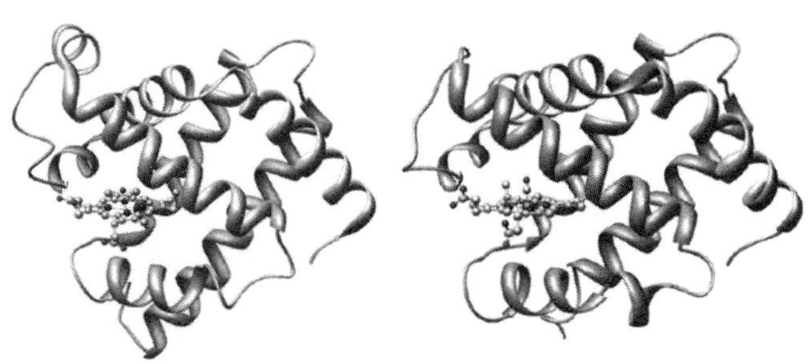

Myoglobin (1MBN) One chain of hemoglobin (1JEB)

FIGURE 1.26 Structural comparison of myoglobin and a single chain of hemoglobin.

```
1MBN    VLSEGEWQLVLHVWAKVEADVAGHGQDILIRLFKSHPETLEKFDRFKHLKTEA
1JEB    SLTKTERTIIVSMWAKISTQADTIGTETLERLFLSHPQTKTYFPHFDLHPGSA

1MBN    EMKASEDLKKHGVTVLTALGAILKKKGHHEAELKPLAQSHATKHKIPIKYLEFI
1JEB    QLRA------HGSKVVAAVGDAVKSIDDIGGALSKLSELHAYILRVDPVNFKLL

1MBN    SEAIIHVLHSRHPGDFGADAQGAMNKALELFRKDIAAKYKELGYQG
1JEB    SHCLLVTLAARFPADFTAEAHAAWDKFLSVVSSVLTEKYR------
```

FIGURE 1.27 Sequence comparison of myoglobin and a single chain of hemoglobin.

Even though there are several amino acids that can be changed while still maintaining the functionality of the globin fold, there is a critical location at which a mutation can produce a significant change in functional behavior. The reader may note that the sixth residue of 1JEB is "*E*" for glutamic acid. Sickle-cell anemia is an inherited disease that leads to single mutational change in which this glutamic acid is replaced by valine. This produces a small hydrophobic patch on the surface of the protein. Unfortunately, this patch is able to bind with a hydrophobic pocket in the deoxygenated conformation of another hemoglobin molecule. This *stickiness* of the hemoglobin proteins causes a clumping that leads to distortion of the red blood cell, which then takes on a *sickle-cell* appearance. These blood cells have a shorter life span than normal cells and so the patient has anemia caused by low levels of hemoglobin. Sickle cells can also become trapped in small blood vessels preventing blood flow.

As a final note on tertiary structure, Figure 1.28 is presented to illustrate a protein that does have a remarkable amount of regular structure. This is from a human placental ribonuclease inhibitor that binds the blood vessel-inducing protein human angiogenin [PS97]. The angiogenin molecule is presented to show the extraordinary repetition of alpha helix and beta sheet structures that combine to form the large *horseshoe* tertiary structure. Although there is an extensive beta sheet present, it does not have the same twist that is often seen for smaller sheets. The beta strands have collected into a cylindrical formation that almost goes full circle. The regularity of their positions seems to be coordinated by the *hair roller* placement of the helices in the structure.

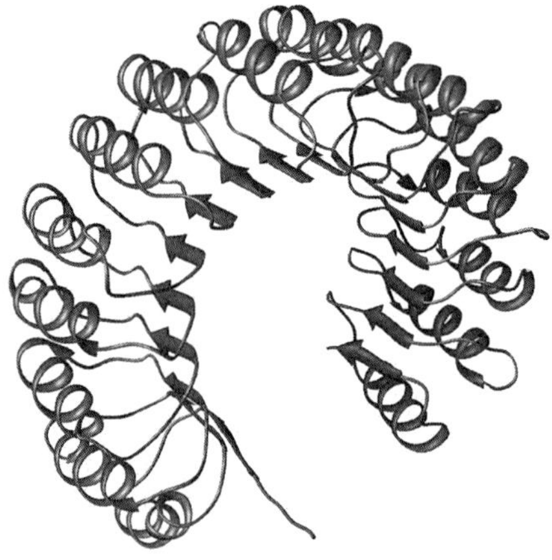

FIGURE 1.28 A tertiary structure involving repeated secondary structures (chain D of 1A4Y).

Many other remarkable folds can be seen in [OM97] and especially [PR04]. Space limits the presentation of these diagrams but the interested reader may fetch and display the ribbon diagrams for the more picturesque titles: β Barrel (2POR), β 7 Propeller (2BBK chain H), β 2 Solenoid (1TSP), and αβ 4-Layer Sandwich (2DNJ chain A). For most of these examples, the regularities in the structure are best seen by using a protein visualizer that allows the viewer to freely rotate the structures.

1.4.4 Quaternary Structure

A protein has quaternary structure when it is an assembly of multiple tertiary structures. We refer to such structures as oligomeric complexes and we describe these structures using labels that start with a prefix representing the number of subunits followed by the suffix *mer*. The progression is monomer, dimer, trimer, tetramer, pentamer, and so forth. If the subunits are different, then an additional *hetero* prefix is used, as in *heterodimer*. If the subunits are the same, then the *homo* prefix is used, as in *homotetramer*. From an efficiency point of view, it is in the best interest of a cell to build complex structures from smaller building blocks that are all the same. Each unit will be prescribed by the same DNA sequence and so there is a parsimonious genetic representation of the full oligomeric structure.

In some cases, the quaternary structure can show a low level of recursive assembly. For example, three heterodimers may be assembled to form a *trimer of dimers*. A beautiful example of this construction is the insulin protein illustrated in Figure 1.32 later in the chapter. The assembly of this hexamer is a bit complicated, so we will explain it as a series of steps starting with the monomer illustrated in Figure 1.29. This is a single molecule made up of two peptide chains that are joined together using two disulfide bridges. There is a third disulfide bridge, but it connects two cysteine residues within the same chain. To see how the three disulfide bridges interconnect the residues in the three chains, consider Figure 1.30.

Two of these monomers make up the insulin dimer by having their beta strands combine in an antiparallel fashion to make a small beta sheet. This is illustrated in Figure 1.31, where the A and B chains appear in the background with C and D chains in the foreground. Chains C and D correspond to chains A and B in the dimer assembly, and they have the same type of disulfide connectivity. In Figure 1.31, all disulfide bridges have been removed for clarity.

Figure 1.32 shows the full insulin hexamer made up of three dimers each one of the same construction as that shown in Figure 1.31. The positions of the three dimers are coordinated by means of two zinc atoms. To see how the zinc atoms are related to a dimer, they have been included in Figure 1.31.

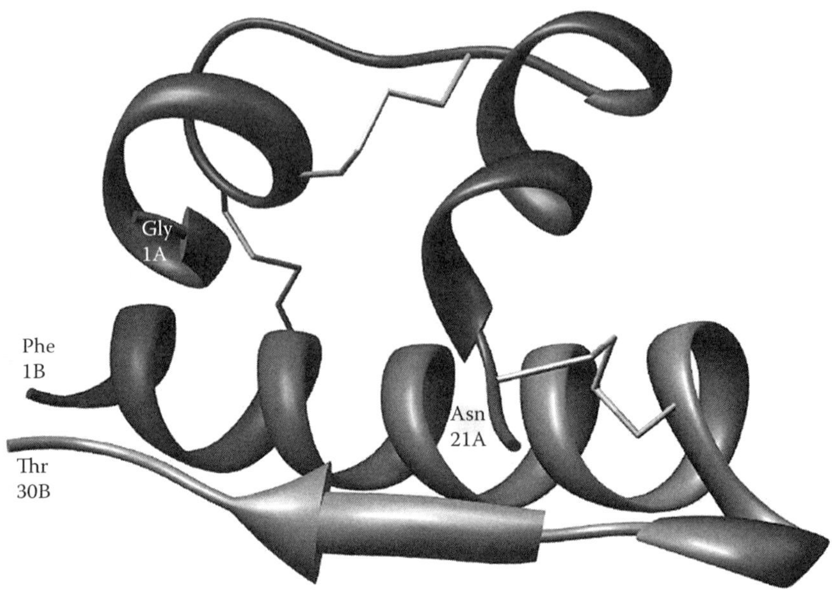

FIGURE 1.29 The insulin monomer has two chains connected by disulfide bridges.

FIGURE 1.30 Six cysteine residues of the monomer are interconnected using disulfide bridges.

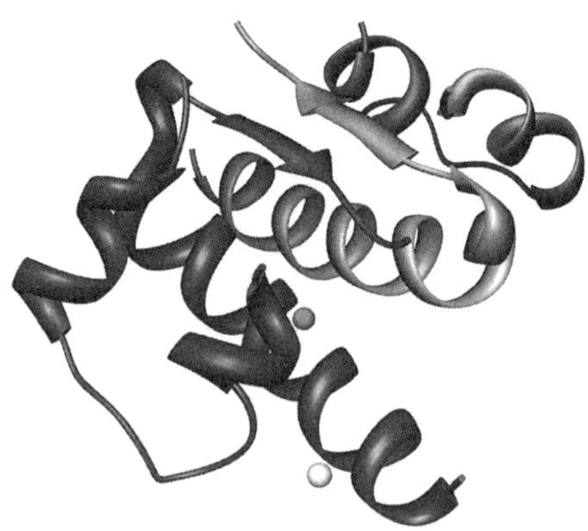

FIGURE 1.31 One of the three dimers making up the hexamer of an insulin structure.

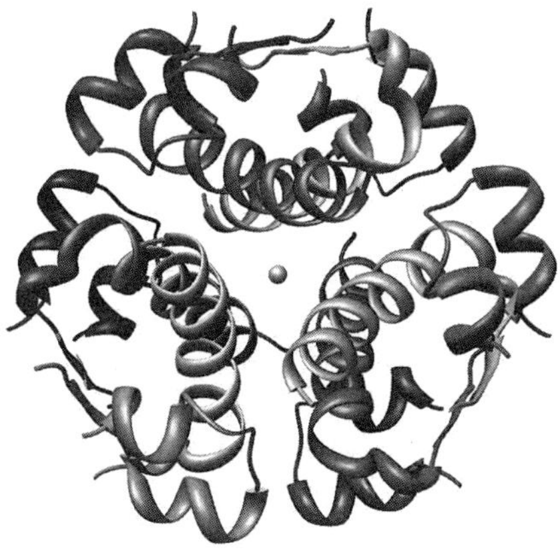

FIGURE 1.32 Hexamer structure of insulin (PDB code 1ZNJ).

To understand how the dimers combine to make a hexamer, imagine taking the dimer of Figure 1.31 and tilting the entire assembly so that your line of view is such that the darker zinc atom is exactly behind the brighter zinc atom. This will put the dimer into the exact orientation that the upper third of the hexamer shows in Figure 1.32. Notice that in this orientation, the beta sheets are seen on edge and so they are somewhat difficult to discern.

There are many other examples of quaternary structure, some of which we will see later.

1.4.5 Protein Functionality

As already stated, protein structure facilitates protein functionality. Even though this book does not consider functionality in any significant depth, we now present a very high level overview of the various processes and structural elements that use proteins.

The study of protein function includes the following:

- Enzymatic catalysis
    - Almost all chemical reactions in a cell are facilitated by a catalytic mechanism involving a protein.
    - Rate of reactions is accelerated, sometimes by as much as a factor of 10^{17}.

- Enzymes may use any of the following strategies:
  - Provide an alternative reaction route.
  - Stabilize reaction intermediates.
  - Facilitate alignment of reacting groups.
- Signal transmission
  - Information can be moved from one part of a biological system to another. This is done, for example, by neurotransmitters, hormones, and cytokines.
    - Cytokines act as signals in immune systems and during embryogenesis.
    - They typically bind to a cell-surface receptor and this initiates a signal cascade resulting in an upregulation or downregulation of gene expression.
- Regulation
  - Protein–protein interactions (PPI): PPI may enable or inhibit one of the proteins involved in an interaction.
  - Transcription regulation: Proteins may assemble to form transcription factors that regulate gene expression in DNA.
- Immune system
  - The immunoglobulins are examples of proteins that play a central role in the defense mechanisms of the body. They recognize foreign molecules and help the body to destroy viruses and bacteria. Other components of the immune system include major histocompatibility complexes and various receptor proteins.
- Molecular transport
  - Some membrane proteins are capable of moving small molecules into or out of a cell. For example, the Na^+/glucose transporter is membrane protein that brings in both sodium ions and glucose molecules with sodium ions moving down their concentration gradient while the glucose molecules are pumped up their concentration gradients.

- Motor elements
    - Changes in protein conformation produce mechanical movement, be it large muscle movement or the small flagellar motors of a bacterium.
- Structural systems
    - Proteins act as the building blocks of living systems.
    - In large complex systems, they contribute to the formation of large structures such as bone, hair (collagen), nails (keratin), and many other tissues.
    - Smaller structures include strands of silk, viral coats, and frameworks for the movement of molecules within cells.

1.4.6 Protein Domains

We have already seen the progression of structural organization: primary sequence, secondary structure, tertiary structure, and quaternary structure. The protein *domain* is another structural concept that is important in the discussion of protein functionality. Although the definition of a domain varies somewhat in protein research papers, here is a reasonable definition:

> A domain is a region of a protein that has more interactions within itself than with the rest of the chain.

These interactions include the formation of a hydrophobic core. Indeed, it has been observed that protein domains can often fold independently. The observation of independent folding is due to experiments that have shown that the subsequence of amino acids in a domain can often fold into the final conformation even when the rest of the protein is absent.

The following points provide more information about the formation of domains:

- Domains can be structurally similar even though they are in proteins that are otherwise quite different. Such domains usually show a high sequence similarity.
- Domains tend to be compact and globular. Linkages between domains are often loop structures and hardly ever helices or beta strands.

- Domains have distinct solvent accessible surfaces and are typically separated by water molecules.
- Residues within a domain will contact other residues in the same domain. There is very little contact, if any, between residues in different domains.
- Usually, a domain is formed from a contiguous residue sequence of the protein. In rare cases, a domain is made up of two or more regions of protein that are subsequences from one or more polypeptide chains. This is why expert visual inspection is often required to determine domains.

Multiple domains within a single protein often cooperate to accomplish related tasks. For example, reverse transcriptase protein of HIV has various domains that work together to convert viral RNA to DNA followed by destruction of the viral RNA. So, by studying the domain structure of a protein we can strive to understand how design principles facilitate particular types of functionality. Although this is an important goal, the mechanisms of cooperative function are not fully understood in many cases.

Figure 1.33 presents a simple example of the 6PAX protein that uses two domains to interact with a segment of DNA. This protein is part of a family called the homeodomain proteins. They interact with DNA to provide transcription regulation and they are important in development and cell differentiation. When the 6PAX protein interacts with DNA, the event initiates a cascade of protein interactions that lead to the development of various eye structures. In the figure it is clear that the protein has two hydrophobic cores each responsible for interacting with a particular DNA segment. The two cores are linked by a rather lengthy loop structure.

FIGURE 1.33 The two domains of 6PAX interact with DNA.

1.5 EXERCISES

Work through the topics in the first part of the tutorial in the User Guide that can be found at (http://www.cgl.ucsf.edu/chimera/current/docs/UsersGuide/frametut.html).

After practicing with menu invocations, you should be able to do the following:

1. Getting familiar with Chimera menus

   Use the Chimera menus to fetch the protein structure with PDB ID 1AY7. Then do the following exercises:

   a. Work with the menus to isolate the three strands of the beta sheet to display them in round ribbon form with the residues in stick form, as in Figure 1.15.

   b. Use the menus to isolate the beta sheet and display the backbone atoms in the strands, as in Figure 1.14.

   c. Use the **Favorites… Sequence** menu to select the amino acid sequence LSALPPEAT within chain A. Proline is usually regarded as a *helix breaker* (Figure 1.34). Use the Chimera menus to show this sequence in three representations: ribbon, ribbon plus atoms, and atoms/bonds only. Avoid any molecular rotations between each of these scenes. Pick a vantage point that best demonstrates the bend in the helix.

   d. Get familiar with the menu operations needed for a hydrogen bond display. You will also need to select the backbone atoms that are confined to the beta sheet as Figure 1.12.

   e. Use the **Actions… Surface…** menu to put a surface over the 1AY7 protein.

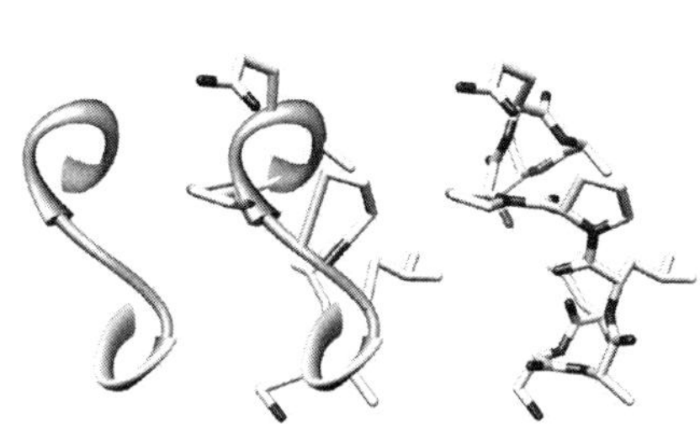

FIGURE 1.34 Coil with proline residue.

2. Structure and functionality

   Figure 1.35 shows the ribbon diagram of rice cytochrome c (PDB ID 1CCR). There are only two histidine residues in this protein and they are shown colored in black. One other residue has been made visible: There is a methionine residue with its sulfur atom hovering just above the Fe atom in the heme group. Do some web searching to discover the biological function of cytochrome c. Compare the structures of myoglobin and cytochrome c by referring to Figures 1.21 and 1.35, respectively. How is the difference in functionality supported by the differences in structure?

3. Searching the PDB

   Figure 1.36 shows the ribbon diagram of a 12-residue fragment of a longer protein. What is the PDB identifier of that protein?

4. Beta sheet topologies

   Taking into account both parallel and antiparallel associations of strands, how many different topologies can we have for a three-stranded beta sheet? How many for a four-stranded beta sheet? Draw each sheet as a set of arrows. Ignore the connectivity between the strands. Be sure that you do not count a sheet twice when it is the same sheet already listed but with a different orientation in three-space.

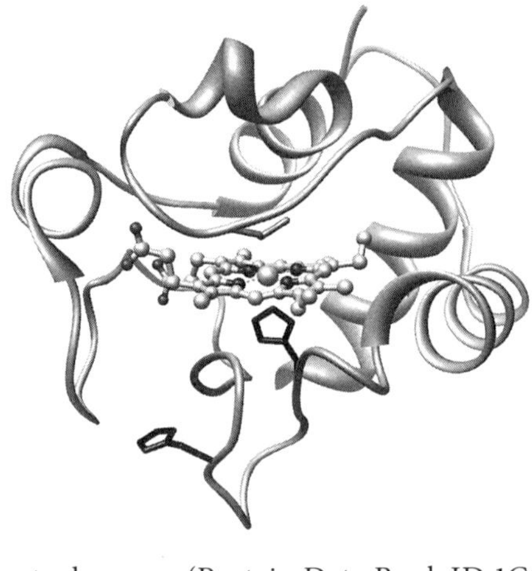

FIGURE 1.35 Rice cytochrome c (Protein Data Bank ID 1CCR).

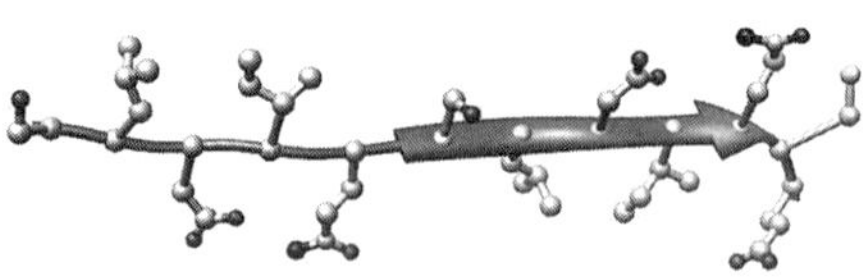

FIGURE 1.36 (**See color insert.**) What is the Protein Data Bank ID?

5. Cyclotides

   Essay assignment: Cyclotides are peptides that have a cyclic backbone. They usually have short primary sequences and often fold into loop structures that are stabilized by cysteine bridges. What are the anticipated uses of cyclotides? How would their structure facilitate these applications? How is the loop closure accomplished?

6. Heme affinity for carbon monoxide

   Essay assignment: It has been observed that the heme prosthetic group shows an affinity for CO that is up to 100,000 times higher than the affinity for O_2 in an unhindered model [JK95]. However, a distal histidine can provide some steric hindrance (discussed in [OM88]). It should be noted that the geometry of the Fe–C–O configuration has been debated. For example, Lim et al. [LJ95] state: "These solution results help to establish how myoglobin discriminates against CO, a controversial issue dominated by the misconception that Fe–C–O is bent." However, a casual inspection of the Fe–C–O triplet in co-crystallizations such as 1VXF would seem to indicate otherwise. You may also find [KP99] to be of interest. The investigation for your essay should cover the various issues that relate to binding affinity of CO relative to a heme group. In particular, how does affinity depend on issues such as hydrogen bonding, electrostatic attraction, Fe–C–O angle, steric hindrance, and the containing protein (hemoglobin vs. myoglobin)?

REFERENCES

[Cr88] F. Crick. *What Mad Pursuit: A Personal View of Scientific Discovery*. Basic Books, New York, 1988 (a reprint edition came out in June 1990).

[Di05] R. E. Dickerson. *Present at the Flood: How Structural Molecular Biology Came About*. Sinauer, Sunderland, MA, 2005.

[Do03] C. M. Dobson. Protein folding and misfolding. *Nature*, **426** (2003), 884–890.

[Do06] N. V. Dokholyan. Studies of folding and misfolding using simplified models. *Current Opinion in Structural Biology*, **16** (2006), 79–85.

[Dy07] R. B. Dyer. Ultrafast and downhill protein folding. *Current Opinion in Structural Biology*, **17** (2007), 38–47.

[EM04] E. G. Emberly, R. Mukhopadhyay, C. Tang, and N. S. Wingreen. Flexibility of β-sheets: principal component analysis of database protein structures. *Proteins: Structure, Function, and Bioinformatics*, **55** (2004), 91–98.

[FB97] A. N. Fedorov and T. O. Baldwin. Cotranslational protein folding. *The Journal of Biological Chemistry*, **272** (1997), 32715–32718.

[FR79] J. S. Fruton. Early theories of protein structure. *Annals of the New York Academy of Sciences*, **325** (1979), 1–20.

[HC02] B. K. Ho and P. M. G. Curmi. Twist and shear in β-sheets and β-ribbons. *Journal of Molecular Biology*, **317** (2002), 291–308.

[IA05] K. Imai, N. Asakawa, T. Tsuji, M. Sonoyama, and S. Mitaku. Secondary structure breakers and hairpin structures in myoglobin and hemoglobin. *Chem-Bio Informatics Journal*, **5** (2005), 65–77.

[JK95] P. Jewsbury and T. Kitagawa. Distal residue-CO interaction in carbonmonoxy myoglobins: a molecular dynamics study of three distal mutants. *Biophysical Journal*, **68** (1995), 1283–1294.

[KB58] J. C. Kendrew, G. Bodo, H. M. Dintzis, R. G. Parrish, H. Wyckoff, and D. C. Phillips. A three-dimensional model of the myoglobin molecule obtained by x-ray analysis. *Nature*, **181** (1958), 662–666.

[Ko01] V. A. Kolb. Cotranslational protein folding. *Molecular Biology*, **35** (2001), 584–590.

[KD99] J. Kuriyan and J. E. Darnell Jr. An SH2 domain in disguise. *Nature*, **398,** Mar. 4 (1999), 22–24.

[KP99] G. S. Kachalova, A. N. Popov, and H. D. Bartunik. A steric mechanism for inhibition of CO binding to heme proteins. *Science*, **284** (1999), 473–476.

[LB98] D. V. Laurents and R. L. Baldwin. Protein folding: matching theory and experiment. *Biophysical Journal*, **75** (1998), 428–434.

[LJ95] M. Lim, T. A. Jackson, and P. A. Anfinrud. Binding of CO to myoglobin from a heme pocket docking site to from nearly linear Fe-C-O. *Science*, **269** (1995), 962–966.

[MG00] T. Mizutani, C. Goto, and T. Totsuka. Mammalian selenocysteine tRNA, its enzymes and selenophosphate. *Journal of Health Science*, **46** (2000), 399–404.

[OM88] J. S. Olson, A. J. Mathews, R. J. Rohlfs, B. A. Springer, K. D. Egeberg, S. G. Sligar, J. Tame, J.-P. Renaud, and K. Nagai. The role of the distal histidine in myoglobin and haemoglobin. *Nature*, **336,** Nov. 17 (1988), 265–266.

[OM97] C. A. Orengo, A. D. Michie, S. Jones, D. T. Jones, M. B. Swindells, and J. M. Thornton. CATH—a hierarchic classification of protein domain structures. *Structure*, **5** (1997), 1093–1108.

[PC53] L. Pauling and R. B. Corey. A proposed structure for the nucleic acids. *Proceedings of the National Academy of Sciences of the United States of America*, **39** (1953), 84–97.

[PO03] O. Poirot, E. O'Toole, and C. Notredame. Tcoffee@igs: a web server for computing, evaluating and combining multiple sequence alignments. *Nucleic Acids Research*, **31** (2003), 3503–3506.

[PR04] G. A. Petsko and D. Ringe. *Protein Structure and Function*. New Science Press, London, UK, 2004.

[PS97] A. C. Papageorgiou, R. Shapiro, and K. Ravi Acharya. Molecular recognition of human angiogenin by placental ribonuclease inhibitor—an x-ray crystallographic study at 2.0 Å resolution. *The EMBO Journal*, **16** (1997), 5162–5177.

[SG99] A. W. Sowa, P. A. Guy, S. Sowa, and R. D. Hill. Nonsymbiotic haemoglobins in plants. *Acta Biochimica Polonica*, **46** (1999), 431–445.

[Sk05] J. Skolnick. Putting the pathway back into protein folding. *Proceedings of the National Academy of Sciences of the United States of America*, **102** (2005), 2265–2266.

[TR01] C. Tanford and J. Reynolds. *Nature's Robots: A History of Proteins*. Oxford University Press, UK, 2001.

CHAPTER 2

Accessing and Displaying Molecular Data with Chimera

2.1 MOTIVATION

Python scripting with Chimera allows the user to design, develop, and execute a variety of algorithms that can work with macromolecular data (e.g., the atom coordinates of these molecules), perform any needed calculations, and subsequently print results to the IDLE Shell window or to a file. A script can also change the molecular scene being displayed.

As described earlier, Chimera provides the user with a wide range of capabilities that can be invoked using menus and command lines. In many situations, these high-level capabilities are sufficient for many of the standard tasks that need to be done by the structural biologist. So, why would we need to work with the lower level computations provided by Python scripting? This is an important question because Python scripting requires a time consuming commitment to develop skills in object-oriented programming.

The following points give three reasons for such an investment of your time:

1. Specialized applications

   In many situations you may need to develop your own specialized applications that are not achievable by using current menus and command lines. For example, you might want to develop a different approach for the visualization of some molecular feature or molecular functionality. In general, the ability to do scripting will reduce your dependency on canned programs and thus avoid the constraints they may impose.

2. Standard procedures might need some customized modifications

   It might be the case that Chimera does provide the basic capability, but you simply need to extend this functionality to meet some extra project requirements. For example, Chimera does provide structure comparison tools but structure comparison is very complicated and there are several different criteria for measuring structural similarity. Your application goals might require a different approach to the problem.

3. Calculations that are repeated across several instances

   Various applications involve data collection to gather statistical information about macromolecules such as RNA or proteins. This might involve the processing of thousands of instances of side chains or even the processing of thousands of files. For these situations, simple menu invocations will not suffice and you will need a script that can automatically sift through the extensive data.

2.2 PYTHON SKILLS

This book will assume an understanding of object-oriented programming concepts. Although it is not necessary to be an expert *Pythonista*, you should be equipped with a basic level of Python programming skills. If you are new to Python, there are several freely available tutorials on the Internet. You can also purchase introductory books on Python, some with an emphasis on bioinformatics. A list of these sources is given in the Reference section. Fortunately, Python is very user friendly and is typically regarded as an easy language to learn. This is especially true if you already have some experience with an object-oriented language such as C++ or Java. Suggested references ([BA10], [BE09], [HM00], and [KI09]) and website links can be found at the end of the chapter. You should have a working knowledge of the following concepts and topics before proceeding:

Numbers: integer and floating point arithmetic
Strings
Collectives: lists, tuples, dictionaries, and sets

Conditionals: `if` statements, `else` statements

The `range()` function

Looping techniques: `for` statements and `while` statements

Other control statements: `break`, `continue`

Defining functions: including default arguments and keyword arguments

Input and Output: including formatting and reading/writing of files

Classes: understanding instances, methods, attributes, and inheritance

A novice Python programmer may find it difficult to follow the meaning and intention of a script and so I have tried to provide extra comments as the need arises. Application functionality has been expressed through function definitions and class definitions when object-oriented encapsulation was considered to be achievable. To make the scripts more comprehensible, variable names have often been given suffixes as follows:

`_A`	Array
`_B`	Boolean
`_D`	Dictionary
`_L`	List
`_M`	Matrix
`_P`	Point
`_S`	Set
`_T`	Tuple

So a variable such as `hydrophobes_L` is a list of objects and we would expect it to be used with methods such as `append`, `count`, or `extend`.

2.3 PYTHON SCRIPTING

Python is part of the Chimera download. To get a Python Shell you can invoke the menu item **Tools... General Controls... IDLE**. You might find it extremely helpful to set up an IDLE button on your Toolbar. This will allow you to get the Python Shell with a simple button click, thus avoiding the menu descent. To set up this, invoke the menu item **Favorites... Add to Favorites/Toolbar...**. You will be presented with a dialog box that can be used to put an IDLE button on your Toolbar.

Considering the material that will be covered in the rest of this book, it is advisable to set up three more buttons in the Toolbar: the Model Panel, Side View, and Sequence buttons. After you complete these preferences, the top-left corner of your Chimera window should look something like that in Figure 2.1.

The Python Shell has multiple uses as follows:

- You can test syntax or execution of short Python scripts.
- You can use the Shell window to get output results from a script that is in execution.
- You can open an editing window that may be used to generate a new Python script or to modify a recently produced script.
- You can also provide input to a running script.

Let us discuss each of these in turn. You can consider the Shell window to be a sophisticated calculator. Type in an expression or small code sequence and it can be immediately executed. The triple ">>>" is a command prompt that awaits input of Python statements. For example, consider the following statements:

```
>>> x = 3.0
>>> x**3 + 5.
32.0
>>>
```

The variable x has been set to 3.0 and the next statement computes the value of an expression that uses x. Note that the result is given on a line without the prompt.

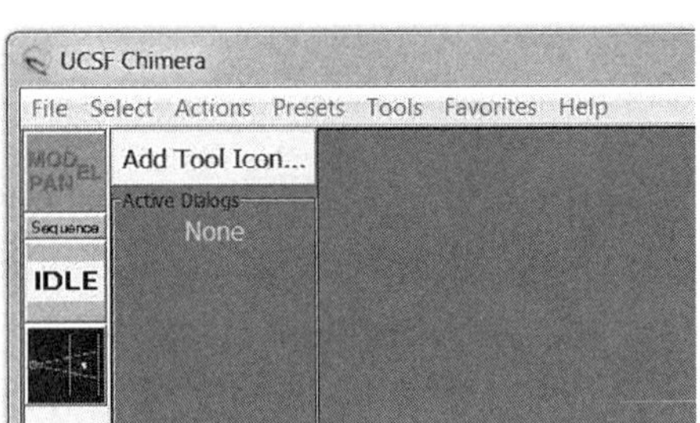

FIGURE 2.1 Recommended toolbar buttons for Chimera.

Let us try something with more relevance to structural bioinformatics. To get a protein from the Protein Data Bank (PDB), we can execute:

```
ourProtein = chimera.openModels.open('1AY7', type = 'PDB')[0]
```

If you execute this script line, Chimera will fetch the 1AY7 PDB file from the RCSB and it will be displayed in the Chimera application window. We will deal with the details of this statement in Section 2.4. Incidentally, it is possible to delete the model holding the protein data and, as a result, clear the contents of this window by invoking the menu item **File... Close Session** in the Chimera window.

Suppose we want to set up this code snippet as a script in a file. We would go through the following steps:

- Use the **File... New Window...** menu item to get a new edit window.
- Type in lines:

  ```
  import chimera
  ourProtein = chimera.openModels.open('1AY7', type = 'PDB')[0]
  ```

- Use the **File... Save As...** menu item to get a Save As dialog. Give the file a relevant name with a ".py" suffix, for example, myTestProg.py. The file can be saved in some directory, which is appropriate for these initial test cases. Now, the edit window can be closed with the assurance that the script has been stored in a.py file.
- Later, after launching Chimera and getting a Python Shell window, we can invoke the menu item **File... Open...** to open an edit window containing the contents of this file. By invoking **Run... Run Module** we can execute the script and it will bring in the 1AX7 protein.

So, that was our first Python script. It was a very short program, but it did have the power to fetch data from the PDB and display it on the screen. If we wanted to fetch a PDB file from a local directory, we would simply replace the PDB ID string with a string that contained the pathname of the target PDB file.

We can add a statement that asks for user input. For example, consider the following script:

```
import chimera
pdbID = raw_input("Type in a PDB ID \n")
ourProtein = chimera.openModels.open(pdbID, type = 'PDB')[0]
for i in range(len(ourProtein.residues))
    print ourProtein.residues[i].type
```

2.3.1 Script 2.1: Using Raw Input to Get a PDB ID

It will ask the user for a PDB ID. When this is typed into the Shell window, the script will fetch the protein file and will iterate through all the residues printing out the type of each residue. For example, if we typed in 1AY7, we would get a list of residue types: ASP, VAL, SER, and so on.

Important notes:

- Be sure to realize that you will often be dealing simultaneously with two windows: the Shell window and the Chimera display window. Your next appropriate interaction will depend on which window is currently active. For example, if the Shell window needs more input from you, but the display window is active because you have done some scene manipulation, such as a rotate or zoom, then you will have to use a mouse click to make the Shell window active before you can type in the needed input.
- When running one script after another, be sure to get the display back to its initial state, thus clearing it for the next script. You will need to do a **File, Close Session** menu invocation before running the next script.

Section 2.4 will provide some details that explain the significance of the Chimera functions used in Script 2.1. After that we can get back to other functions that can be used to extract data from a PDB file.

2.4 CHIMERA OBJECT HIERARCHY

Python is an object-oriented programming language that supports the use of data structures and their associated methods. By starting our scripts with the statement

```
import chimera
```

we immediately gain access to all the objects that are available within Chimera. For example, the Chimera object openModels will track the molecular models that have been extracted from a file that has been opened. For example,

```
model_L = chimera.openModels.open('1AY7', type = 'PDB')
```

will set up a list of models that we have designated as `model_L`. The list contains only one model because that is all there in this PDB file. This can be verified by looking at the Model Panel (brought to your screen by the menu invocation **Tools... General Controls... Model Panel** or by simply clicking on the Model Panel button if you have that set up on your Toolbar). We can get the single protein model by using an index of 0, as in

```
ourProtein = model_L[0]
```

that accomplishes the same thing as our earlier statement defining `ourProtein`.

It is possible to have a PDB file with several models. This is often the case with PDB files that give us the results of nuclear magnetic resonance (NMR) experiments. For example, fetching the file 1ZNF will give you 37 models. Look at the Model Panel to verify this. Any one of these models can be accessed by using an index in the range of integers running from 0 to 36 inclusive.

Chimera takes full advantage of Python's object-oriented provisions by creating an object hierarchy that reflects the hierarchy of constituents in chemical structures. For example, a protein is a molecule that would have a list of residues, atoms, or bonds. This hierarchy is illustrated in Figure 2.2.

Each of the objects just mentioned (`chimera.Molecule`, `chimera.Residue`, `chimera.Atom`, `chimera.Bond`) has a set of methods. An important aspect of object-oriented programming is that computations dealing with an object are constrained and must use one or more of the methods that are *encapsulated* with the object. In Figure 2.2, each box contains a set of attributes for the object named in the header of the box. For the sake of simplicity, many of the attributes have been left out. To view a full list for a molecule object, type the following line in the Python Shell:

```
>>> ourProtein = chimera.openModels.open('1AY7', type = 'PDB')[0]
```

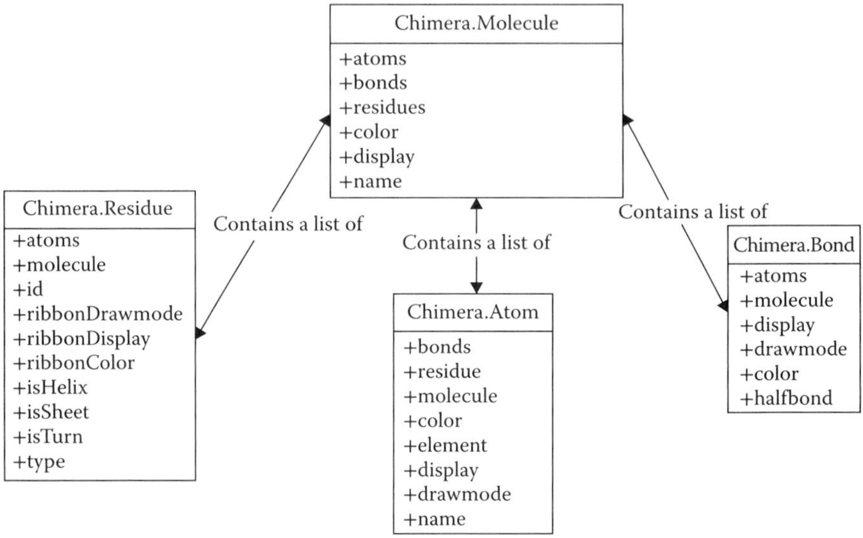

FIGURE 2.2 Chimera's object model. (From http://www.cgl.ucsf.edu/chimera/docs/ProgrammersGuide/Examples/Main_ObjectModel.html.)

In response to the next prompt that appears after the file has been fetched, type in "`ourProtein`" immediately followed by a period. You should get a list box containing all the attributes for that molecule, as in Figure 2.3. Do not use the mouse with the scroll bar; the list will simply disappear. Instead, use your up/down arrow keys to go through the list. You will see that the list of attributes is far longer than those given in Figure 2.2. Using the arrow keys will move the highlighting band to the attribute that you want to use. It will be selected as an autocompletion append to your statement if you simply resume typing.

Attributes include methods for the object and various data descriptors. For example, `ourProtein.findAtom(33)` is a method that would get the atom corresponding to index 33 in the list of atoms that can be accessed as an object by typing: `ourProtein.atoms`. So, we could get the name of atom 33 by indexing a list of atoms or by simply using the findAtom method:

```
>>> ourProtein.atoms[33].name
u'CA'
>>> ourProtein.findAtom(33).name
u'CA'
>>>
```

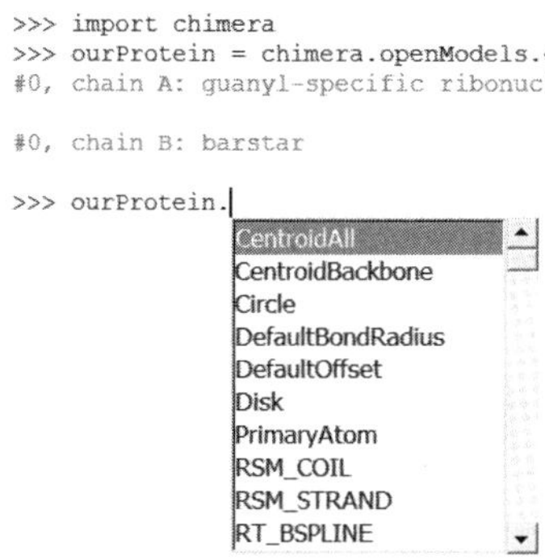

FIGURE 2.3 Autocompletion list for a molecule.

How many atoms are in the 1AY7 molecule?

```
>>> len(ourProtein.atoms)
1678
```

You can get a full list of the attributes and more details about these attributes by using the `help` function in the Python Shell. For example,

```
Help(ourProtein)
```

prints text in the Shell window describing all the attributes that are associated with the Molecule `ourProtein`.

Now that we have the basic techniques for dealing with Python scripts in the Chimera environment, we can review some of the attributes that will prove to be useful for building larger programs.

In Sections 2.5, we will assume that the following statement has been executed:

```
qorProt = chimera.openModels.open('1qor', type = 'PDB')[0]
```

This protein with PDB ID 1QOR is a quinone oxidoreductase from *Escherichia coli*. It will be a good running example for the next sections because it has two chains, a good representation of secondary structures and two ligands. Although somewhat complicated, it has a rather elegant structure involving an extensive beta sheet and other *barrel like* structures.*

* Let us pause to appreciate the intricate molecular beauty of a protein that does its work within the fecal coliform bacteria living in our intestines.

2.5 ATTRIBUTES FOR MOLECULE OBJECTS

As described at the end of Section 2.4, we are assuming that `qorProt` is a molecule object. In this case, the molecule is a protein so we can also reference its residues as well as atoms and bonds. We now consider some attributes that will prove useful later in the text.

Each of the following statements will generate a list of objects:

```
qorProt.atoms
qorProt.bonds
qorProt.residues
```

Their meaning should be obvious. We can access any object in a list by specifying an index. For example, we can access the atom object at index 44 by using `qorProt.atoms[44]`.

Each of the following is a function that can be used to get an object specified by an index:

```
qorProt.findAtom()
qorProt.findBond()
qorProt.findResidue()
```

For example, we can access the residue object at index 21 by using the construct `qorProt.findResidue(21)`. The following represents two more approaches for referencing residues:

```
qorProt.residueAfter()
qorProt.residueBefore()
```

In these cases, the argument for the function is a residue object. For example, we could find the first residue in the protein by defining:

```
resObj = qorProt.findResidue(0)
```

Then inside a while loop we can progress through the sequence of residues by using a replacement statement such as:

```
resObj = qorProt.residueAfter(resObj)
```

For example, the following script will print the types of all the residues in 1QOR:

```
resObj = qorProt.findResidue(0)
while resObj:
    print resObj.type
    resObj = qorProt.residueAfter(resObj)
```

If you execute this code in the Python Shell, you will notice that final "residues" in the list are all "HOH." If the programmer wants a residue list that is free of water, then the code should be modified to avoid the "HOH" entries.

To get the name of the molecule, we can use:

```
qorProt.name
```

In this case, the Python Shell would return the character string "1qor."

Finally, to delete the model we can use:

```
chimera.openModels.close([qorProt])
```

This removes the molecule object and clears it from the display.

Chimera does not explicitly support a Chain class. However, it does have a Sequence class that essentially covers the same functionality.

2.6 ATTRIBUTES FOR SEQUENCE OBJECTS

We can extract a list of sequences from a protein molecule by working with the `sequences()` function. Consider the following Python Shell interaction after defining `qorProt`:

```
>>> sequences_L = qorProt.sequences()
>>> len(sequences_L)
2
```

This is expected because 1QOR has two chains. This can be verified by checking the menu item **Select... Chain**, which will drop a submenu containing the two chain identifiers "A" and "B."

2.6.1 Referencing a Sequence Object

Suppose we continue by extracting the first member of this list, calling it `seqA`:

```
>>> seqA = sequences_L[0]
>>> seqA
<chimera.Sequence.StructureSequence object at 0x000000000A4E14A8>
```

If you happen to know the chain identifiers for the chains (not always "A," "B," ...) you can get the sequence by using the `sequence()` function:

```
seqA = qorProt.sequence('A')
```

To get a list of chain identifiers, you can demand that the `sequences()` function provides its results as a dictionary and then you may ask for the keys to the dictionary:

```
chainId_L = qorProt.sequences(asDict = True).keys()
```

2.6.2 Identification of a Sequence

If the sequence was extracted from the list of sequences, then you might not have the chain identifier. This is available with the `chain` attribute:

```
>>> seqA.chain
u'A'
```

There are various other attributes that will provide more information describing the sequence:

```
>>> seqA.fullName()
u'1qor (#0) chain A'

>>> seqA.molecule.name
u'1qor'

>>> seqA.name
u'chain A'
```

2.6.3 Sequence Entries

To get the single letter residue types that are in the sequence, we can work with the `sequence` attribute:

```
>>> seqA.sequence
['M', 'A', 'T', 'R', ...., 'L', 'I', 'P']
```

To conserve space on this page, most of the sequence entries have been replaced with "...."

2.6.4 Getting Residue Objects from a Sequence

Finally, we can get a list of the residue objects within the chain by using:

```
seqA.residues
```

You have to exercise a bit of care here. The length of this list is the same as the length of the sequence but some of the residue entries may be Python "NoneType" objects. Consider the following code snippet that accesses the protein with ID = "2FIF," builds a dictionary of sequences and then attempts to put all the alpha carbon atoms of chain "A" into an atom list:

```
alphaCarbs_L = []
prot = chimera.openModels.open("2FIF", type="PDB")[0]
chains_D = prot.sequences(asDict = True)
for r in chains_D['A'].residues:
    alphaCarbs_L.append(r.findAtom('CA'))
```

The snippet produces the following error display in the Python Shell window:

```
Traceback (most recent call last):
  File "<pyshell#30>", line 2, in <module>
    alphaCarbs_L.append(r.findAtom('CA'))
AttributeError: 'NoneType' object has no attribute 'findAtom'
```

Run the same code again with a slight change in the last line:

```
  if r: alphaCarbs_L.append(r.findAtom('CA'))
```

Now the code works. If you use the menu invocation: **Favorites... Sequence** and then show the sequence for chain "A," you can see why this happens. The sequence ends with the residues ...RLRGG. However the last three entries (RGG) do not have structure information. The PDB file says that the protein has these amino acids in the sequence, but the x-ray analysis of the protein does not give coordinate information for the final three residues in the chain. The change in the code snippet uses an `if` statement to check whether the residue object actually exists before attempting to access any of its structural attributes.

2.6.5 Residue Maps for Sequences

The attribute `resmap` is a dictionary that maps residue objects to a zero-based sequence position. Consider the following Shell interactions:

```
>>> resAtIx2 = qorProt.findResidue(2)
>>> resAtIx2.type
u'ARG'
>>> seqA.resMap[resAtIx2]
3
>>> seqA.sequence[3]
'R'
```

You might see this as a curious inconsistency because there are two different indices associated with the same object. The PDB file for 1QOR states that the residue sequence starts with MATRIEF... (see Figure 2.4b). However, residues corresponding to the 3D structure (complete with all atomic coordinates) start with ATRIEF... (see Figure 2.4a). The `findResidue(2)` expression is finding the residue with index 2 in the 3D structure as displayed. This is zero-based indexing, so ALA is at index 0, THR is at index 1, and ARG is at index 2.

So, `resAtIx2`, the residue object at index 2 is ARG. The `resMap` attribute is reporting that this residue object is at index 3 *in the sequence*. This is reaffirmed by the last expression that assures us that `seqA.sequence[3]` is indeed "R" (the single letter specification for ARG). This is also zero-based indexing. Of course, zero-based indexing is what programmers expect. A biologist looking at the sequence will want to see the numbering start at 1. This is what we see in Figure 2.4b. In this situation the ARG residue has position 4. This can be determined in a script by using the `id.position` attribute:

```
>>> resAtIx2.id.position
4
```

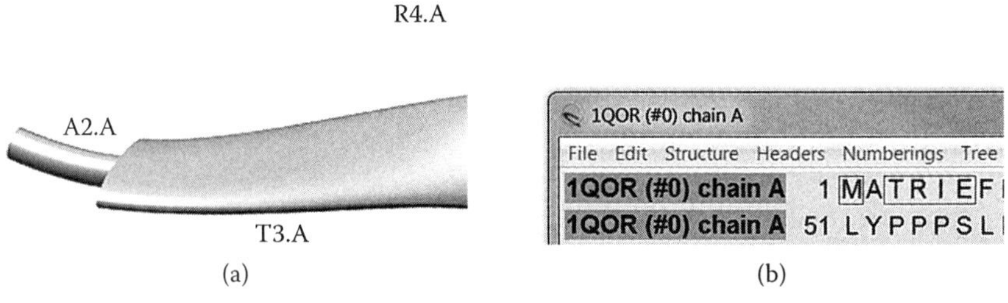

FIGURE 2.4 (a) Labels on the first three. (b) Start of sequence for 1QOR.

So, we have the cautionary note: Be careful! Depending on the context, a residue can be located using any one of three different numbers. In our running example with 1QOR, the ARG residue is located using 2, 3, or 4.

These issues have implications for the execution of code that is working with the `residues` attribute of a `sequence` object. You might not get a residue object if the index corresponds to a residue that is missing from the 3D structure. This is illustrated in the following Shell interactions:

```
>>> seqA.residues[1].type
u'ALA'
>>> seqA.residues[0].type
Traceback (most recent call last):
  File "<pyshell#5>", line 1, in <module>
    seqA.residues[0].type
AttributeError: 'NoneType' object has no attribute 'type'
>>>
```

In this last example, we have only one residue missing from the structural display, namely, the methionine at the beginning of chain A. It is also possible to have missing structural information within a chain. If you inspect the sequence for 3ZY7, you will see that the residues NTNP have been outlined in a red rectangle to indicate that they are missing from the 3D representation of the structure. In the display, the missing backbone is represented by a straight dashed line. To see a more severe example of missing residues, look at Exercise 2.

2.7 ATTRIBUTES FOR RESIDUE OBJECTS

2.7.1 Referencing a Residue Object

In this section we assume that `resObj` is a residue object that has been taken from our protein molecule. For example,

```
resObj = qorProt.findResidue(555)
```

As described earlier, it is also possible to generate a residue object by using molecule methods such as `residueAfter()` and `residueBefore()`. We can also extract a residue object by affixing an index to a list of residue objects generated by using the attribute `residues` on a protein molecule object.

2.7.2 Identification and Characterization of Residues

The following Python Shell interactions show the results for various useful attributes of a residue object:

```
>>> resObj.type
u'ARG'
>>> resObj.isHelix
False
>>> resObj.isSheet
True
>>> resObj.isStrand
True
```

The previous three statements produce Boolean values that specify the type of secondary structure containing this residue. Note that `isSheet` and `isStrand` produce the same results.

If a residue is in a coil secondary structure (e.g., the residue with index 111) then all three of these Boolean results will be `False`.

```
>>> resObj.isHet
False
```

The attribute `isHet` will be `False` if the residue is a *standard* residue (one of the basic 20 amino acids). The `isHet` attribute is useful for finding ligands in the protein. For example, the following code:

```
for res in qorProt.residues:
    if res.isHet and not res.type == 'HOH':
        print res.type
```

will print out the names of the 4 molecules that accompany the protein. This includes the two NAP (nicotinamide adenine dinucleotide phosphate) molecules that are ligands in the binding pockets of 1QOR.

```
SO4
NAP
SO4
NAP
```

If the exclusion of "HOH" was removed from the `if` statement, we would get the previous 4 lines, followed by 432 lines, each containing "HOH."

2.7.3 Location of a Residue

The `id` attribute has two useful subattributes:

```
>>> resObj.id.chainId
u'B'
>>> resObj.id.position
231
```

Note: `resObj.id.position` has the value 231, which is the position of the residue within chain "B." We would expect this to differ from 555, which is the index in the entire set of residues. We can also reference the molecule object that contains this residue:

```
>>> resObj.molecule
<_chimera.Molecule object at 0x00000000047C21C0>
```

Using this we can determine the name of the molecule containing this residue:

```
>>> resObj.molecule.name
u'1qor'
```

The `ssId` attribute provides an integer ID for secondary structure in a succession of secondary structures. A change in `ssId` can help to distinguish two helices that are distinct secondary structures but are adjacent in the sequence (there is no intervening residue with a different secondary structure characterization). For example, the snippet:

```
myoProt = chimera.openModels.open('1JEB', type = 'PDB')[0]
for r in myoProt.residues:
    print r.type, r.id.position, r.id.chainId, r.ssId, r.isHelix
```

produces a lengthy list that specifies, for each residue, its type, position, chain ID, ssId, and a Boolean that is `True` if the residue is in a helix. This includes the six lines:

```
GLY 16 B 8 True
LYS 17 B 8 True
VAL 18 B 8 True
ASN 19 B 9 True
ALA 20 B 9 True
ASP 21 B 9 True
```

There is a change in `ssID` in going from `VAL 18 B` to `ASN 19 B`. An inspection of 1JEB (see Figure 2.5) shows why this happens. There is sudden bend in the structure consisting of two consecutive helices meeting at an angle that is approximately 90 degrees and so these two secondary structures are considered to be different helices.

2.7.4 Deriving Substructure

The atoms attribute provides a list of the atoms in the residue:

```
resObj.atoms
```

For example, with the current definition of `resObj`, the following snippet:

```
outStr = ""
for atm in resObj.atoms:
    outStr += " " + atm.name
print "Residue ", resObj.type
print "has the following atoms: " , outStr
```

produces:

```
Residue ARG
has the following atoms: C CA CB CD CG CZ N NE NH1 NH2 O
```

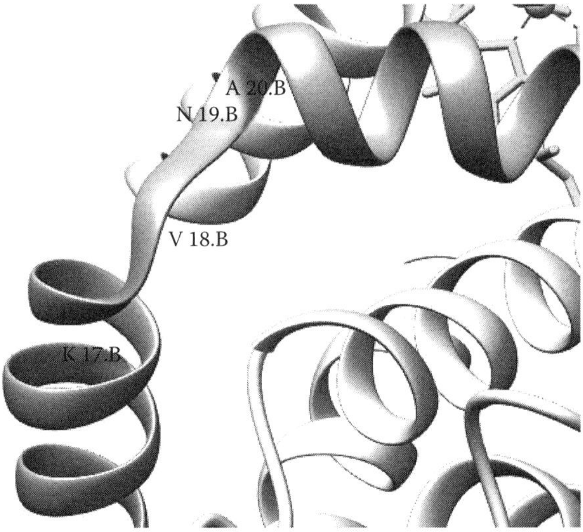

FIGURE 2.5 Change in secondary structure ID within 1JEB.

If you know the name of the atom, then you can get the atom object by using the `atomsMap` attribute:

```
resObj.atomsMap
```

This is a dictionary that produces a list of atoms. Here is a Shell script example:

```
>>> resObj.atomsMap['N'][0].name
u'N'
```

An alternative strategy is to use the `findAtom()` method.

```
>>> resObj.findAtom('N')
<_chimera.Atom object at 0x000000000497C6C0>
```

To get the number of atoms in the residue we can use:

```
>>> resObj.numAtoms()
11
```

2.7.5 Other Useful Attributes

To get the Kyte-Doolittle hydrophobicity value, try:

```
>>> resObj.kdHydrophobicity
-4.5
```

To list the names of the atoms in the residue use the `atomNames()` function. It will return a set of strings:

```
>>> resObj.atomNames()
set([u'C', u'CB', u'CA', u'CG', u'NE', u'O', u'CD', u'CZ', u'NH1',
     u'NH2', u'N'])
```

To determine whether two residues are bonded, you can use the `bondedTo()` function. It returns a Boolean value. In general, you would expect the following expression to be `True` (unless the residue after `resObj` is in the next chain).

```
resObj.bondedTo(qorProt.residueAfter(resObj))
```

It should be noted that `bondedTo` will return `True` if the two residues are cysteines linked by a disulfide bridge.

Chi angles and phi, psi angles:

```
>>> resObj.chi1
173.89880802506943
>>> resObj.chi2
155.88664384279298
>>> resObj.chi3
43.69074443200619
>>> resObj.chi4
78.94742632085887
>>> resObj.phi
-58.981261074117576
>>> resObj.psi
134.32458920108093
```

2.8 ATTRIBUTES FOR ATOM OBJECTS

2.8.1 Referencing Atom Objects

We can get an atom object by indexing the atom list for the molecule. For example,

```
>>> qorProt.atoms[4128]
<_chimera.Atom object at 0x000000000497C6C0>
```

This atom has the name:

```
>>> qorProt.atoms[4128].name
u'N'
```

In the remainder of this section we assume that `atomObj` is an atom object that has been taken from our 1QOR protein. We use the following example of an atom object taken from the `resObj` defined as in Section 2.7.1:

```
atomObj = resObj.findAtom('N')
```

2.8.2 Identification of Atoms and Atom Properties

As seen earlier, we can get the name of an atom by using the `name` attribute:

```
>>> atomObj.name
u'N'
```

The `idatmType` is an attribute that is adapted from the program IDATM. This is described in a research paper of Meng and Lewis.*

For example, the current `atomObj` has `idatmType` given by

```
>>> atomObj.idatmType
u'Npl'
```

In this case, `Npl` corresponds to the following description:

sp^2-hybridized nitrogen bonded to three other atoms, formally neutral (trigonal planar; amide, aniline)

A complete list of IDATM atom types can be found at http://www.rbvi.ucsf.edu/chimera/current/docs/UsersGuide/idatm.html

To get the radius of the atom, use the `radius` attribute:

```
>>> atomObj.radius
1.6399999856948853
```

To get the B-factor of the atom use the `bfactor` attribute. As stated in the pdbwiki, the B-factor is a mix of real thermal displacement, static disorder (multiple but defined conformations), and dynamic disorder.

```
>>> atomObj.bfactor
22.91
```

Finally, we note that the attribute `serialNumber` gives the serial number for the atom. This is the entry in the second column of the ATOM record for this atom in the PDB file for the protein. For our example, atom:

```
>>> atomObj.serialNumber
4130
```

2.8.3 Working with Atom Coordinates

In our algorithms dealing with protein structure, it will be necessary to get the atomic coordinates of any atom in the protein. This is done using the `coord()` method. For example,

```
>>> atomObj.coord()
chimera.Point(38.57699, 48.56300, 8.41300)
```

---

* E. C. Meng and R. A. Lewis. Determination of molecular topology and atomic hybridization states from heavy atom coordinates. *Journal of Computational Chemistry*, **12**(7), Sept. (1991), 891–898.

Note that the result of this operation is a Chimera Point object. If need be, you can easily convert a point object to a tuple or a list:

```
>>> tuple(atomObj.coord())
(38.577, 48.563, 8.413)
>>> list(atomObj.coord())
[38.577, 48.563, 8.413]
```

To extract a particular ordinate from the point, you can use an indexing operator:

```
>>> atomObj.coord()[0]
38.577
```

We can also set the coordinates of an atom by using the `setCoord()` method as shown in the next code sample. The argument must be a Chimera Point object. To build a point object you will have to import the Point class:

```
from chimera import Point
atomObj.setCoord(Point(33.,55.,16))
```

Caution: The display software in Chimera does not check for the reasonableness of atomic positions. It simply does its best to draw the scene. In Figure 2.6, we see that an arbitrary setting of the atomic coordinates for our `atomObj` can produce a molecule with unrealistic bond lengths.

2.8.4 Atom Attributes Related to Chemical Structure

To determine the residue or molecule to which an atom belongs, use the `residue` and `molecule` attributes. In the following examples, these attributes have been followed by `type` and `name` attributes, respectively, to show how they work:

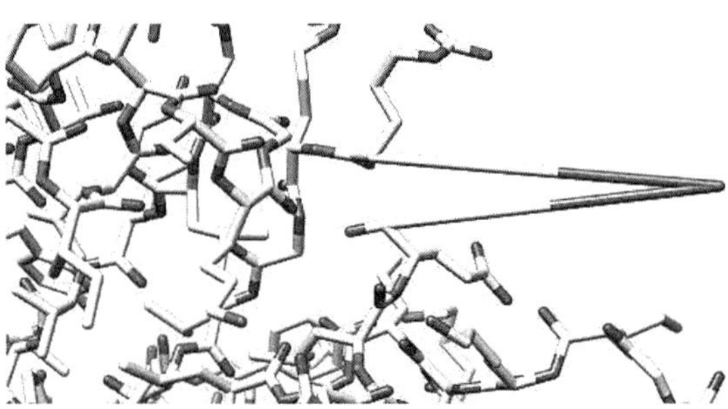

FIGURE 2.6 Arbitrary movement of an atom can stretch bonds!

```
>>> atomObj.residue.type
u'ARG'
>>> atomObj.molecule.name
u'1qor'
```

2.8.5 Atoms and Bonds

To get the neighboring atoms bonded to the current atom, we can use the neighbors attribute:

```
>>> atomObj.neighbors
[<_chimera.Atom object at 0x000000000497C7D8>,
<_chimera.Atom object at 0x000000000497C670>]
```

or use the primaryNeighbors() method:

```
>>> atomObj.primaryNeighbors()
[<_chimera. Atom object at 0x000000000497C7D8>,
<_chimera.Atom object at 0x000000000497C670>]
```

We can index these lists to pull out a desired neighbor:

```
>>> atomObj.neighbors[1].name
u'CA'
>>> atomObj.neighbors[0].name
u'C'
```

The bonds attribute will provide a list of bonds and by using an index we can extract a particular bond, for example,

```
>>> atomObj.bonds[1]
<_chimera.Bond object at 0x000000000CF9E490>
```

Or we can simply use the `primaryBonds()` method to get a particular bond:

```
>>> atomObj.primaryBonds()[1]
<_chimera.Bond object at 0x000000000CF9E490>
```

See Section 2.9.2 for a discussion of attributes for bonds.

Suppose we define `caNbr` to be the alpha carbon neighbor of our "N" `atomObj`:

```
caNbr = atomObj.neighbors[1]
```

There are three strategies for referencing the bond object connecting `atomObj` to this neighboring atom:

```
>>> atomObj.connectsTo(caNbr)
<_chimera.Bond object at 0x000000000CF9E490>
>>> atomObj.findBond(caNbr)
<_chimera.Bond object at 0x000000000CF9E490>
>>> atomObj.bondsMap[caNbr]
<_chimera.Bond object at 0x000000000CF9E490>
```

The methods `connectsTo()` and `findBond()` take the atom neighbor as an argument and return the bond object that connects them. Note that `bondsMap` does essentially the same functionality but it is a dictionary.

2.8.6 Atom Attributes Related to the Display

The following atom attributes will be useful for getting information from the display:

```
>>> atomObj.color.rgba()
(0.18823, 0.31372, 0.97254, 1.0)
```

This four component tuple gives the RGB and transparency properties of the color assigned to the atom.

Finally, we can determine whether an atom is currently being displayed:

```
>>> atomObj.display
False
```

In this example, we got a `False` result because the structure was displayed entirely with ribbons and no individual atoms were visible.

2.9 ATTRIBUTES FOR BOND OBJECTS

2.9.1 Referencing Bond Objects

Chimera maintains a bond list for the entire molecule. So, we could reference these bond objects by indexing the `bonds` attribute:

```
>>> qorProt.bonds[0]
<_chimera.Bond object at 0x0000000009F8E3A0>
```

Most algorithms will get bond objects that are associated with specific atoms. The following Shell interaction shows how the `bonds` attribute for an atom object can be used to reference a list of bonds associated with `atomObj`.

```
>>> atomObj.bonds[1]
<_chimera.Bond object at 0x000000000CF9E490>
```

In the next examples, `atomObj` will be the same as that used in earlier sections:

```
atomObj = qorProt.findResidue(555).findAtom('N')
```

The running example for the bond will be that between `atomObj` and the alpha carbon neighbor with index 1:

```
>>> atomObj.neighbors[1].name
u'CA'
```

The bond was referenced using:

```
>>> alphaC_nbr = atomObj.neighbors[1]
>>> bondToCA = atomObj.bondsMap[alphaC_nbr]
```

2.9.2 Bond Attributes

We can get the molecule object that contains this bond, illustrated here with a snippet that also gets the name of the molecule:

```
>>> bondToCA.molecule.name
u'1qor'
```

The `atoms` attribute gives a tuple containing the two atoms at either end of this bond.

```
>>> bondToCA.atoms[0].name
u'CA'
>>> bondToCA.atoms[1].name
u'N'
```

The method `contains(an_atom)` returns `True` if and only if the `an_atom` object is at either end of the bond. For example,

```
>>> bondToCA.contains(atomObj)
True
```

Similar in use to the atoms attribute, the method findAtom(atomIx) returns the atom object indexed by atomIx. Because only two atoms specify a bond, atomIx must be 0 or 1:

```
>>> bondToCA.findAtom(0).name
u'CA'
>>> bondToCA.findAtom (1).name
u'N'
>>> bondToCA.findAtom (2).name
Traceback (most recent call last):
    File "<pyshell#161>", line 1, in <module>
      bondToCA.findAtom (2).name
IndexError: index out of range
```

The bond method otherAtom(an_atom) returns the atom that is at the other end on the bond containing the atom designated by an_atom.

```
>>> bondToCA.otherAtom(atomObj).name
u'CA'
>>> bondToCA.otherAtom (alphaC_nbr).name
u'N'
```

To get the length of a bond, use the length method:

```
>>> bondToCA.length()
1.4503137591569617
```

The method sqlength() returns the square of this distance.

```
>>> bondToCA.sqlength()
2.1034099999999976
```

2.10 WORKING WITH A BATCH OF FILES

Later chapters will deal with calculations involving molecular structure (atomic coordinates, bond angles, and so on), but for now, we will concentrate on scripts that work with attributes of atoms, for example, the type of residue containing the atom and the position of that residue within the protein chain.

Before covering this material, we will go through some elementary scripts that can be incorporated into various larger scripts to be covered in subsequent chapters.

2.10.1 Dealing with Files

It is common for a programmer to write a script that deals with several protein files. To avoid the delays that occur when a script must access these files from the PDB, it is convenient to have the data files stored on your own disk. By accessing all the files and putting them on your drive, these delays are encountered once and not during the repeated execution of your script (as might happen when you are debugging your script).

Script 2.2 will read a simple text file containing the PDB ID's of several proteins. Then it will use Chimera to fetch the files and then store them into the directory containing the script. The text input has a simple format, for example,

Myoglobins

1a6m Sperm whale

1mba Slug sea hare

1mbs Common seal

1mwc Pig

1ymb Horse

2mm1 Human

1myt Yellow fin tuna

The first line is to explain the significance of the list and will be ignored by the script. In each succeeding line, only the first four characters will be read.

The following code will do the job:

```
import chimera, urllib2, tkFileDialog
# Launch dialog to get path of PDB ID list:
filePathName = tkFileDialog.askopenfilename()
print "Path name for PDB file list is: ", filePathName

# Read file containing list of PDB identifiers:
listFileHandle = file(filePathName)
listFileHandle.readline() # Skip first line
pdbID = listFileHandle.readline()
while pdbID:
    print pdbID[0:4]
    request = urllib2.Request('http://www.rcsb.org/pdb/files/%s.pdb'
                              % pdbID[0:4])
    try:
        fileHandle = urllib2.urlopen(request)
    except urllib2.URLError, e:
        print 'Failure to fetch file'
        if hasattr(e, 'reason'):
            print 'Reason: ', e.reason
```

```
        elif hasattr(e, 'code'):
            print 'Error code: ', e.code
    else:
        pdbFileContents = fileHandle.read()
        fileHandle.close()
        outFile = file(pdbID[0:4] + '.pdb', 'w')
        outFile.write(pdbFileContents)
        outFile.close()

    pdbID = listFileHandle.readline()
    # Go for next pdbID in the list

listFileHandle.close() # Close the list file.
```

Script 2.2: Getting a Batch of Files from the Protein Data Bank

Note the use of the `tkFileDialog.askopenfilename()` method to launch a dialog window that will allow the user to specify the pathname of the file containing the PDB ID list. After that, the code uses the basic file-handling statements that should be understandable to most programmers with an introductory knowledge of Python. A more esoteric statement involves the `urllib2.Request()` method from the `urllib2` module. Its use requires that we know the exact URL and its appropriate format to access a PDB file.

A variation of the above script uses Chimera to fetch the file and then write it out as a PDB file:

```
import chimera
from chimera import runCommand
import tkFileDialog

# Launch dialog to get path of PDB ID list:
filePathName = tkFileDialog.askopenfilename()
print "Path name for PDB file list is: ", filePathName

# Read file containing list of PDB identifiers:
listFileHandle = file(filePathName)
listFileHandle.readline() # Skip first line
pdbID = listFileHandle.readline()
while pdbID:
    print pdbID[0:4]
    prot_L = chimera.openModels.open(pdbID[0:4], type = "PDB")

    # Model 0 can be now be modified in preparation
    # for the PDB file write.
    # Insert code here to do modifications.
    outFilePathName = pdbID[0:4] + ".pdb"
    runCommand("write format pdb 0 " + outFilePathName)
    pdbID = listFileHandle.readline()
    # Go for next pdbID in the list.
    chimera.openModels.close(prot_L)

listFileHandle.close() # Close the list file.
```

Script 2.3: Getting a Batch of PDB Files by Using Chimera

The script is slightly shorter but more versatile, because the molecule can be changed before the file write. For example, the `runCommand` line can be altered to incorporate a `selected` option:

```
runCommand("write format pdb selected 0 " + outFilePathName)
```

Before this command, you can insert statements that select those atoms that will be written out to the PDB file. The selection can be accomplished by means of other `runCommand` statements or by using a more sophisticated selection procedure that might involve calculations. In effect, this allows the user to remove atoms that are *not* to be included in the PDB file that is to be written (e.g., see Exercise 3).

Further options for the `write` command can be found at http://www.cgl.ucsf.edu/chimera/docs/UsersGuide/midas/write.html.

Another useful variation of this script would be to access files containing a protein co-crystallized with a ligand. By selecting the ligand before the `write` command, we can generate a file that contains only that molecule. Note that the format for such a write could be `mol2`. This would give us the coordinates of ligand atoms when the ligand is docked into the binding site of the protein (see Exercise 4).

2.11 HILITER CLASS

Scripting with Chimera is particularly enjoyable when the output of your program produces a visual change in the display. These changes can also be used to provide feedback when testing or debugging a lengthy script.

The `StructBio` package, available from the book's website, contains a class called `HiLiter` that will be useful to illustrate the behavior of various scripts in later chapters. We will not discuss the details of the script because it is reasonably straightforward. Instead, we will describe the typical use of `HiLiter`. The class can be accessed by doing the following import:

```
from StructBio.Utilities.hiLite import HiLiter
```

Suppose that `prot` is a molecular object that has been previously defined. The statement

```
hiLiter = HiLiter(prot, "white")
```

creates the object `hiLiter` as an instance of the `HiLiter` class. The initialization routine for the class will cause the molecule to be displayed in the *wire* representation and the background will be colored white. If this second argument is omitted, the default color is black. After the object is created, we can use the `emphasize` method to highlight various atoms in the molecule:

```
hiLiter.emphasize(primaryList, secondaryList)
```

The `primaryList` is a list of atom objects that will have their representation changed from wire to ball & stick. The `secondaryList` is a list of atom objects that will have their representation changed from wire to stick. All atoms not in these lists go back to the wire representation if not already present in that representation. Figure 2.7 (produced by the script for Exercise 11) shows the typical appearance of these changes. In this figure, the larger ball is an alpha carbon that was the single entry in the primary list, while the two smaller spheres (ball & stick representation) are the carbon and nitrogen atoms that were the two entries in the secondary list.

The function `clearEmphasis()` can be used to bring the entire scene back to a wire representation without any emphasis. The function call `setVerbose(True)` can be used to put `HiLiter` instance in verbose mode. In this mode, the Python Shell window will get a print out of all atom names in both primary and secondary lists whenever the emphasis changes.

The use of `HiLiter` during debugging and testing will help you to settle questions such as: "Did my last computation correctly identify the atoms that my algorithm needs for further computation?" Putting the `emphasize` method in the body of a loop that is subject to a stepping interaction (such as waiting for a user keystroke) generates an animation that helps you understand the behavior of the program.

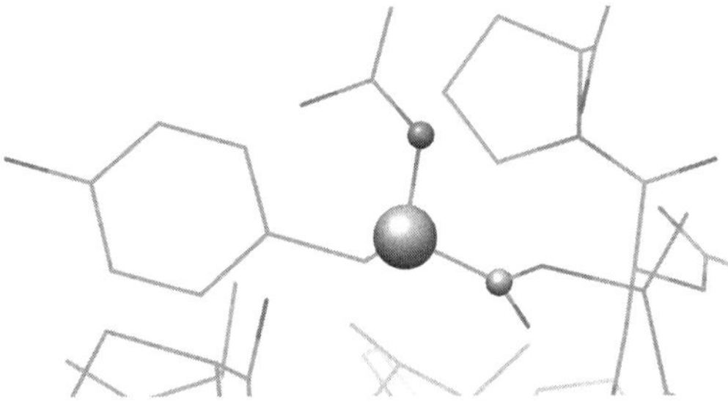

FIGURE 2.7 Example of a HiLiter instance putting emphasis on three atoms.

2.12 HBONDDICTS CLASS

In a protein, it is possible for bond lengths and bond angles to change but typically with very small alterations. Protein flexibility is mainly achieved by changes in the phi and psi dihedral angles. However, these changes are not arbitrary and are subject to various constraints mostly present due to the existence of hydrogen bonds. For example, beta strands tend to stay in *sheet* structures that are held in particular formations by the existence of hydrogen bonds.

The `StructBio` package, available from the book's website, contains a class called `HBondDicts` that is useful for the creation of dictionaries that describe the hydrogen bonds in a protein. Use the following to access the class:

```
from StructBio.HBonds import HBondDicts
```

The interested reader can view the source code to get more details about its operation. Instead of a line-by-line analysis of the program, we will provide a higher level description of both its activities and the dictionaries that have been provided for user applications.

When an `HBondDicts` object is instantiated, the `__ini__` method will execute the `findHBonds` method in the `FindHBond` class that is part of the Chimera installation. This will generate a list of tuples that specify all the hydrogen bonds. The initialization method then builds the `hBond_D` dictionary to be described in Section 2.12.1. Some private support functions such as `__SScode` and `__categoryCode` are used to help in the construction of these dictionaries.

2.12.1 Using the HBondDicts Class

After fetching a protein using a statement such as:

```
prot = chimera.openModels.open(protID, type="PDB")[0]
```

The protein object can be passed to the class `__init__` function by instantiating an `HBondDicts` object, for example,

```
hBds = HBondDicts(prot)
```

This instantiation builds all the required dictionaries for further use in the program. Generation of hydrogen bonds is actually a complicated procedure that involves several issues. If you pull down the Chimera menu item **Tools... Structure Analysis... FindHBond** you will get a dialog window that shows several of the options that are available in the creation of hydrogen bond representations. In the `HBondDicts` initialization just shown, default values are used for the geometric constraints placed on the bond. The definition line for the `__init__` function is

```
def __init__(self, prot, dist_Slop = 0.4, angle_Slop = 20.0):
```

so, you can use constraints that are different from the default values if desired. Going along with our example, let us describe the dictionaries that would be available for the `hBds` object:

- The dictionary `hBds.hBond_D` provides a mapping from donor atoms to acceptor atoms. The keys for the dictionary are atom objects representing the donor atoms (usually a subset of the nitrogen atoms in the protein backbone). The dictionary value corresponding to a key is a list of acceptor atoms with each atom having a single hydrogen bond going to the donor atom specified by the key. Usually, a list will have only one entry, but occasionally one may encounter a situation involving more than one acceptor for a single donor.
- The dictionary `hBondSameChain_D` provides some descriptive information about the hydrogen bonds. Keys for this dictionary are (donor, acceptor) tuples. Each entry in the tuple is an atom object. Naturally, each tuple specifies a hydrogen bond and the value corresponding to a key is *True* if and only if the donor and acceptor are in the same chain.
- The dictionary `hBondCategory_D` provides more information about the hydrogen bond. The keys are the same as for the `hBondSameChain_D` dictionary. The value for a key is a string that holds a code describing the significance of the bond:
    - "HE": The donor and acceptor atoms are both in the same helical secondary structure.
    - "PL": The donor and acceptor atoms are in beta strands that are running in parallel.

- "AP": The donor and acceptor atoms are in antiparallel beta strands.
- "NA": "None of the above," this code is a catchall for any other H-bond scenario.

See Exercises 13 and 15 for problems related to this class.

2.13 EXERCISES

1. Starting with a simple Python script

   Get a Python Shell and use it to get a new window in which you will write a Python script. The script should read a PDB file when given the PDB ID of a protein (the ID can be a parameter string in the program). The script should fetch the PDB file and then print (to the Shell window) a table with a single row for each residue in the protein. Each line should contain the residue type, followed by the residue position and chain identifier (the latter two entries separated by a period). A final column should print out a code that specifies the secondary structure containing this residue: H for helix, S for strand, and C for coil. Here is the first output line for the protein with PDB ID = 1QU9:

   ```
   SER 2.A C
   ```

   Some miscellaneous observations related to 1QU9: There are no atom coordinates for the Met residue in position 1. Consequently, it is not in the residues list for the protein. Many of the residues have type "HOH."

   Do some Internet searching to get information about the CSP residue at position 107.

2. Mapping atom labels to atom objects

   In later chapters, we will be doing exercises that involve finding atoms that are specified by an atom name within a residue that is identified by position and chain identifier. Although it is easy to find a residue if you know its index in the protein's residue list, it is a bit more difficult to locate a residue by using its position in the chain. The goal of this exercise is to program a short function that will generate a dictionary that maps atom labels to atom objects. If you had an atom object (call it a), you could map the atom object to an atom label by using the string:

```
':' + str(a.residue.id.position) + '.' +
a.residue.id.chainId + '@' + a.name
```

The generated dictionary will allow you to reverse this mapping: When such a string is given to the dictionary as a key, it will return the atom object as the corresponding dictionary value.

When your function is given a protein molecule object, it should scan the atoms in the protein and generate the dictionary. Test the dictionary with a short mainline program that fetches the PDB file "1BRF" and then calls the function. To show that the dictionary works, print out the names of the atoms with labels `":54.A@FE"` and `":8.A@SG"`. They should be `FE` and SG. Further testing should print out the residue positions and residue types for all keys in the dictionary.

3. A Python script that deals with atom coordinates

   This exercise will get you started with Python scripts that work with protein files and atom coordinates: Write a script that reads a PDB file when given the PDB ID of a protein (the ID can be a parameter string in the program). The script should access the atomic coordinates in the PDB file and then print the following:

   a. The centroid of all atomic positions given in the file.
   b. The centroid of all alpha carbon atoms in the protein.
   c. The centroid of all atoms in each chain.
   d. The centroid of all alpha carbon atoms in each chain.

   Test your script on the PDB file 1HIV. Parts (c) and (d) should be applied to chains A and B only. Chain I is a ligand and you can ignore it for parts (c) and (d).

4. Listing bond pairs and geminal pairs

   Write a script that accesses the protein with ID 1OOH. It should build a *bond dictionary* `bondDict` with 20 keys. A key will be a string with three characters specifying the type of one of the 20 amino acids. The entry corresponding to a key should be a list of pairs of atom names that specify the bonds that can be found in that residue. For example, `bondDict["CYS"]` should return the list:

```
['C-CA', 'C-O', 'CA-N', 'CA-C', 'CA-CB', 'CB-CA', 'CB-SG',
'N-CA',  'O-C', 'SG-CB']
```

The script should also build a geminal dictionary `gmnlDict` with 20 keys. The keys have the same significance as just described, but now the entry corresponding to a key should be a list of pairs of atom names that specify the geminal pairs in that residue. Two atoms form a geminal pair if they are both bonded to a third atom. For example, `gmnlDict["MET"]` should return the list:

```
['CA:O', 'O:CA', 'N:C', 'N:CB', 'C:N', 'C:CB', 'CB:N', 'CB:C',
'CA:CG', 'CG:CA', 'CB:SD', 'SD:CB', 'CG:CE', 'CE:CG']
```

The script should write out the dictionaries as two files: `bondsInResidues.txt` and `geminalsInResidues.txt`.

5. 90° rotation

   Read the document at: http://www.rcsb.org/pdb/education_discussion/molecule_of_the_month/download/PotassiumChannels.pdf. Use Chimera to get a display of the protein 1K4C. You will notice that the PDB file contains all the atoms for a single protein molecule. However, the potassium channel is actually comprised of four such molecules. The objective of this exercise is to compute the coordinates of the other three proteins so that when you look down the *z* axis, the display will contain a scene that is similar to that shown in the second figure of the PDF document PotassiumChannels.pdf. Each of the four proteins will be stored as a Chimera *model*. When your Python script is running properly, you will see the four models listed in a small dialog box that can be launched by the menu invocation: **Favorites/Model Panel**. The display of the four models can be controlled by checking the appropriate boxes.

   How do we build the other three protein models?

   If you look carefully at the 1K4C display you will notice that it has 7 potassium atoms within the *het chain*. Furthermore, if you inspect the PDB file, you will discover that the atomic coordinates of these potassium atoms are specified in the HETATM records as follows:

```
HETATM 4021 K   K 3001  155.336 155.342 -30.553 1.00 14.49     K
HETATM 4022 K   K 3002  155.331 155.331 -33.953 1.00 15.51     K
HETATM 4023 K   K 3003  155.341 155.323 -37.162 1.00 15.69     K
HETATM 4024 K   K 3004  155.330 155.324 -40.505 1.00 16.84     K
HETATM 4025 K   K 3005  155.327 155.335 -47.577 1.00 24.94     K
HETATM 4026 K   K 3006  155.339 155.327 -22.975 1.00 47.51     K
HETATM 4027 K   K 3007  155.343 155.330 -26.017 1.00 65.44     K
```

The interesting observation here is that the (x, y) coordinates of all seven atoms are almost the same. In fact, the average x coordinate has a value of 155.335 and the average y coordinate has a value of 155.330. Allowing for a small amount of *noise* in the (x, y) coordinates, we see that the potassium atoms are all arranged along a line that is parallel to the z axis. Consider what happens if the 3D position of *every* atom in a molecule is modified by subtracting (155.335, 155.330, 0.0) from the coordinates. Note that the z coordinate does not change. Each potassium atom will end up on the z axis.

To get the four models, you should do the following:

a. Generate four models each a copy of 1K4C. Consider the models to be indexed as 0, 1, 2, and 3.

b. Move the atoms in all molecules so that the potassium atoms line up on the z axis.

c. Rotate model 1 about the z axis through an angle that is equal to 90°, rotate model 2 about the z axis through an angle that is equal to 180°, and finally, rotate model 3 about the z axis through an angle that is equal to 270°.

Relevant notes:

a. The rotation does not require any sophisticated use of matrices or trigonometric calculations. Chimera has a function called `xform` that provides an easy way to do translation and rotation, but you are not allowed to use this approach. Instead, you should note that because the angles are all multiples of 90°, there is a very simple relationship between the initial coordinates of an atom and the final coordinates after such a rotation operation. Your script should work with the actual coordinates to achieve such a rotation. Why avoid xform if it is available? The goal of this question is for you to learn scripting techniques that deal with the access of atomic coordinates and their use in subsequent calculations. These techniques will be very important for structural calculations in later exercises.

b. During the initial load of the proteins, they will be established in a viewing frustum in front of the Chimera camera. When your script modifies atom coordinates, it may happen that the molecules are shifted outside the frustum and the display is empty. To recover the scene, you should do the menu invocation: **Favorites... Side View**. This will get you a Viewing dialog. You can then click on **View All** to get the proteins back into the frustum.

Your display should be similar to that of Figure 2.8. Note the orange *girdle* around the lower part of the channel. This part of the surface is above residues that are generally hydrophobic. This is the region of the channel that would be embedded in the cellular membrane (note that there are no water molecules around this part of the channel).

6. Indices and positions

   Consider the following Shell interactions:

```
>>> opkProt = chimera.openModels.open('1opk', type='PDB')[0]
>>> resAtIx0 = opkProt.findResidue(0)
>>> resAtIx0.type
u'ASN'
>>> seqA = opkProt.sequence('A')
>>> seqA.resMap[resAtIx0]
43
>>> resAtIx0.id.position
83
```

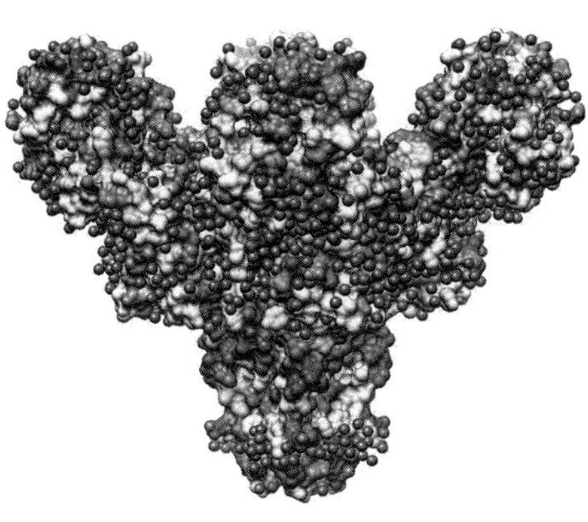

FIGURE 2.8 (**See color insert.**) Potassium channel 1K4C with all chains in place.

This code fetches the PDB file with ID = "1OPK." This protein (c-Abl tyrosine kinase) has been studied because of its relationship to chronic myelogenous leukemia. Unfortunately, the first 82 residues of the protein are missing in the structural x-ray analysis (they are in a long loop that is difficult to localize during crystallization). In the preceding code, we have three numbers: 0, 43, and 83. Explain the significance of each. How do they relate to the sequence display (menu item **Favorites... Sequence**) and the label of the first residue in the display of the protein. Explain the results that you get when the following two expressions are executed in the Shell:

```
seqA.residues[43].type
seqA.residues[42].type
```

7. Writing PDB files each with a single model

   Write and test a function `writeModels(prot)` described as follows:

   Input: A protein model containing several submodels (typical of PDB files for protein structures derived from NMR experiments).

   Output: The function should write out one or more files in PDB format, each file containing only one submodel from the file. The file name should have an appropriate suffix to distinguish it from the original PDB file. For example, the first submodel obtained from `1AH1.pdb` would have file name `1AH1_01.pdb`.

   Test the function on any of the following PDB entries: `1BMR`, `1DW5`, `1QDP`, `1D1H`, `1N4N`, `1IJC`, `1PFD`, `2LI3`, `1AH1`.

8. Writing PDB files each with a single chain

   Write and test a function `writeChains(prot, unique = False)` described as follows:

   Input: A protein model and a Boolean flag `unique` with default value `False`.

   Output: The function should write out one or more files in PDB format, each file containing only one chain taken from the protein. If the `unique` parameter is set to `True`, then the function is to avoid the repeated writing of chains that are the same. In other words, a homodimer will generate only one file. Test the function on the following PDB entries: `3MI1` (should generate 2 files) and `3M1I` (should generate 3 files). To distinguish these single chain files from regular PDB files, you should append the chain ID as a suffix in the file name. For example, the first file written out for `3MI1` would have file name: `3MI1A.pdb`.

9. Writing PDB files each with a single protein substrate

   Modify Script 2.3 so that it accesses the PDB files specified in the following text file. It should extract the substrate peptide from each file and save it to your directory as a PDB file. Each substrate peptide can be identified as chain P.

```
HIV Protease with Substrate Complexes
  1KJF
  1KJG
  1KJH
  1KJ4
  1KJ7
  1F7A
```

10. Writing MOL2 files each with a single ligand molecule

    Modify Script 2.3 so that it accesses the PDB files specified in the following text file. It should extract the inhibitor from each file and save it to your directory as a MOL2 file. Each inhibitor can be identified as a residue with type given by the three character code in the second column.

```
HIV Protease Inhibitors
  3NDW  RIT  Ritonavir
  2NNP  ROC  Saquinavir
  2AVO  MK1  Indinavir
  3EL5  1UN  Nelfinavir
  3SM2  478  Amprenavir
  2Q5K  AB1  Lopinavir
  3EL9  DR7  Atazanavir
  2O4P  TPV  Tipranavir
  3SO9  017  Darunavir
  2FDE  385  Brecanavir
```

11. B-factors and flexibility

    The B-factor of an atom is linearly related to the mean square displacement of an atom from its average position. Various research papers have used the B-factor of an alpha carbon atom as a first approximation for the assessment of the flexibility of the protein backbone at that position: a larger B-factor indicates more flexibility. Note that `bfactor` is an attribute for an atom in the Chimera object model.

A word of caution: The B-factor is influenced, to some extent, by the effects of crystal packing before x-ray analysis. Consequently, the use of B-factors does not necessarily give us a completely accurate assessment of the protein's flexibility as it would occur in the cytosolic environment.*

The goal of this exercise is to assess the change in flexibility when HIV protease is co-crystallized with Tipranavir (one of the protease inhibitors currently available as an anti-HIV drug). The exercise will also give you some experience with the use of plotting facilities that are part of `matplotlib` module.

Steps for this exercise are as follows:

a. Get familiar with the plot facilities available with Chimera.

b. Generate the data.

c. Generate a bar chart for the data and make appropriate observations.

Start by reviewing Script D.1 in Appendix D. This will show you how `matplotlib.pyplot` can be imported and used to generate bar charts.

To generate the data, your script should access two PDB files: 1HHP (the "apo" or ligand-free form of HIV protease) and 2O4P (HIV protease in complex with Tipranavir). For each protein, you should generate an array containing the B-factors of the 99 alpha carbon atoms in chain A.

A direct comparison of the B-factors corresponding to two different crystallizations is not advisable because each x-ray analysis may go through a refinement procedure that leads to different scales for the B-factors.† So, each data set should be modified by a normalization calculation:

$$\text{B-factor} = \frac{\left(Bf - \langle Bf \rangle\right)}{\sigma\left(Bf\right)}$$

where Bf is the B-factor that Chimera gets from the PDB file, $\langle Bf \rangle$ is the mean B-factor computed across all the alpha carbon atoms of

* G. Song and R. L. Jernigan. vGNM: A better model for understanding the dynamics of proteins in crystals. *Journal of Molecular Biology*, **369** (2007), 880–893.

† Z. Yuan, J. Zhao, and Z. X. Wang. Flexibility analysis of enzyme active sites by crystallographic temperature factors. *Protein Engineering*, **16** (2003), 109–114.

the protein chain (except for the first three and last three residues), and $\sigma(Bf)$ is the standard deviation computed across all the alpha carbon atoms of the protein chain (except for the first three and last three residues). The first three and last three residues are usually omitted because chain termini are typically very flexible and may lead to bias in the computation of the mean and standard deviation values. The mean and standard deviation functions are available in the Python `numpy` module.

Use a modified version of the Bar Chart Example script to show how the two proteins differ in their B-factors as you move along the 99 alpha carbon atoms in the backbone (Figure 2.9). (Use green for the apo form and red for the inhibited form of the protein.)

Now answer the question: How is the information that you see in the bar chart consistent with the behavior of the Tipranavir inhibitor acting as an anti-HIV drug? Hint: Residues 46–56 are usually considered to be the so-called *flap* region of the protease.

12. HiLite class test

    To get familiar with use of the `HiLiter` class, available from the book's website (Chapter 2 scripts), write a small mainline program that reads a protein file and then uses the `HiLite` class to sequentially emphasize the backbone atoms in the residues. As you step from one residue to the next, let the alpha carbon atom be the single entry in the primary list while the adjacent nitrogen and carbon atoms define two entries for the secondary list. Step pacing can be achieved by waiting for the user to depress the enter key.

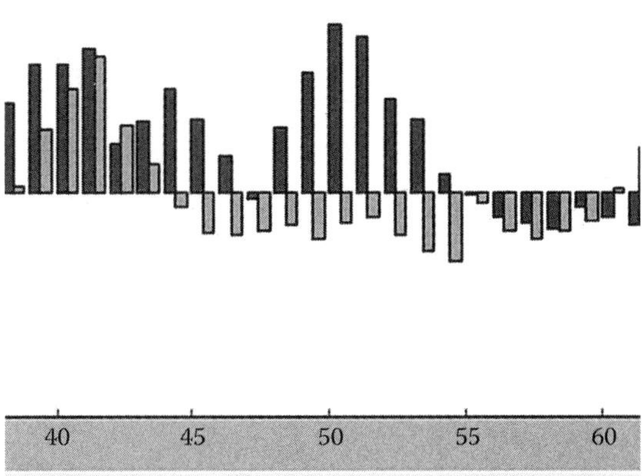

FIGURE 2.9 (**See color insert.**) A section of the plot comparing the B-factors.

13. HBondDicts class test

Write a script that sequentially highlights the atoms involved in hydrogen bonding. Use the `HiLiter` class and the `HBondDicts` class. In response to the user pressing the enter key, the animation should move to the next donor, acceptor pair with the donor atom in the primary list and the acceptor atom in the secondary list of the `HiLiter` object. Use the verbose mode of `HiLiter` and, for each hydrogen bond, you should also print out the dictionary values for `hBondSameChain_D and hBondCategory_D.`

14. Introducing the Surfaces class

The next exercise makes use of the `Surfaces` class (see Appendix B) that can be used to display various surfaces in your molecular scene. For this exercise, you should experiment with the `addPolygon` function that can be used to display convex polygons. Each polygon is specified by an ordered list of vertex coordinates. A polygon is drawn as a set of triangles. Each triangle is comprised of three vertices: two consecutive vertices in the polygon vertex list and the centroid of all vertices in the list. Note that the vertices do not have to be coplanar, but in most applications they will be either coplanar or fairly close to being coplanar.

The class can be instantiated with an empty argument list, for example,

```
pDisp = Surfaces()
```

After this, we can bring polygons into the display by adding them to the instance, for example,

```
pDisp.addPolygon(vertices_L, polyColor, transparency)
```

Here, `vertices_L` is a list of `numpy` arrays specifying the 3D coordinates of the vertices of a polygon in the display space. The color is designated by `polyColour`. This is a three component tuple specifying the color in RGB format. The `transparency` parameter is a value between 0.0 and 1.0 specifying the transparency of the surface.

```
pDisp.display()
```

For example, running the next script will produce the two polygons in Figure 2.10.

FIGURE 2.10 Displaying two polygons.

```
import chimera, numpy
from chimera import runCommand
from numpy import array
from StructBio.Scenographics.surfaces import Surfaces

runCommand("set bg_color white")

pd = Surfaces()

v1 = array([ 0., 4., 0.0])
v2 = array([ 2., 1., 0.0])
v3 = array([ 1., -3., 0.0])
v4 = array([-1., -3., 0.0])
v5 = array([-2., 1., 0.0])
pd.addPolygon([v1,v2,v3,v4,v5], (0.0, 1.0, 0.5), 0.9)

v1 = array([ 0., 2., 2.0])
v2 = array([ 2., 1., 2.0])
v3 = array([ 1., -3., 2.0])
v4 = array([-1., -3., 2.0])

pd.addPolygon([v1,v2,v3,v4], (0.0, 0.5, 1.0), 0.9)

pd.display()
```

To get familiar with this class, try the following problem: Generate the display of a cube with the eight vertex coordinates specified as $(\pm1,\pm1,\pm1,)$. Choose any random coloring scheme (perhaps with some bias to the primary colors: red, green, and blue).

15. Drawing *H-Bond Plates*

A worthwhile objective in structural bioinformatics is to represent complicated structures with simplified visual components. The classic ribbon diagram is an example of this goal. In this exercise, we will generate rectangular plates that bridge neighboring beta strands. The visual objective is to get a beta sheet to look like a *sheet*. In Section 3.7, we will provide a more compelling argument for the construction of

H-bond plates. This visualization is especially useful for larger proteins that have the *beta meandering motif*, for example, PDB ID = 1FJ1.

To give these plates some biological relevance, we will stipulate the following rules:

a. Each plate corresponds to a single hydrogen bond involving a backbone nitrogen donor and its associated oxygen acceptor. Both donor and acceptor atoms must be in a beta strand. The strands need not be in the same chain.

b. The four vertices of a plate will be determined by the alpha carbon atoms on either side of the donor and acceptor pair.

c. Each of the *short sides* of a plate will have vertices that are consecutive alpha carbon atoms and in the same chain.

For example, Figure 2.11 shows three strands in a parallel beta sheet. The hydrogen bonds are represented by thin gray lines. In this figure, the four sided *plate* outlined with dashes corresponds to the hydrogen bond donor labeled with N_i where the subscript i denotes the index for the residue containing that atom. The short sides of the plate are defined by the pairs (CA_j, CA_{j+1}) and (CA_{i-1}, CA_i). The final rule stipulating that each member of this pair must be in the same chain is necessary to avoid peculiar situations near the start and end residues of a chain. Although the pairs can be in different chains, we want to avoid the undesired situation in which a short side jumps from one chain to another.

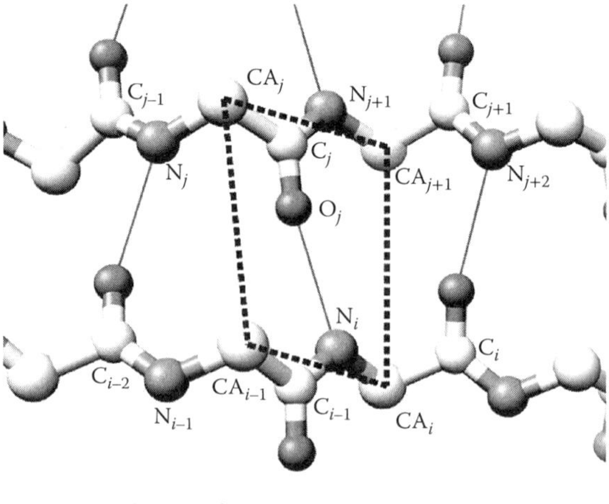

FIGURE 2.11 Atom numbering for parallel strands.

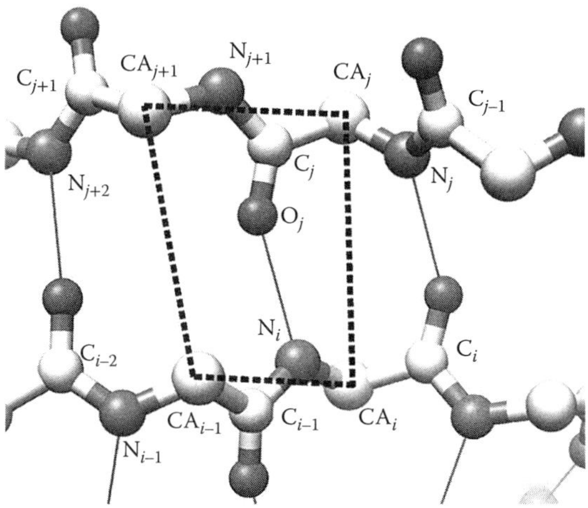

FIGURE 2.12 Atom numbering for antiparallel strands.

Figure 2.12 shows a plate that is in an antiparallel beta sheet.

To give these plates additional structural significance they are to be colored according to the following rules:

Color	RGB Code	Significance
forest green	034 139 034	H-bond linking parallel strands, same chain
cornflower blue	100 149 237	H-bond linking antiparallel strands, same chain
spring green	000 255 127	H-bond linking parallel strands, different chains
sky blue	135 206 235	H-bond linking antiparallel strands, different chains

Let us look at an example. The trimeric autotransporter adhesin protein from Bartonella henselae (PDB ID = 3D9X). It is a beautiful structure with three chains that are intertwined to form various beta sheets. Figure 2.13 shows a ribbon representation with H-bonds present. Ribbons have been colored as follows: Chain A (goldenrod), Chain B (aquamarine), and Chain C (medium purple).

Incidentally: If you inspect this structure, it is clear that any one of the three chains cannot adopt its final structure when it has just been generated by the ribosome. It is likely that the antiparallel sheets form early, but the final structure can only be formed when all three chains meet up to intertwine with each other.

With all the plates drawn, you should get a scene as in Figure 2.14. In this scene, the backbone has been drawn as a *chain trace* in stick form (no side chains visible). Note that each plate has two edges that are coincident with two alpha carbon to alpha carbon line segments of the chain trace.

This exercise is somewhat complicated, so we will break it up by considering the steps that will be necessary:

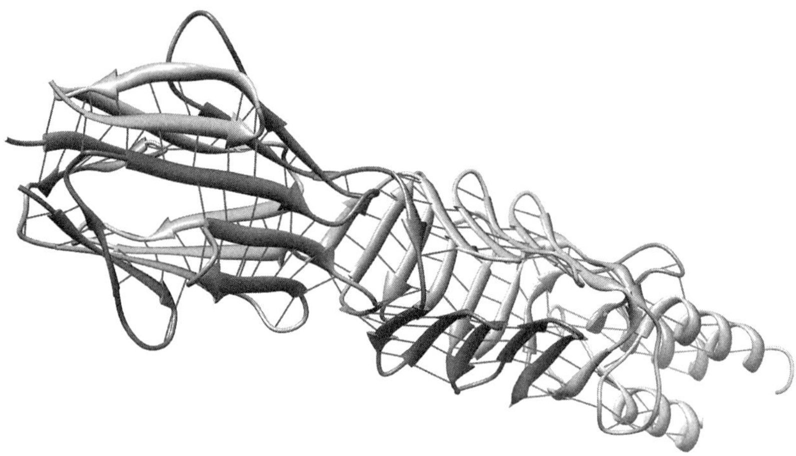

FIGURE 2.13 (**See color insert.**) The autotransporter adhesin protein from Bartonella (PDB ID = 3D9X).

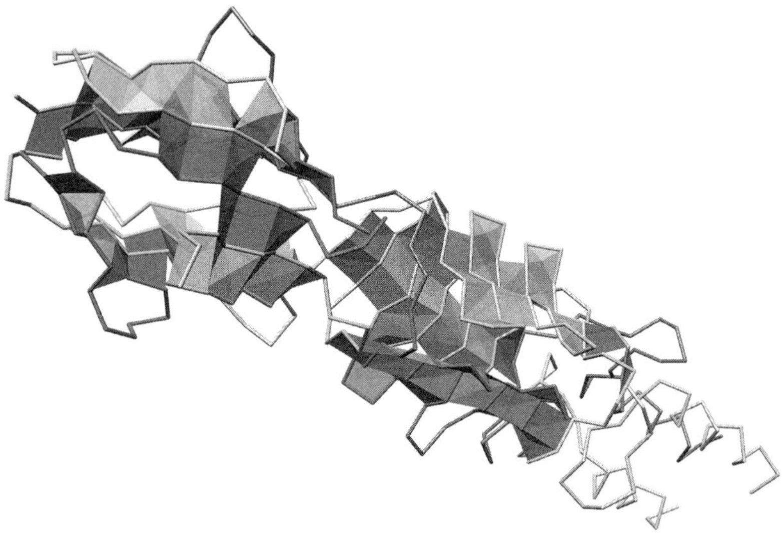

FIGURE 2.14 3D9X drawn with H-bond plates.

a. After fetching the PDB file, generate an `HBondDicts` instance.
   Suggestion: Use `dist_Slop = 0.5` and `angle_Slop = 35.0`.

b. Use the `hBondCategory_D` dictionary of the `HBonds` instance to get the hydrogen bonds and their categorization.

c. Determine the residues of the donors and acceptors and then deduce the locations of the alpha carbon atoms that flank the donors and acceptors. Be careful with *end of chain* issues: You want to avoid having a plate that goes from one chain to another.

d. Using the `Surfaces` class, construct the plates and subsequently display them.

More details for these steps are as follows:

Your script should go through all the H-bond donors and for each donor it should get the coordinates of the four alpha carbons defining the corners of the plate (as described in Figures 2.11 and 2.12). Use the `addPolygon` function of the `Surfaces` class to generate the plate in the display. After all the plates have been added to the surface, as a collection of polygons, the script can execute the `display()` function to get the surface model displayed in the Chimera window.

You should be able to see the model as an entry in the Model Panel named `"H-Bond Plates"` if you instantiate the `Surfaces` object using the function call:

```
Surfaces(modelName = "H-Bond Plates").
```

Recall that the Model Panel can be accessed using the menu invocation: **Favorites... Model Panel**.

You should now have enough information to write your script. Cautionary note: If you are referencing residues in a protein, using the `residueBefore` or `residueAfter` method, be careful to take the appropriate action when you cross from one chain to the next. If you are not careful about this, it is possible to get long skinny triangles that have no relevance to our problem.

Show how your script works with the following four proteins (as usual, be sure to invoke the menu **File... Close Session** between each test):

a. 3D9X (do you get the same result as in Figure 2.13?)

b. 2F0C (note the beta prism in the middle of the protein.)

c. 1EMA (a nice example of a beta barrel.)

d. 3ZY7 (see next paragraph.)

e. A protein of your own choice (try to find a protein that has a beta sheet configuration different from the test cases above.)

Note: The 3ZY7 protein is a computational design of a symmetric homodimer. The research paper at: http://www.pnas.org/content/108/51/20562.abstract discusses how the designers got the two chains to *stick* together along two beta strands. Your output should show this bridging as a sequence of sky blue plates (everything else colored in cornflower blue). Another note: The H-bond file for this protein contains extra lines for *alternate* positions of the donor and acceptor atoms. Use the alternates labeled "A" (e.g., "N.A") and ignore lines designated with "B" alternates.

REFERENCES

[BA10] S. Bassi. *Python for Bioinformatics*. Chapman & Hall/CRC, Boca Raton, FL, 2010.

[BE09] D. M. Beazley. *Python Essential Reference*. Fourth edition, Addison-Wesley, Upper Saddle River, NJ, 2009.

[HM00] D. Harms and K. McDonald. *The Quick Python Book*. Manning Publications, Greenwich, CT, 2000.

[KI09] J. Kinser. *Python for Bioinformatics*. Jones and Bartlett, Sudbury, MA, 2009.

WEBSITES FOR PYTHON

The Python Tutorial: http://docs.python.org/tutorial/

TP (tutorials point): http://www.tutorialspoint.com/python/index.htm

A Beginner's Python Tutorial: http://www.sthurlow.com/python/

Python via YouTube: http://www.youtube.com/playlist?list=PLEA1FEF17E1E5C0DA

CHAPTER 3

Algorithms Dealing with Distance

3.1 MOTIVATION

Distance measurements are fundamental operations in the analysis of molecular structure. Most of these measurements are done to compute interatomic distances that are part of a more extensive analysis such as studies dealing with protein flexibility, docking of ligands, or evaluation of energy functions. All these measurements assume that the coordinates of the atoms are accessible. For example, coordinates would be available when a Protein Data Bank (PDB) file has been accessed and processed. This chapter also considers other calculations dealing with atomic coordinates such as the computation of the centroid of a collection of atoms. More esoteric computations would involve the perpendicular distance between an atom and a plane determined by other atoms in the system.

3.2 CALCULATING INTERATOMIC DISTANCES

There are several ways to calculate interatomic distances. One can use the following menu invocation: **Tools... Structure Analysis... Distances.** This will launch the Structure Measurements dialog window, and one can then designate atom pairs for the distance calculation. Since our emphasis is on Python scripts, we cover the techniques that allow one to calculate distances that can be used for further calculations in a script. All distance calculations will be based directly or indirectly on the familiar Euclidean formula given by Equation 3.1:

$$d(a,b)=\sqrt{(a_x-b_x)^2+(a_y-b_y)^2+(a_z-b_z)^2} \tag{3.1}$$

where atom a has coordinates (a_x, a_y, a_z) and atom b has coordinates (b_x, b_y, b_z).

To illustrate the various techniques, we repeat some of the distance calculations that were done by Watenpaugh and others [WS80] in their study of rubredoxin (PDB identifier: 4RXN). Rubredoxin is a metalloprotein containing a single iron atom coordinated by four cysteine sulfur atoms. In this beautiful arrangement, the cysteine residues act like the prong setting of a diamond ring holding the iron atom at the center of a tetrahedral structure (Figure 3.1).

Calculations done by Watenpaugh and others [WS80] included the computation of distances between the iron atom and each of the four coordinating sulfur atoms. As a running example, we pose the following simple question: what is the interatomic distance between the iron atom and the sulfur atom of CYS 6 in 4RXN?

We start with the following code snippet, which derives the coordinates of the two atoms:

```
p = chimera.openModels.open("4RXN", type = "PDB")[0]
for r in p.residues:
    if r.type == "FE":
         feCoords = r.findAtom("FE").coord()
    if r.id.position == 6:
         sgCoords = r.findAtom("SG").coord()
```

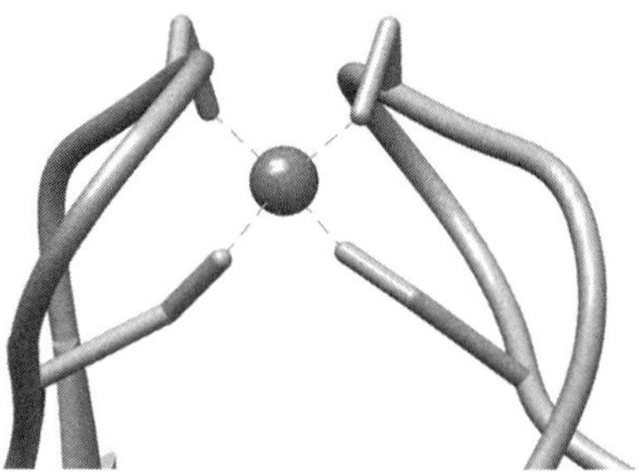

FIGURE 3.1 An iron atom held by cysteine residues in rubredoxin (Protein Data Bank identifier: 4RXN).

This script finds the target atoms, and for each atom the atomic coordinates are accessed as an attribute of the corresponding atom object. Note that the coordinates are returned by the `coord()` function as a `Point` object. With the `Point` objects stored in the variables `feCoords` and `sgCoords`, we can compute the interatomic distances using a straightforward application of Equation 3.1. This formula uses the individual components of the coordinates of the atoms. In our case, a_x, a_y, and a_z correspond to `feCoords`[0], `feCoords`[1], and `feCoords`[2], whereas b_x, b_y, and b_z correspond to `sgCoords`[0], `sgCoords`[1], and `sgCoords`[2], so calculation of the distance becomes

```
s = 0.0
for i in range(3):
    s += (feCoords[i] - sgCoords[i])**2
distance = s**0.5
```

Experienced Python programmers will typically avoid loops whenever possible. To get a more concise implementation of the distance function, we note that some calculations can be done on `Point` objects. For example, subtracting a `Point` object from another `Point` object will produce a `chimera.Vector` object. So the previous four lines of code can be replaced by the following single line:

```
distance = ((feCoords - sgCoords)*(feCoords - sgCoords))**0.5
```

where the "*" operation is computing the inner product of the two vectors.

Another approach is to use the `norm` function within the numerical Python module. After importing the `linalg` module (which may be needed for other calculations as well),

```
import numpy
from numpy import linalg
```

we can derive the distance using the following simple calculation:

```
distance = linalg.norm(feCoords-sgCoords)
```

We should also note that Chimera has a distance function for coordinate objects:

```
distance = feCoords.distance(sgCoords)
```

In all four cases, the value of the distance will be 2.333. This is consistent with the observations made in [WS80].

3.3 APPLICATIONS

Distance calculations are often the first steps performed in more extensive calculations, for example, the evaluation of potential energy between two atoms that share some sort of bond. For example, the hydrogen bonds discussed in Chapter 2 will have a bond strength that is dependent on interatomic distance. Chimera works with this dependency when it has to determine whether a hydrogen bond has enough significance to be displayed. The threshold for this decision can be altered by the user: the menu invocation **Tools... Structure Analysis... FindHBond** will launch a dialog window that allows you to relax these distance constraints by an amount of your choosing.

Many algorithms in structural bioinformatics rely on the collection of statistical information related to interatomic distances. For example, Zhang and others [ZL04] use a database of collected statistical information to derive a distance-dependent, pairwise statistical potential $\bar{u}(i,j,r)$ for the atom type pair specified by indices (i, j). They refer to this as the distance-scaled finite ideal-gas reference state energy function. Their analysis uses the inverse-Boltzmann formula:

$$\bar{u}(i,j,r) = -RT\,\text{In}\frac{N_{obs}(i,j,r)}{N_{\text{exp}}(i,j,r)} \tag{3.2}$$

where R is the gas constant, T is the temperature, $N_{\text{obs}}(i,j,r)$ is the observed number of atom type pairs (i, j) within the distance shell specified by the range $r-\Delta r/2$ to $r+\Delta r/2$ in the database, and $N_{\text{exp}}(i,j,r)$ is the expected number of atom type pairs (i, j) in the same distance shell if there are no interactions between the atoms.

In building such a database, it would be very useful to have an algorithm that can rapidly determine the atoms that are contained within a given distance shell. This algorithm is the topic of Section 3.4.

3.4 RAPID DETERMINATION OF ATOM MEMBERSHIP IN SHELLS

For this application, we are given a radius value (`maxR`) and we wish to construct a data structure that will facilitate rapid access of nearby atoms. Specifically, when a center atom is specified, the data structure should help us to generate a list of all atoms that are within `maxR` of the center atom. A slight modification of the algorithm will allow us to avoid nearby atoms that are closer than a radius value (`minR`).

The algorithm is to be implemented as a Python class called `Shell` and stored in a module called `shellNeighbors`. The instantiation of the `Shell` object will set up a data structure that is suitable for shells with arbitrary centers but with prespecified radii (`minR` and `maxR`). In practice, `maxR` will be some small floating point value such as 5.0. To get the neighboring atoms within this distance shell centered on a particular atom, we will invoke a `Shell` method called `getAtomsInShell`. More precisely,

Input for `Shell` object initialization:

A list of atom objects

Two nonnegative numbers `minR` and `maxR`, with `minR` < `maxR`

The `Shell` class should have a method called `getAtomsInShell`:

Input for `getAtomsInShell()`:

An atom object acting as the center atom

Output for `getAtomsInShell()`:

A list of atoms such that the distance between any atom in the list and the center atom is a value that falls in the range defined by the interval [`minR`, `maxR`]

The initialization of the `Shell` object should execute in O(n) time, where n is the number of atoms in the list passed to the `__init__` function of the `Shell` object. The execution time for the `getAtomsInShell` query function should be $O(1)$, that is, constant time. This requirement will demand a strategy that is much faster than the straightforward approach that involves a simple scan of all the other atoms in the list, this strategy taking $O(n)$ time to satisfy the query.

3.4.1 Implementation Strategy

The algorithm will behave as if the atoms in the atom list are contained in a rectangular three-dimensional (3D) *bounding box*. To be more specific, we want the smallest bounding box with sides having lengths that are exact multiples of maxR. We will consider the inside of the box to be a 3D array of cubes, each cube having a length, width, and depth equal to maxR. Each cube in the box will have 3D **integer** *grid* coordinates (c_x, c_y, c_z) that identify the cube's position within the bounding box. Figure 3.2 shows the front of the bounding box as a red grid, the bottom of the bounding box as a blue grid, and the left side of the box as a green grid. The other three sides of the box are not shown, and (as we will see) they are not needed in the computations.

These three sides will be determined by the minimum values of the atomic coordinates. They can be computed using a small function that is private for the Shell class:

```
# Function to compute the three minimum values of
# the coordinate entries.
def __minValsOfCoords(self):
    # Initialize to infinite values:
    xMin = yMin = zMin = float("inf")
    for atm in self.atoms_L:
        cc = atm.coord()
        if cc[0] < xMin: xMin = cc[0]
        if cc[1] < yMin: yMin = cc[1]
        if cc[2] < zMin: zMin = cc[2]
    return [xMin, yMin, zMin]
```

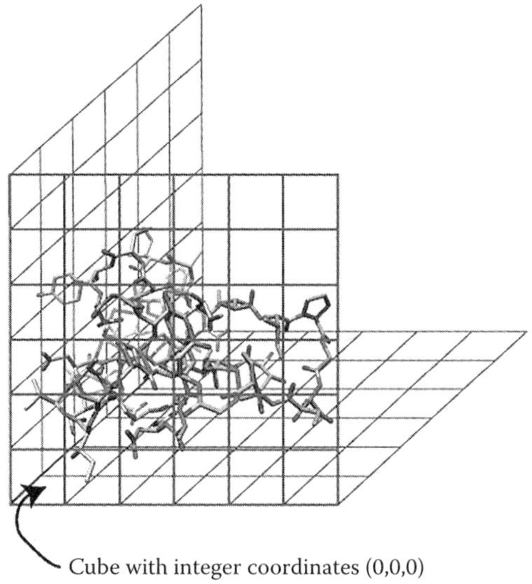

FIGURE 3.2 **(See color insert.)** Three sides of the bounding box.

Having these minimum values and knowing the dimensions of a cube, we can compute the integer coordinates of a cube containing an atom when we are given the coordinates of the atom:

```
# Function to convert 3D coordinates into
# grid coordinates.
def __convCoordsToGridCoords(self, coords):
    fV = floor((numpy.array(coords) - self.minVals)/self.maxR)
    return (int(fV[0]), int(fV[1]), int(fV[2]))
```

How are we going to use the grid coordinates of the atoms? A possibility is to have them act as indexes for a huge 3D array, each entry of the array containing a list of protein atoms that are in the cube with these grid coordinates. There is a big drawback to this scheme: it would require a lot of space and many of the entries would be empty because the protein typically does not intersect all the cubes in the bounding box. Instead, consider a code that builds a dictionary. This dictionary will map grid coordinates to lists of atoms. More specifically, a dictionary key will be a 3-tuple of integers representing the grid coordinates of a cube inside the bounding box. The dictionary contents associated with a key will be a list of atoms that can be found in that cube.

The dictionary can be built as part of the initialization for the `Shell` object:

```
self.gridDict = {}
for atm in self.atoms_L:
    # Generate integer coordinates for the cube
    # containing this atom.
    gridCoords = self.__convCoordsToGridCoords(numpy.array(atm.
    coord()))
    if not self.gridDict.has_key(gridCoords):
        self.gridDict[gridCoords] = []
    self.gridDict[gridCoords].append(atm)
```

How do we use the dictionary? If you have a *query* atom object, you can easily compute a *near neighbor list* specifying all the atoms that are located within the distance `maxR` from the query atom. This can be done in the following way: first convert the 3D coordinates of the query atom to its grid coordinates. This will designate a *central cube* that is surrounded by 26 other cubes having grid coordinates that are easily computed. Central cubes near the sides of the 3D grid will need special consideration. We can use `gridDict` to get all atoms in the $3 \times 3 \times 3$ set of cubes. Some of these atoms will be within the distance `maxR` of the query atom and others will be beyond this threshold, so we will have to do a final check to verify that an atom is within threshold. We can also eliminate any neighboring atoms that are beyond the `minR` constraint. If the distance between an atom and the query atom is less than or equal to `maxR`, it is put into the return list. The return list should *not* include the query atom. Note that the amount of computation that is done to handle a query is not related to the total number of atoms because (realistically) there is a small upper bound on the number of atoms that can be found in a set of 27 cubes. Consequently, a query runs in constant time. We have just described the operation of the following function:

```
def getAtomsInShell(self, centerAtom):
    # This function accepts an atom and returns a list
    # of all atoms that are within the shell surrounding the
      centerAtom.
    # The shell is specified by minimum radius minR and
    # maximum radius maxR.

    atomsInShell_L = []

    # Generate integer coordinates for the cube containing
    # the center atom.
    gridCoords = self.__convCoordsToGridCoords(centerAtom.coord())
    xI = gridCoords[0]
    yI = gridCoords[1]
    zI = gridCoords[2]

    # Inspect the 27 cubes to get the atoms that are within maxR.
    for i in range(xI - 1, xI + 2):
        for j in range(yI - 1, yI +2):
            for k in range(zI - 1, zI + 2):
                coordTuple = (i, j, k)
                if self.gridDict.has_key(coordTuple):
                    nbrList = self.gridDict[coordTuple]
                    for atm in nbrList:
                        interAtomDistance = \
                        numpy.linalg.norm(centerAtom.coord() - atm.
                        coord())
                        if interAtomDistance <= self.maxR:
                            # Do not include the center atom.
                            if interAtomDistance < 0.01: continue
                            # Eliminate atoms closer than the
                            # minimum radius.
                            if interAtomDistance < self.minR:
                                continue
                            atomsInShell_L.append(atm)

    return atomsInShell_L
```

The `Shell` class is in the `StructBio` package and can be imported using

```
from StructBio.Utilities.shellNeighbors import Shell
```

The module will prove to be useful for various other applications in the remainder of this book.

3.5 CONTACT MAPS FOR PROTEINS

If we have the coordinates of all the atoms in a protein, then we can compute a *distance matrix*. This is a square symmetric matrix of cells that are indexed by the residues of the protein being studied. In simple versions of a distance matrix, the cell $D[i, j]$ at row i and column j is the distance between alpha carbon $[i]$ and alpha carbon $[j]$. Depending on the requirements of the application, other definitions may be used, for example, working with the beta carbons in lieu of the alpha carbons with the alpha carbon atoms being used only for glycine. Another possibility is to measure distances between the centers of mass of the various side chains.

A contact map is essentially a visual discretization of the distance matrix. A cell is colored black if the distance between alpha carbon $[i]$ and alpha carbon $[j]$ is less than some particular threshold (say, 10 Å); otherwise, it is left white. Note that cell $D[i, j]$ will have the same coloration as cell $D[j, i]$, and so the matrix is symmetric.

Contact maps provide a reduced two-dimensional (2D) representation of a protein and have the advantage of being invariant to rotations and translations of the coordinates. Contact maps may be used to describe similarity between protein structures. An example of a contact map is shown in Figure 3.3.

The protein used here is chain A of a bovine Rabex-5 fragment cocrystallized with ubiquitin (PDB identifier: 2FIF). It was chosen because it is a small protein that has a good selection of secondary structural elements: helices and a beta sheet made up of both parallel and antiparallel strands.

Note that row numbering starts at the top and increases as you go down to the bottom of the figure. This is the simplest type of contact map with only two levels of distance information. In Figure 3.4, we show the top half of the contact map, but this is using five levels of distance information. As values from the distance matrix are being mapped to a range of gray levels, we will refer to this type of representation as a *distance map*. By comparing Figures 3.3 and 3.4, you can see the similarities of the main structural elements with Figure 3.4 giving a finer representation of the same data.

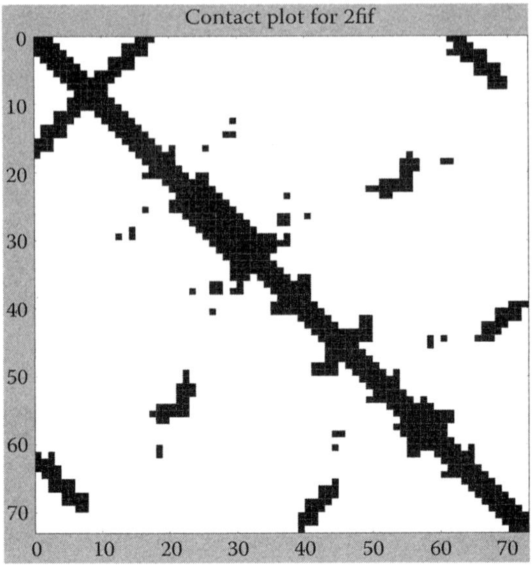

FIGURE 3.3 Contact map for chain A of 2FIF.

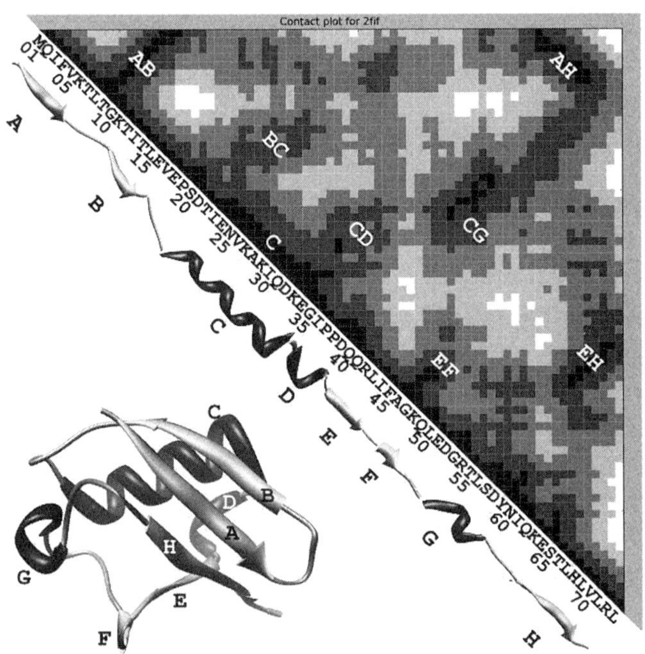

FIGURE 3.4 **(See color insert.)** Distance map for chain A of 2FIF.

Below the top half of the distance map, there is the *strung out* version of 2FIF chain A with the Chimera visualization of the protein in the lower left corner. The secondary structure elements (SSEs) have been labeled as **A**, **B**, **C**, …, **H**. Notice the contributions made by these secondary structures in the distance map. Dark areas in the map correspond to secondary structure pairs that are near to each other. The areas labeled as **AB**, **EF**, and **EH** represent antiparallel beta strands. The area labeled as **BC** is caused by the **B** strand coming near the **C** helix. Note that this is also seen as an antiparallel interaction, and if you follow the direction of increasing alpha carbon indices in the helix you will see why this is so. There is a sudden backbone turn between helix **C** and helix **D** to the extent that **D** comes up against **C**, thus producing the darker area in the distance map labeled as **CD**. As seen in the structure image, strands **A** and **H** run parallel to one another and this is seen in the map as area **AH**, which runs parallel to the main diagonal but some distance away from it. Finally, it should be noted that the helices **C**, **D**, and **G** produce dark areas along the diagonal, a typical distance map signature for a helical structure where residues are interacting with other nearby neighboring residues. Below the diagonal, we have inserted the primary sequence of the protein along with the corresponding residue positions ranging from 1 to 73 inclusive.

3.5.1 Script for Generating Contact and Distance Maps

Suppose we have a 2D array of numbers with each number in the range [0, 1]. We wish to display a 2D *dot plot* of these numbers subject to the following rule: if the matrix number at row i and column j is between 0 and 0.5, then the corresponding dot plot cell $[i, j]$ is colored black; otherwise, it is colored white. The following function will draw a 2D plot that follows this rule:

```
import matplotlib.pyplot as plot
from matplotlib.colors import ListedColormap

def generate2Dplot(zG_A, plotTitle):

    # [0,.5) black;  [.5, 1.] white:
    cmap = ListedColormap(['#000000', '#FFFFFF'])

    n = int(zG_A.shape[0])
    xSeq = arange(0, n+1, 1)
    ySeq = arange(0, n+1, 1)
    xG, yG = meshgrid(xSeq, ySeq)
```

```
plot.ion()
plot.pcolor(xG, yG, zG_A, cmap = cmap)
plot.axis([0, n, n, 0])
plot.axes().set_aspect('equal')

plot.title(plotTitle)

raw_input("Hit Enter to dismiss the plot.")
plot.close()
```

This function is an adaptation of the density plot code (Script D.2 in Appendix D). Here is an explanation of the code. The first argument for the function call is the array of numbers, and the second argument is a title that will be displayed at the top of the plot. The first executable statement defines a list of colors that will be used in the plot. Hexadecimal numbers "#000000" and "#FFFFFF" designate black and white, respectively. This color map will be used in the `pcolor` function. The value of n is the number of atoms in the atom list, and the statements defining `xSeq`, `ySeq`, `xG`, and `yG` are used to set up a grid of coordinates that will act as the independent variables specifying locations for the cells in the dot plot.

The `plot.ion()` function call establishes interactive mode. With this established, each drawing command will immediately update the plot, so there is no need for a draw command. After all the updates are displayed in the plot, the script can issue a `close()` command so that control returns to the Python shell window.

The `pcolor` function works with the mesh grid and the distance map matrix to color the plot according to the color map specified by `cmap`. Since two colors are specified, the distance range [0.0, 1.0] is split into two equal subranges, and colors from the color list are given to cells according to the following rule: distance maps in the range [0.0, 0.5] color a cell black, whereas distance maps in the range [0.5, 1.0] color a cell white. If you wish to make a correspondence between a color and the interatomic distance, you need only multiply the range corresponding to the color by the value of the `distanceThreshold`. For example, with a `distanceThreshold` value of 16, any plot cell colored black corresponds to an interatomic distance between 0.0 and 8.0 Å.

The `axis` function establishes tick values for the horizontal and vertical axes of the plot, whereas the command `axes().set _ aspect('equal')` makes each of the plot cells a perfect square. The remaining lines have obvious functionality.

Script 3.01, available from the book's website (www.structuralbioin formatics.com), contains the `generate2Dplot` function along with a mainline script that fetches a PDB file, extracts the alpha carbon atoms from a specified chain, and then displays a contact or distance map plot depending on the choice of `cmap` statements. The mainline works with the `Shell` class to illustrate the use of this class, and it does save some time because interatomic distances are only evaluated for atom pairs with an interatomic distance that is less than the `distanceThreshold` value. This allows us to compute all needed distances in $O(n)$ time because the number of atoms that are within the distance threshold from some specific alpha carbon is bounded above by some small constant. Strictly speaking, the entire algorithm is $O(n^2)$ because the initialization of the distance map array [done by the `ones((n,n), float)` call] is intrinsically $O(n^2)$.

To adapt the script to your own requirements, it is expected that you would choose the appropriate definition of `cmap` and also establish the value of `distanceThreshold`.

3.6 INERTIAL AXES

3.6.1 Motivation

Very often in a structural analysis, we want to approximate a secondary structural element with a single straight line. For example, Figure 3.5 shows a straight line that acts as the longitudinal axis of a helix.

Figure 3.6 shows the alpha carbon atoms of the same helix and gives us an indication of a reasonable strategy for the determination of the axis: we should have the straight line positioned among the atoms so that it is closest to all these atoms in a least-squares sense. This is called the *inertial axis*.

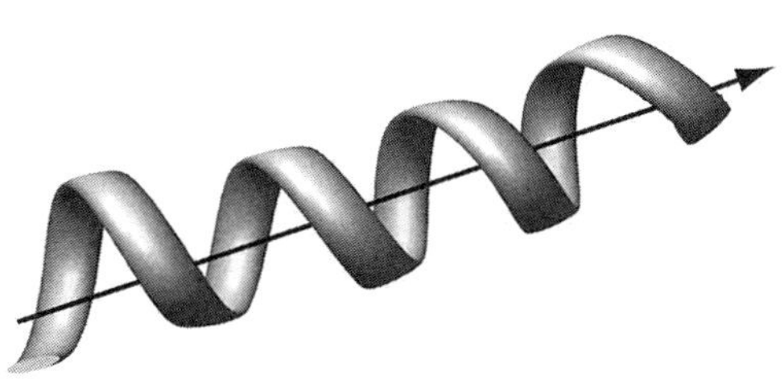

FIGURE 3.5 Longitudinal axis for a helix.

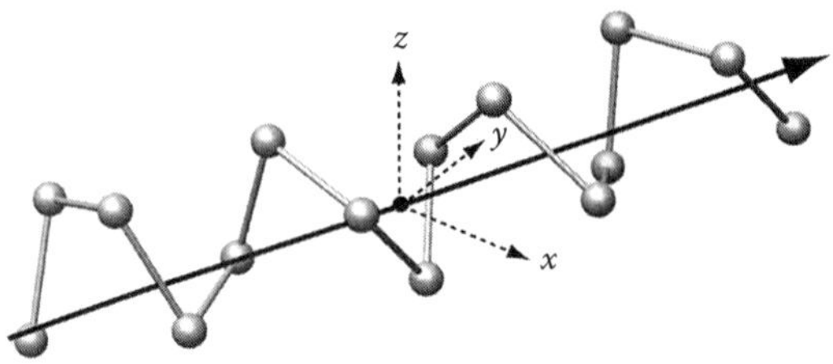

FIGURE 3.6 Longitudinal axis defined relative to the alpha carbon atoms of the helix.

Various research studies have used the inertial axis as a simplifying geometric construct to represent secondary structural elements. This is especially important when one wishes to compare the tertiary structures of related proteins. The following references are good starting points: [Aq86], [AM88], [OB92], and [YS99].

Finally, here is an interesting note related to the principal axes of inertia computed for an entire protein (taking into account atomic masses): some researchers have observed that, with high frequency, at least one principal axis penetrates the surface of a protein in a region used for ligand binding (see [FR00]).

3.6.2 Mathematical Analysis

Our strategy for computing the inertial axis is to have the axis pass through the centroid of the atoms with a direction chosen to minimize the sum of squares of the perpendicular distances from the atoms to the helix axis.

More precisely, if $\|d_i\|$ represents the perpendicular distance between atom $\mathbf{a}^{(i)}$ and the helix axis, then we choose the axis direction so that the sum S is minimized with S defined as

$$S = \sum_{i=1}^{N} \|d_i\|^2 \tag{3.3}$$

Depending on the needs of the application, it is possible to define the position of this axis using a different set of atoms, for example, all the atoms in the backbone or even all the atoms in the helix (backbone plus side chains). Before going further in this discussion, it should be stated that such an axis could also be computed to go among the atoms of a beta strand or, in fact, any arbitrary set of atoms.

We now provide an analysis that shows how the axis is dependent on the coordinates of the chosen atoms. To make the analysis simpler, we first consider one atom; call it **a**, with position vector $a=\left(a_x, a_y, a_z\right)^{\mathrm{T}}$. As noted earlier, this vector is using a frame of reference that is an x, y, z coordinate system with origin positioned at the centroid of the atoms under consideration (Figure 3.6). The perpendicular distance from the atom to the axis is illustrated in Figure 3.7. We are trying to determine the values of w_x, w_y, and w_z that define the directions of the inertial axis. These are set up as components of a unit vector w, and so $w=\left(w_x, w_y, w_z\right)^{\mathrm{T}}$ with $w^{\mathrm{T}}w=1$. The required axis is essentially a scalar multiple of w, and d represents the perpendicular vector going from atom **a** to this axis.

Some trigonometry defines the square of the norm of d as

$$\|d\|^2=\|a\|^2\sin^2\theta=\|a\|^2\left[1-\cos^2\theta\right]=\|a\|^2\left[1-\frac{\left(a^{\mathrm{T}}w\right)^2}{\|a\|^2\|w\|^2}\right] \tag{3.4}$$

We have replaced $\cos\theta$ with its equivalent formulation in terms of a dot product. As $w^{\mathrm{T}}w=\|w\|^2=1$, we can remove $\|w\|^2$ from the denominator and replace the 1 in the square brackets in Equation 3.4 with $\|w\|^2$. Then, multiplying the terms in the large square brackets by $\|a\|^2$ yields

$$\|d\|^2=\|a\|^2\cdot 1-\left(a^{\mathrm{T}}w\right)^2=\|a\|^2\|w\|^2-\left(a^{\mathrm{T}}w\right)^2 \tag{3.5}$$

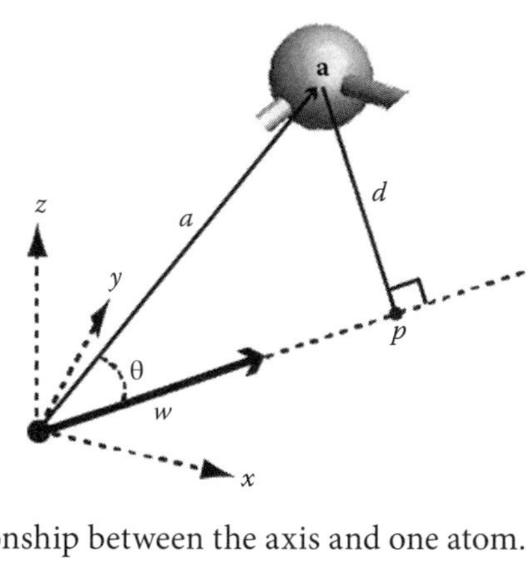

FIGURE 3.7 Relationship between the axis and one atom.

The replacement of 1 with $\|w\|^2$ might seem a strange step in this derivation, but it is extremely useful because we now have a formula for $\|d\|^2$ in which w appears in each term as a quadratic. We will see that this leads to an elegant matrix formulation for $\|d\|^2$ and lays the groundwork for a subsequent strategy to solve our minimization problem. Deriving the matrix formulation for $\|d\|^2$ is straightforward but somewhat daunting due to the lengthy expressions that are involved. We continue as follows:

$$\begin{aligned}\|d\|^2 &= \left(a_x^2 + a_y^2 + a_z^2\right)\left(w_x^2 + w_y^2 + w_z^2\right) - \left(a_x w_x + a_y w_y + a_z w_z\right)^2 \\ &= Aw_x^2 + Bw_y^2 + Cw_z^2 - 2\left[Fw_y w_z + Gw_z w_x + Hw_x w_y\right]\end{aligned} \tag{3.6}$$

where

$$\begin{aligned} A &= a_y^2 + a_z^2 \qquad & F &= a_y a_z \\ B &= a_z^2 + a_x^2 \qquad & G &= a_z a_x \\ C &= a_x^2 + a_y^2 \qquad & H &= a_x a_y \end{aligned} \tag{3.7}$$

Now the quadratic formulation pays off and we can write as follows:

$$\|d\|^2 = \begin{pmatrix} w_x \\ w_y \\ w_z \end{pmatrix}^{\mathrm{T}} \begin{bmatrix} A & -H & -G \\ -H & B & -F \\ -G & -F & C \end{bmatrix} \begin{pmatrix} w_x \\ w_y \\ w_z \end{pmatrix} \tag{3.8}$$

This gives us an elegant and concise representation for the $\|d\|^2$ value related to a single atom a; but you will recall that we want to minimize S, which is the sum of all such squared norms going across all the N atoms in the set. We let the coordinates of the ith atom $\mathbf{a}^{(i)}$ be represented as $a^{(i)} = \left(a_x^{(i)}, a_y^{(i)}, a_z^{(i)}\right)^{\mathrm{T}}$; then

$$S = \sum_{i=1}^{N} \|d_i\|^2 = \begin{pmatrix} w_x \\ w_y \\ w_z \end{pmatrix}^{\mathrm{T}} \begin{bmatrix} D_{xx} & -E_{xy} & -E_{xz} \\ -E_{yx} & D_{yy} & -E_{yz} \\ -E_{zx} & -E_{zy} & D_{zz} \end{bmatrix} \begin{pmatrix} w_x \\ w_y \\ w_z \end{pmatrix} \tag{3.9}$$

where

$$
\begin{aligned}
D_{xx} &= \sum_{i=1}^{N} \left(a_y^{(i)}\right)^2 + \left(a_z^{(i)}\right)^2 & E_{yz} &= E_{zy} = \sum_{i=1}^{N} a_y^{(i)} a_z^{(i)} \\
D_{yy} &= \sum_{i=1}^{N} \left(a_z^{(i)}\right)^2 + \left(a_x^{(i)}\right)^2 & E_{xz} &= E_{zx} = \sum_{i=1}^{N} a_z^{(i)} a_x^{(i)} \\
D_{zz} &= \sum_{i=1}^{N} \left(a_x^{(i)}\right)^2 + \left(a_y^{(i)}\right)^2 & E_{xy} &= E_{yx} = \sum_{i=1}^{N} a_x^{(i)} a_y^{(i)}
\end{aligned}
\tag{3.10}
$$

This gives a concise representation for S:

$$S = w^{\mathrm{T}} T w \tag{3.11}$$

where

$$T = \begin{bmatrix} D_{xx} & -E_{xy} & -E_{xz} \\ -E_{yx} & D_{yy} & -E_{yz} \\ -E_{zx} & -E_{zy} & D_{zz} \end{bmatrix} \tag{3.12}$$

Equation 3.12 is dependent only on the components of the vectors holding the atomic positions of all atoms in the given set. In physics, T is called the *inertial tensor* and it often appears in discussions involving studies of rotational inertia.

We can now formulate our minimization problem as a Lagrange multiplier problem in which we minimize S subject to the constraint $w^{\mathrm{T}} w = 1$. The Lagrangian will be as follows:

$$L(w, \lambda) = w^{\mathrm{T}} T w - \lambda\left(w^{\mathrm{T}} w - 1\right) \tag{3.13}$$

where λ is the Lagrange multiplier. Proceeding with the usual differentiation, we get

$$\frac{\partial L}{\partial w} = 2Tw - 2\lambda w \tag{3.14}$$

Setting this to 0 gives us the following eigenvalue problem:

$$Tw = \lambda w \tag{3.15}$$

The 3 × 3 tensor matrix T is symmetric, and considering that S is always positive for any vector w we see that T is also positive definite. Consequently, all three eigenvalues are positive. Note that once we have an eigenvalue–eigenvector pair we can write

$$S = w^{\mathrm{T}} T w = \lambda w^{\mathrm{T}} w = \lambda \tag{3.16}$$

Therefore, we have the solution to our problem. Once T is calculated, we compute its eigenvalues and the required w (specifying the directions for our axis) will be the eigenvector corresponding to the smallest eigenvalue. It should be noted that three eigenvectors are produced by this procedure and, because of the symmetry of T, they are mutually orthogonal and can be used as an orthogonal basis for the set of atoms under consideration. In some applications, this *inertial frame of reference* is a useful construct and the eigenvectors are called the principal axes of inertia.

3.6.3 Script for Computing the Inertial Axis

Chimera already has a menu item (**Tools...Depiction...PipesAndPlanks**) that provides this functionality, but it is instructive to develop the code that computes the inertial axis and, of course, having your own code means that you can modify the algorithm to meet any individual needs. The algorithm has been implemented in the class called `InertialAxes`, which is imported as follows:

```
from StructBio.LeastSquares.inertialAxes import InertialAxes
```

Specifications for the class are as follows:

Input for `InertialAxes` object initialization:

A list of atom objects

Input for `projectionOnAxis()` an `InertialAxes` method:

An atom object

Input for `distanceToAxis()` an `InertialAxes` method:

An atom object

Let us go over some of the essential core functions that are needed for the problem. We start with the routines that are to generate the inertial axis for a list of ordered atoms, for example, the N, CA, and C atoms in a protein backbone. In the code that follows, we assume that `atoms_L` is a list of atoms prepared by the mainline script.

As seen in Figure 3.6, we need to compute the centroid for atoms in the list:

```
def centroid(self):
    sum = zeros(3, float)
    for a in self.atoms_L:
        sum += a.coord()
    return sum/float(len(self.atoms_L))
```

Code for computing the centroid is quite straightforward and is presented mainly to illustrate the utilization of the collection of atoms that will determine the inertial axis. In the discussion that follows, it is important to realize that atoms will have coordinates defined in one of two frames of reference. These are the original PDB frame of reference and a frame of reference established by shifting the origin to the centroid. The latter is needed when doing computations for the inertial axes corresponding to a particular atom list. The next function, `tensorMatOneAtom`, will be responsible for the calculation of the matrix in Equation 3.8. First, note the following identity:

$$\begin{bmatrix} A & -H & -G \\ -H & B & -F \\ -G & -F & C \end{bmatrix} = \begin{bmatrix} a_y^2+a_z^2 & -a_xa_y & -a_za_x \\ -a_xa_y & a_z^2+a_x^2 & -a_ya_z \\ -a_za_x & -a_ya_z & a_x^2+a_y^2 \end{bmatrix} =$$

$$\begin{bmatrix} a_x^2+a_y^2+a_z^2 & 0 & 0 \\ 0 & a_x^2+a_y^2+a_z^2 & 0 \\ 0 & 0 & a_x^2+a_y^2+a_z^2 \end{bmatrix} - \begin{bmatrix} a_x^2 & a_xa_y & a_za_x \\ a_xa_y & a_y^2 & a_ya_z \\ a_za_x & a_ya_z & a_z^2 \end{bmatrix} =$$

$$\left(a_x^2+a_y^2+a_z^2\right)I - \begin{bmatrix} a_x \\ a_y \\ a_z \end{bmatrix} \begin{bmatrix} a_x & a_y & a_z \end{bmatrix}$$

So, given the coordinates of an atom, we can do some simple calculations on vectors to derive the matrix result. Note that `coordArray` is used to pass the coordinates of the atom to the function. If you are not familiar with the various array calculations available in numpy, then it may be of benefit to review some of the statements in this script:

```
# This computes the contribution made by one atom to the tensor
# matrix.
# The return value is a matrix that will be added to other matrices
# corresponding to other atoms.
# The 3D coordinates of the atom are provided by coordArray.
# This routine relies on the observation that this contribution is
# equal to:
# the identity matrix multiplied by the norm squared of the
# coordinates minus the outer product of the coordinate vector
# with itself.
# This routine assumes that coordArray holds the coordinates
# of the atom relative to the centroid.
# Note: array([coordArray]) is a 1 by 3 array.
def __tensorMatOneAtom(self, coordArray):
      sumSquares = sum(coordArray*coordArray)
      return sumSquares * numpy.identity(3) - \
            array([coordArray])*array([coordArray]).transpose()
```

With these two functions being available, generation of the inertial axes can be done in the _init_ function for the class:

```
def __init__(self, atoms_L):
    self.atoms_L = atoms_L
    self.centroid = self.centroid()

    # List of vectors that are the inertial axes:
    self.inertialAxes_L = []

    # The primary axis will be at index 0.
    self.primaryAxis = None

    tensorMat = numpy.zeros((3,3))
    for atom in self.atoms_L:
        tensorMat += self.__tensorMatOneAtom(numpy.array(atom.coord()) -
                                                  self.centroid)

    # With the tensor matrix now computed, we extract the eigenvector
    # corresponding to the smallest eigenvalues:
    eigvals, eigvecs = linalg.eig(tensorMat)
    ind = eigvals.argsort()    # Sort the eigenvalues.
                    # Be careful here to extract eigenvectors
                    # from the transpose of eigvecs!
    for i in range(3):
        self.inertialAxes_L.append(eigvecs.transpose()[ind[i]])

    self.primaryAxis = self.inertialAxes_L[0]
```

The initialization function sums the matrix contributions made by all the atoms in the atom list to get the matrix in Equation 3.12. It then solves the eigenvalue problem described in Equation 3.15 and defines the primary axis (the inertial axis that we want) to be the eigenvector corresponding to the minimum eigenvalue.

On completion of the initialization function, we have the primary axis defined as a direction vector that originates from the centroid. Referring to Figure 3.7, the direction vector is shown as w and the origin of the x, y, and z axes are positioned at the centroid. For many applications, we will need some extra functionality, which is provided by the following two functions:

```
# Given an atom object this function gets the coordinates of
# the atom and projects these coordinates onto the primary axis.
# It returns the coordinates of this projection.
# The coordinates are given in the PDB frame of reference.

def projectionOnAxis(self, atom):
    # Compute atom coords relative to centroid.
    atmCoords = numpy.array(atom.coord()) - self.centroid
    return sum(self.primaryAxis * atmCoords) * self.primaryAxis + \
         self.centroid
    # The centroid was added to convert back to the PDB
    # frame of reference.

# -------------------------------------------------------
# Given an atom object this function computes
# the perpendicular distance from the atom to the primary axis.
def distanceToAxis(self, atom):
    atmCoords = numpy.array(atom.coord())
    return numpy.linalg.norm(atmCoords - self.projectionOnAxis(atom))
```

The first function `projectionOnAxis` computes the coordinates of the projection of an atom onto the primary axis. Referring to Figure 3.7, we are computing the coordinates of the point labeled p. This is done by shifting the coordinates of the atom so that its frame of reference has origin at the centroid. This essentially gives us the vector going from the origin to the atom labeled **a** in Figure 3.7. Computing the inner product of this vector with the vector $\mathbf{w}$ will give the distance from the origin to the point p. This distance can be used as a scaling factor that stretches vector $\mathbf{w}$ out to point p. Then, adding the centroid to this p vector gives us the coordinates of p in the PDB frame of reference.

The function `distanceToAxis` computes the distance between an atom and the primary axis. The computation is very straightforward. We use the `norm` function to calculate the distance between the atom and its projection on the axis.

The `InertialAxes` class will be used in Exercises 7, 8, and 9 at the end of this chapter.

3.7 DEHYDRONS

3.7.1 Motivation

As discussed in [CH05], hydrophobic attraction between two nonpolar groups involves a major electrostatic contribution because it increases the hydrogen bonding among the surrounding water molecules. As a reminder on the strength of hydrophobic interactions, recall from Section 1.4.3.3 that a disease state (sickle-cell anemia) can be induced by the creation of a hydrophobic surface patch due to a mutational change in a single amino acid. It is well known that proteins fold due to electrostatic forces [AN73], [FS02], [CH05], [JP03]. In particular, hydrophobic collapse into the folded state is likely coupled with hydrogen bond formation [FK03]. This is supported by recognizing that the backbone of an unfolded protein is highly polar due to the amide and carbonyl groups in each residue. In an unfolded state, these groups are typically surrounded by polar water molecules, so the hydrophobic collapse and its concomitant removal of water (dehydration) leads to the thermodynamically favored formation of hydrogen bonds between the carbonyl and the amide groups.

Empirical evidence has shown that the strength and stability of a hydrogen bond depends on various structural attributes such as the atom types acting as donor and acceptor; distance between donor and acceptor; and angle formed by the three atoms (donor, hydrogen, and acceptor). But this is not the full story. The strength and stability of the hydrogen bond also depends on the local *microenvironment* of the bond. In particular, the strength of the bond is best preserved when it is shielded from the weakening effects of nearby polar water molecules.

Consequently, a *sustainable* hydrophobic collapse involves the *formation* and *protection* of hydrogen bonds occurring between the various backbone amides and carbonyls.

Using the terminology and procedures introduced by Fernández and Scott [FS03], we consider a hydrogen bond to be protected when it is sufficiently surrounded or *wrapped* by hydrophobic carbonaceous nonpolar groups having the form CH_n, $n = 0, 1, 2, 3$. The microenvironment containing these groups will be defined as the interior of two intersecting *desolvation* spheres with radii 6.5 Å and centered at the alpha carbons of the residues containing the hydrogen bonded amide and carbonyl groups.

Using the guidelines stipulated in [FS03], we consider a hydrogen bond to be *deficiently* wrapped, or *under wrapped*, if the number of carbonaceous wrapping atoms is less than or equal to 19. Deficiently wrapped hydrogen bonds are referred to as dehydrons. As noted in [FS03], these simple rules provide a fast but not exact strategy for the determination of deficiently wrapped hydrogen bonds. Figures 3.8 and 3.9 show well-wrapped and deficiently wrapped hydrogen bonds. In both figures, the atoms and bonds are represented in wire form with the following exceptions: dehydrons appear as green lines, well-wrapped hydrogen bonds appear as gray lines, the very large translucent desolvation spheres are colored blue (red) if the central alpha carbon is in the residue providing the hydrogen bond donor (acceptor), the smallest spheres designate the alpha carbon centers of the desolvation spheres, the middle-sized spheres represent the carbonaceous

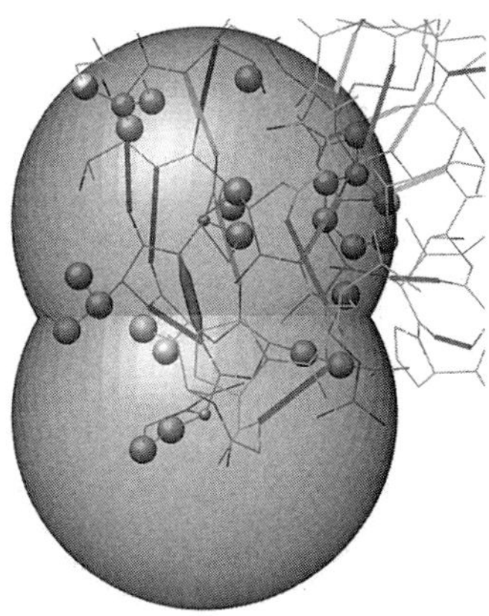

FIGURE 3.8 **(See color insert.)** Well-wrapped hydrogen bond.

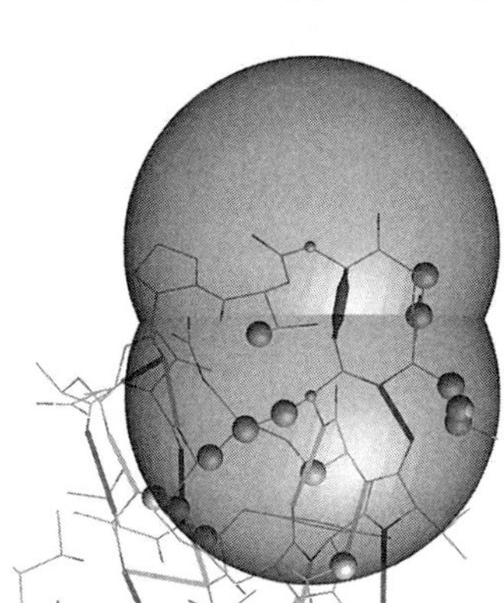

FIGURE 3.9 **(See color insert.)** Dehydron: a deficiently wrapped hydrogen bond.

wrapping groups, and the spindle designates the hydrogen bond under consideration (colored green if it is a dehydron, otherwise gray).

Counting the carbonaceous atoms (25 in Figure 3.8 and 14 in Figure 3.9) reveals the categorization of the hydrogen bond when each count is compared with the value 19. Figure 3.8 shows the CYS 26–PRO 22 hydrogen bond of 1CRN, whereas Figure 3.9 shows the GLY 20–ARG 17 hydrogen bond of 1CRN.

Why are dehydrons important? There is mounting evidence that wrapping deficiencies constitute areas on the protein surface that are essentially an invitation for other molecules to form a local binding that reduces or eliminates the wrapping deficiency. Consequently, the wrapping defect is a *sticky* substructure that can play a role in various protein associations such as protein–protein interactions, ligand binding, and deleterious aggregations such as the formation of amyloid plaques.

Over the last decade, several papers have been written dealing with dehydrons and their relevance to various protein topics. The reader may consult the following references: dehydrons and their utility for drug design [FE10], [FC08], [FR04], protein evolution and complexity [FS04], and protein–protein interfaces [FSS03].

Work done by Fernández and Berry [FB10] indicates that an abundance of dehydrons often represents a structural deficiency that is compensated by an inordinately large number of disulfide bonds. We return to this discussion in Exercise 14.

3.7.2 Script for Identifying Dehydrons

Figures 3.8 and 3.9 were created by a program (presented as Exercise 11). It makes use of the `Dehydrons` class, which is contained in the `StructBio` package. The class can be accessed by using the following import statement:

```
from StructBio.HBonds import Dehydrons
```

The arguments for the instantiation of the `Dehydrons` class are as follows:

`prot`: the protein molecule object that will be processed to find dehydrons.

`distSlop`: the distance slop used when this instantiation calls `FindHBond.findHBonds` (default value 0.4).

`angleSlop`: the angle slop used when this instantiation calls `FindHBond.findHBonds` (default value 20.0).

`sphereRadius`: the radius used for the desolvation spheres (default value 6.5).

`wrapThreshold`: if the number of carbonaceous atoms in the desolvation spheres is less than or equal to this threshold, then the hydrogen bond is considered to be a dehydron (default value 19).

`lineWidth`: the width of the lines used in the wire representation of `prot` when designating a hydrogen bond (default value 5.0).

After the instantiation of the `Dehydrons` class, we can use the following attributes to access data related to the hydrogen bonds:

`hBonds_L`: this is a list containing 2-tuples, each tuple holding a pair of atom objects corresponding to the donor and acceptor atoms for a hydrogen bond. As the description of a dehydron involves alpha carbon atoms, the `hBonds_L` list will only contain hydrogen bonds between standard amino acid residues. Consequently, hydrogen bonds between standard residues and water molecules will have been filtered out.

`dehydrons_L`: this is a list of dehydrons each designated by a 2-tuple. It is a subset of the entries that are listed in `hBonds_L`.

`wrapper_D`: this is a dictionary that allows one to determine all the carbonaceous atoms in the microenvironment of a hydrogen bond. A key will be a (donor, acceptor) atom pair that defines a hydrogen bond, and the corresponding dictionary value will be a `Set` that holds the atom objects designating all the carbonaceous atoms in the microenvironment.

See the final exercises of this chapter to find examples of scripts that use the `Dehydrons` class.

3.8 EXERCISES

The following exercises will deal with distance calculations and will include problems that incorporate the various classes covered in this chapter:

1. Testing the `Shell` class

   To demonstrate the use of the `Shell` class, write a mainline program that performs the following animation, subject to the following:

   a. Use the `raw_input()` function to get a PDB identifier from the user.

   b. Fetch the PDB file from the PDB (or from your local disk).

   c. Create a `Shell` object using all the atoms in the protein. Let `minR` = 2.0 and `maxR` = 4.0.

   d. Run a loop that goes over all the alpha carbon atoms in the protein, and for each such atom use a `HiLiter` object to emphasize the alpha carbon (a single member of the primary list) and its shell neighbors (set up in the secondary list). Use the verbose mode to print atom identifiers for both lists, under appropriate headings. For example, the first alpha carbon of 1HIV with `minR` = 2.0 and `maxR` = 4.0 should produce the following:

```
Alpha Carbon:
PRO 1.A@CA

Near neighbors in shell (r = 2.0, R = 4.0):
PRO 1.A@CG
PRO 1.A@CD
PHE 99.B@O
GLN 2.A@CA
GLN 2.A@N
HOH 353.A@O
PRO 1.A@O
```

2. Cysteine tetrahedrons

   Consider the rubredoxin protein from *Clostridium pasteurianum* (PDB ID 4RXN). This structure supersedes an earlier structural study reported by Watenpaugh and others [WS80] in 1980 when the PDB had only 70 entries (current content is over 102,000 structures). The protein has four cysteine residues (positions 5, 8, 38, and 41) surrounding an iron atom to make up the FeS_4 complex. In their paper, the authors report on various geometric properties of the complex including the Fe–S bond distances. Recall that Figure 3.1 shows this complex. If you fetch the protein and view it with Chimera, you will see that the sulfur atoms seem to be at the vertices of a tetrahedron. This is to be expected because of the coordination forces of the iron atom. A perfect tetrahedron is undoubtedly compromised by the extraneous forces placed on the sulfur atoms by their cysteine residues that have to meet the positional constraints imposed by the protein backbone.

In the spirit of this investigation described in [WS80], we pose the following additional question: how well do the sulfur atoms in these cysteine residues coordinate with the iron atom to form a tetrahedron? To answer this question, compute all six S–S distances and calculate the average distance. Then, calculate the percentage deviation of each distance from this average. The output format should be three columns with appropriate headings:

Column 1: a pair of integers identifying the two chain positions for the CYS residues

Column 2: the distance between the sulfur atoms in these two CYS residues

Column 3: the deviation of this distance from the average (expressed as a percentage)

See also 1RB9 and 1BRF. Note: to make the code simpler, you can take advantage of the fact that the structures each contain only four cysteine residues.

3. Finding disulfide bridges

Various researchers have been involved in the protein fold prediction problem. In some studies, algorithms can predict the conformation of the backbone but without specification of the disulfide bridges between cysteine residues. As a final step in the refinement of positions for the backbone atoms, a different algorithm will then try to predict the occurrence of these bridges and make adjustments to the backbone atoms so that the bridges can be put into the final predicted model. To do this, it is usually necessary to collect statistics that characterize the geometry of disulfide bridges. This seems to be a simple task, but it can be complicated if you have to deal with redundancy in the data. For example, if a protein is a dimer comprising two structurally identical chains, do you include disulfide bridges from both the chains in the collection of statistics? We will consider these issues to be beyond the scope of the question. To make the exercise more straightforward, our goal will be to write a Python script that deals with all the sulfide bridges in the entire protein.

Your script should fetch a protein file specified by user input typed into the Python Shell. It should find all cysteine residues, and for each such residue it should determine whether or not that residue is involved with a disulfide bridge. Be careful, not all cysteine residues are involved with bridges (e.g., the D chain of 1A4Y has several cysteine residues that are not involved with disulfide bridges). In Figure 3.10, we can see a disulfide bridge linking CYS 191.A with CYS 220.A in chain A of 1BQY. For each bridge found in the protein, your script should print out the following information:

a. Residue names and positions of the cysteine residues that are involved (e.g., CYS 191.A and CYS 220.A).

b. Distance between the alpha carbon atoms in the cysteine residues involved with the bridge.

c. If a cysteine residue is in a backbone but does not participate in a disulfide bridge, then it should be part of the output with a note indicating that there is no bridge structure associated with this residue.

Use the following as input for your program: 1BQY (all CYS involved with disulfide bridges), 1TCR and 1A4Y (a mix of bridged and non-bridged CYS residues), and 1ZNJ (note the bridges *between* chains).

Note: if you are studying cysteine bridges and need data, it might help to start with toxins from various species of scorpions, snails, and

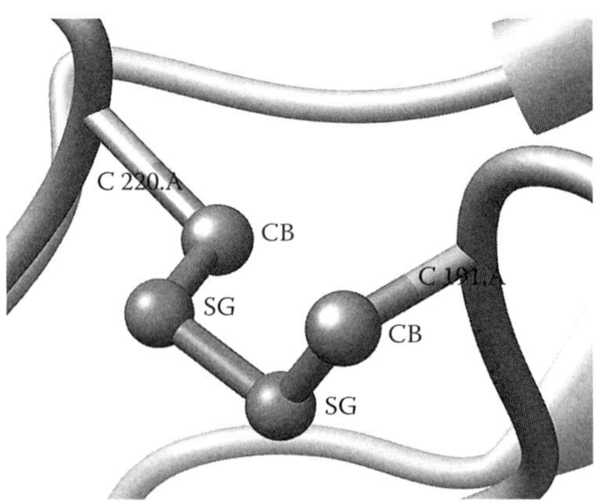

FIGURE 3.10 **(See color insert.)** Disulfide bridge in chain A of 1BQY.

snakes. Your investigation could start with the studies by Saucedo and others [SF12], Jin and others [JN07], and the references therein.

4. Distances between beta carbons in beta sheets

When considering beta sheets, are consecutive residues of a strand closer than neighboring residues on different strands? To answer this question, let us assume that the coordinates of a residue are approximated by the coordinates of the beta carbon atom within a residue. Write a program that reads a PDB file and calculates the average distance between C_β atoms of consecutive residues in a beta strand. See the dotted line in Figure 3.11 for an example of one such distance. The program should also calculate the average distance between neighboring C_β atoms of hydrogen bonded residues that are in different strands of the beta sheet (see the dashed line in Figure 3.11).

Test your program on the parallel and antiparallel beta sheets of a glycosyl hydrolase (PDB ID 1QBA). To test your program on other *beta sheet–rich* proteins, you can consult [ZK00] for some interesting examples.

Note: finding C_β atoms of consecutive residues in a beta strand is fairly easy. Finding neighboring C_β atoms of parallel and antiparallel strands will be more of a challenge, but you can apply the following strategy: generate an instance of the `HBonds` class and pull out all the hydrogen bonds that are in the category "PL" or "AP." Each donor and acceptor pair will correspond to atoms N_i and O_j in Figures 2.11 and 2.12. Now you can compute each of the needed distances (once and only once) by finding the residues containing N_i and O_j. For example, the neighboring C_β atoms in consecutive residues are in the donor residue and the residue just before the donor residue. The C_β atoms for the residues in different strands can be determined by studying Figures 2.11 and 2.12. One of them will be in the donor residue, whereas the partner will be in the residue holding the acceptor atom if the strands are running antiparallel or in the residue just after the acceptor residue if the strands are running parallel. Before

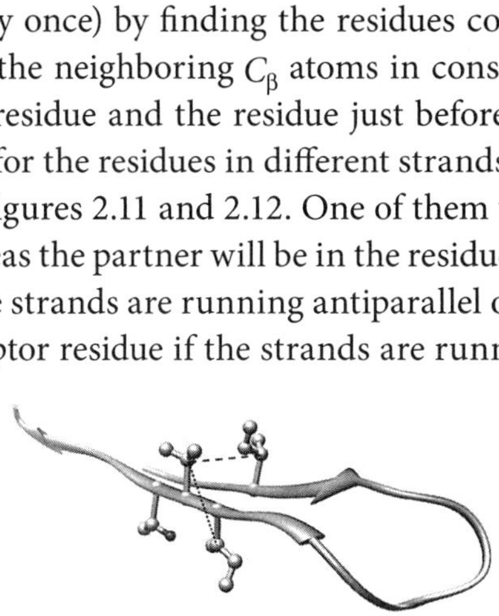

FIGURE 3.11 Distances between beta carbon atoms.

doing any computations, it will be necessary to ensure that all the C_β atoms actually exist. For example, glycine residues will be ignored because they do not have C_β atoms.

You can use the `HiLite` class to put visual emphasis on the atoms involved, and this will aid in debugging. The verbose mode can also be used for output of atom labels. In addition, your output should report on all averages and related statistics. The following output snippets provide an example of the output when the protein 1QBA was processed. Output for a typical beta carbon:

```
Beta carbon:
THR 72.A@CB
Consecutive and adjacent strand beta carbons:
PHE 71.A@CB
ASN 41.A@CB
Distance between consecutive beta carbons: 5.862
Distance between beta carbons across anti-parallel strands: 4.214
```

Final output lines are as follows:

```
Average distance between consecutive beta carbons: 5.719
Number of these distances: 96
Average distance between beta carbons in parallel strands: 5.181
Number of these distances: 31
Average distance between beta carbons in anti-parallel
strands: 5.007
Number of these distances: 65
```

5. Oxygen to imidazole ring distances

In a research study by Chenoweth and Dervan [CD09], they consider an eight-ring cyclic Py/Im polyamide molecule (1P1) that is bound to the central six base pairs of a short DNA structure (Figure 3.12). In this figure, the 1P1 ligand has a ball and stick display representation, whereas the DNA is shown in the pipes and planks depiction.

The study considers various issues related to the interaction of 1P1 and the DNA segment. This includes the measurement of distances between the O4′ oxygen atoms in the DNA backbone and the centroids of nearby imidazole rings (Figure 3.13). All the imidazole atoms are in a ligand accessible with Chimera as *residue type* 1P1 with position 500 in chain B. The following table provides the names of atoms within the 1P1 imidazole rings and for each ring the specification of the nearby oxygen atom (its position, chain identifier, and name). All these oxygen atoms have the name O4′.

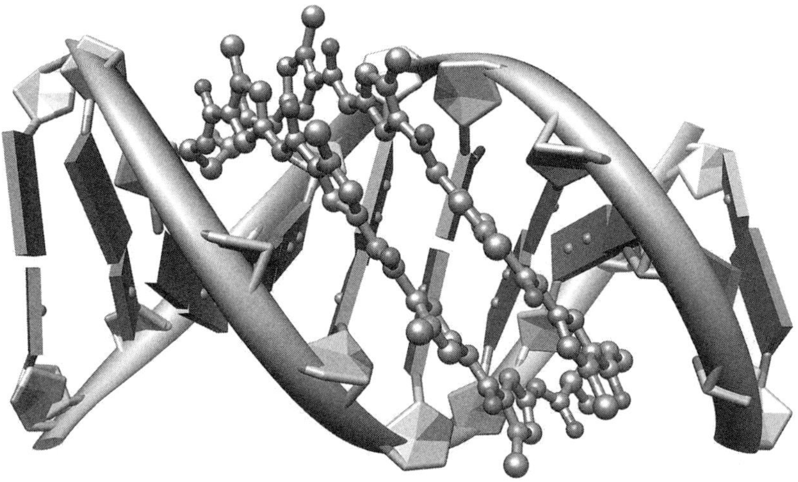

FIGURE 3.12 **(See color insert.)** DNA segment and polyamide ligand (Protein Data Bank identifier: 3I5L).

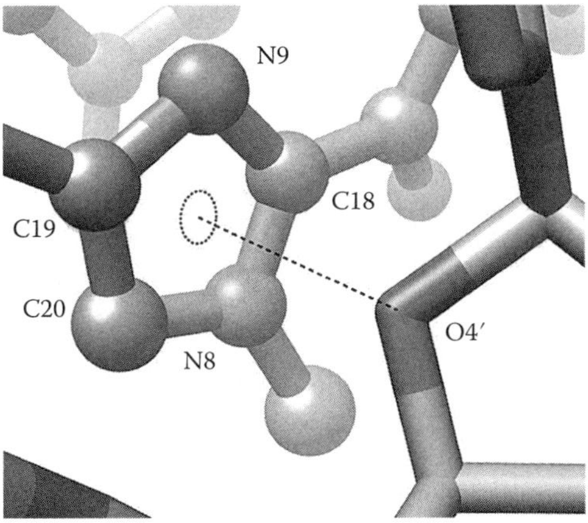

FIGURE 3.13 **(See color insert.)** Distance from centroid of imidazole ring to O4′.

(Chain B) 1P1	**DNA**
C13 C14 N5 C16 N6	6.B O4′
C18 N9 C19 C20 N8	5.B O4′
C39 C40 N17 C42 N18	6.A O4′
C44 N21 C45 C46 N20	5.A O4′

Your script should get all needed atomic coordinates and, for each of the four imidazole rings, it should calculate the centroid of the atoms making up an imidazole ring and then print out the distance from that centroid to the nearby O4′ atom. Figure 3.13 shows this centroid to O4′ distance as a dashed line.

6. Generating test data for distance geometry algorithms

   The distance geometry problem can be stated as follows: can you derive the coordinates of all the atoms in a molecule if you are given the distances between pairs of atoms? It can be shown that the problem is fairly easy if all possible $O(n^2)$ distances are accurately known. The problem becomes much more difficult if only a subset of the distances is given, for example, if we only get as input the distances between any atom and its nearby neighbors. Nearby in this context might mean within 5 Å. Researchers developing algorithms to solve this problem typically build input data sets that are used to test the capabilities of the algorithm. They start with a protein file from the PDB, generate the input test data, apply the distance geometry algorithm to calculate the coordinates of the atoms, and then measure the agreement between the calculated coordinates and the original coordinates from the PDB file.

   In this exercise, you are to write a script that asks the user for a PDB file identifier and a threshold. The script should fetch the PDB file and generate a file that lists all atom pairs such that the distance between the atoms in a pair is less than the specified threshold. A line in the output file should fully identify the atoms in an atom pair followed by the distance and then a code that indicates the type of interaction between the atoms. For example, the following are four lines taken from the test data file generated for 1CRN with threshold = 3.0 Å:

```
THR 1.A O     ILE 35.A N    2.859 H
THR 1.A O     THR 1.A N     2.668 N
THR 1.A O     THR 1.A CA    2.355 G
THR 1.A O     THR 1.A C     1.237 B
```

   An output line contains the residue type, residue position, residue chain identifier, and atom name within the residue for both atoms in the pair. This is followed by the distance (less than the threshold, of course) and the atom pair code. Use the following codes:

   B: bonded (the two atoms are covalently bonded)

   G: geminals (the two atoms are both bonded to a third atom)

H: hydrogen bonded (the two atoms have a hydrogen bond between them)

N: none of the above

The codes are of value if the test data distances are to be subsequently modified by adding some small amount of *noise*. Usually, the amount of noise to be added will depend on the nature of the atom pair. This exercise should make use of the `HBonds` and `Shell` classes. The test data should be written out to a text file that has a file name indicating that it contains near neighbor data, the threshold value, and the PDB identifier. For example, processing 1CRN with threshold 3.5 Å uses the file name NN_3_5_1CRN.txt.

7. Introducing the `Solids` class

   Exercise 8 makes use of the `Solids` class,* which can be used to display a set of elongated *bullet-like* structures that we will call tubes (shapes are similar to test tubes). In the next exercise, they will be used to represent the inertial axes of helical secondary structures. The class can be instantiated with an empty argument list, for example,

   ```
   tubeScene = Solids()
   ```

   After this, we can bring tubes into the display by adding them to the instance, for example,

   ```
   tubeScene.addTube(startPoint, endpoint, tubeRadius, tubeColour)
   ```

   Here, `startPoint` and `endPoint` are two `numpy` arrays specifying the 3D coordinates of two points in the display space. The radius of a tube is defined by the floating point number `tubeRadius`, and its color is given by `tubeColour`, which is a three-component tuple representing the color in RGB format. When all the tubes are added into the scene, we can display it using the following function call:

   ```
   tubeScene.display()
   ```

   To get familiar with this class, you should try it with various arguments. If you run Script 3.08, available from the book's website (www.structuralbioinformatics.com), you should get a scene that looks like Figure 3.14.

* A full description of this class can be found in Appendix B.

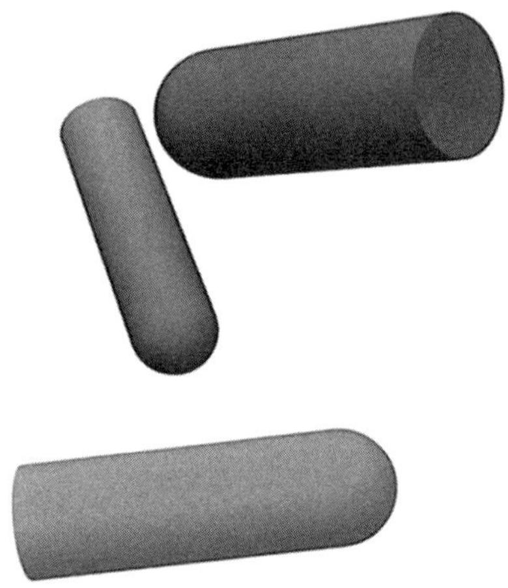

FIGURE 3.14 Testing addTube in the Solids class.

8. Using the InertialAxes class: inertial axis tube visualization

   Very often in a structural analysis, we want to approximate a secondary structural element with a single straight line or some type of visual structure that is positioned using such a line. In this exercise, you are to calculate the inertial axes for the helical secondary structures and then use these axes to define the end points of tubes (see previous Exercise 7) that are drawn within the helix structures. If the secondary structure ribbons are drawn in licorice form, your final display should look something like Figure 3.15.

   Note: the rounded end of a tube corresponds to the residue of the secondary structure that is farther along in the primary sequence. This exercise involves the following steps:

   a. Generate a tube display: write a Python script that opens a user-specified PDB file and subsequently generates a list of inertial axes, one for each SSE of the protein. Do this with a class called AxesForSSEs. Suggestion: this class should use the InertialAxes class. Caution: be careful when handling protein chains that might possibly end and start within a secondary structure; you do not want an axis that starts in one chain and continues into the next chain. The atoms contributing to the computation of an inertial axis will be alpha carbon atoms only. Test the AxesForSSEs class by displaying tubes in helices.

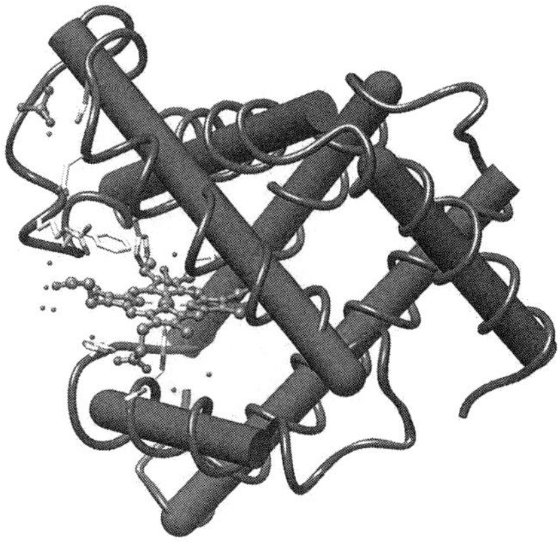

FIGURE 3.15 Inertial axes illustrated with tubes positioned by the `TubesForSSEs` class.

The code should then generate the end points defining the 3D positions of tubes that are to be drawn in the display. This can be done by using the `projectionOnAxis` function of the `InertialAxes` class. Assuming that the atom list is ordered, you can get the start point by projecting the coordinates of the first atom onto the primary axis and then get the end point by projecting the coordinates of the last atom onto the primary axis. After all the tube points are set up in a list of pairs, they can be used to generate tubes in the display by making use of the `addTube` function.

b. Recursive tubing: although the inertial axis provides a simple approximation for a secondary structural element, it does not do well when the secondary structure involves a sudden bend. In the 1MBO display in Figure 3.15, you will notice that there is a tube that is, at times, outside its helix structure. In this part of the exercise, you are to modify the script generated for part (a) of this exercise so that it uses a divide and conquer approach to the placement of tubes within secondary structures. The goal is to get a display such as that in Figure 3.16.

Note that the frontmost helix now has two tubes in place of the single tube shown in Figure 3.15. There are several ways to achieve this, and you should use your own ingenuity to devise a reasonable strategy. The simplest approach is to use a divide and conquer strategy that successively splits the atom lists into halves until all smaller lists produce axes that satisfy a tolerance criterion described next.

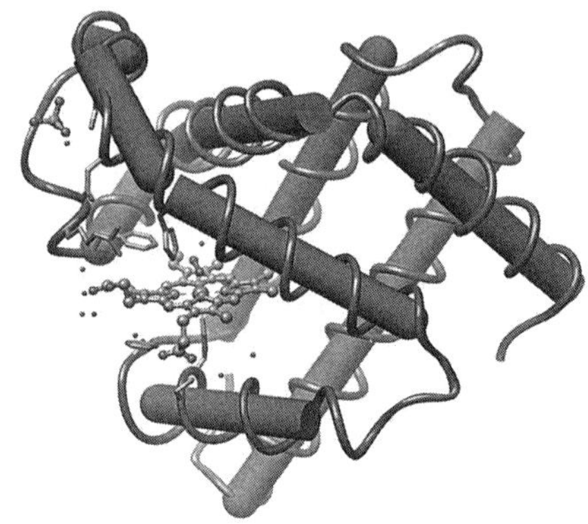

FIGURE 3.16 Using recursive tubes.

For each secondary structure, compute the distances between the alpha carbon atoms and the inertial axis. If the maximum distance is less than some prespecified tolerance (say 3.5 Å), then the axis is considered to be a faithful representation of the secondary structure. If the maximum distance is greater than the tolerance, then the set of residues in the secondary structure is split into two shorter helices and the end point generation function is called for each one. This exercise should not require any modification of the `InertialAxes` and `Solids` classes. Your goal is to modify the `AxesForSSEs` class so that it uses recursive calls to generate a list of inertial axes that meet the tolerance requirements.

Try your script on 1MBO, 2FIF, and 2BG9.

9. Combining hydrogen bond plates with inertial axis tubes

   Write a script that displays a protein using both hydrogen bond plates and inertial axis tubes. Test your script on various proteins including 2BYS.

   For 2BYS, the results should be similar to that displayed in Figure 3.17. This display of the nicotinic acetylcholine receptor (from the snail *Aplysia californica*) shows both the top and bottom views of the structure (both views can be seen simultaneously because of the orientations of the molecules in the crystal). The lobeline ligand is represented in sphere form.

10. Computation of helical hydrophobic moment

    The papers by Eisenberg and others [EW82a], [EW82b] deal with the spatial distribution of hydrophobic side chains in a protein. At this point in our studies, we have an excellent opportunity to extend this

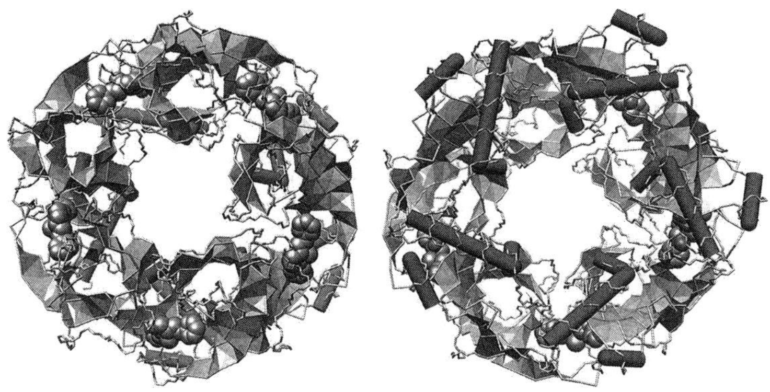

FIGURE 3.17 **(See color insert.)** Plates and tubes for 2BYS.

work so that computation of the hydrophobic moment is accompanied by a visual representation of the hydrophobic moment as an arrow that starts at the inertial axis of the helix and goes through the center of the side chain. There are several ways to define the hydrophobic moment of the side chains in a helix. Let us go with the following procedure:

a. After fetching a protein file, use the `TubesForSSEs` class to construct tube solids for all the helices and strands of the protein.

b. Note that you can access attributes of an object of this class to derive the primary axis of each SSE. The attribute `axes.inertialAxes_L` will be a list of objects instantiated from the `InertialAxes` class. Attributes of this class will give you access to the atom list (`atoms_L`), the centroid of these atoms (`centroid`), and the primary axis (`primaryAxis`).

c. For each residue R_i associated with an axis, compute s_i, defined as the center of the atoms in the side chain (exclude backbone atoms N, Ca, C, and O). Skip any glycine residues.

d. For each helix residue, project s_i onto its inertial axis. Designate the projection point on the axis as p_i. Compute N_i, which is a unit normal vector rooted at p_i with the same direction as the vector going from p_i to s_i. Define $H_i = h_i N_i$, where h_i is the Kyte–Doolittle hydrophobicity value for residue R_i. Note that H_i will point away from s_i if the residue is hydrophilic. Use the `Solids` class to visually display the normal H_i as an arrow. For a helix, you should get a display that looks something like that seen in Figure 3.18. Note that the coloring is consistent with the hydrophobicity of the corresponding residue.

FIGURE 3.18 Arrows for hydrophobic and hydrophilic moments.

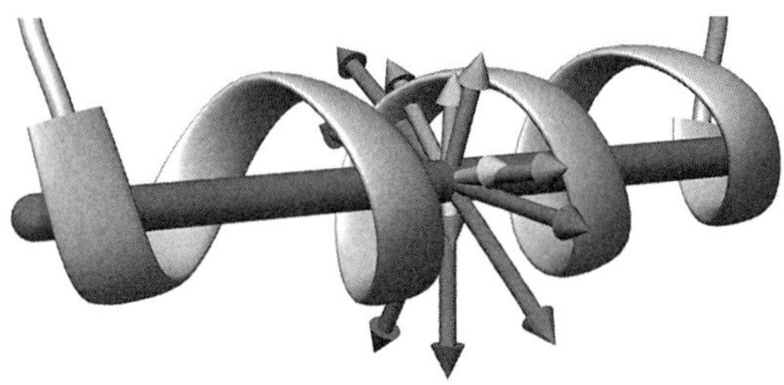

FIGURE 3.19 Arrows in a helical wheel.

e. At the request of the user, the display should change so that all the arrows are translated along the principal axis so that they are rooted at the centroid (Figure 3.19). This produces the *helical wheel* seen on page 372 of [EW82a]. In our case, the arrows in the wheel can be related to the surrounding structure. In particular, one can often see a clear indication of amphiphilicity with the hydrophobic arrows pointing to the core of the protein.

f. For each axis, compute the hydrophobic moment vector M_i as the vector sum of all V_i vectors, where each V_i is the H_i vector translated from p_i to c_i. Compute the sum of all V_i vectors that correspond to hydrophilic residues. Display both arrows. This should produce a result such as that shown in Figure 3.20.

g. Finally, the user should have a fourth display option that generates the previous vector sums divided by the number of residues involved. These average vectors will have an appearance that is similar to Figure 3.20 (the arrows are simply shorter).

Since the script involves a more extensive interaction with the user, it is best to implement two dialogs to facilitate user interaction. Test the script on various helix-rich proteins (e.g., 1JEB).

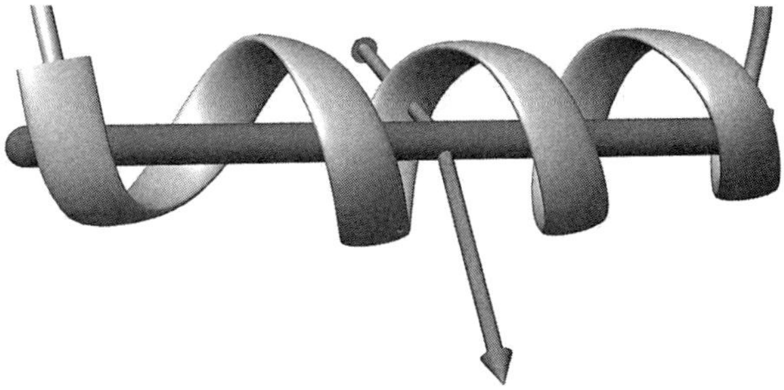

FIGURE 3.20 Sum of normal vectors.

11. Visual constructs for dehydrons (starter script)

    Write a script that does the following:

   a. Ask the user to type in a PDB identifier.

   b. Fetch the protein file and uses the following lines to display hydrogen bonds as gray-colored lines:

```
runCommand("select #" + str(prot.id)+ " & backbone.full")
runCommand("findhbond selRestrict both lineWidth " +
            str(self.line_width) +
            " color gray distSlop %4.1f" % self.dist_Slop +
            " angleSlop %4.1f" % self.angle_Slop)
runCommand("~sel")
```

   c. Invoke the function `FindHBond.findHBonds` to get a list of hydrogen bonds. Be sure to use the same `distSlop` and `angleSlop` parameter values as in the previous step.

   d. Use a `for` loop to go through the list of hydrogen bonds, and for each hydrogen bond generate the visual elements seen in Figures 3.8 and 3.9. Successive iterations of the loop can be paced by asking the user to hit the "Enter" key (use the `raw_input` function). You should use the `Dehydron` class both to find the carbonaceous atoms around each hydrogen bond and to determine whether it is a dehydron. As in Figures 3.8 and 3.9, use the following visual constructs:

      i. The hydrogen bond is emphasized by surrounding it with a spindle (radius = 0.25). Use the `Solids` class within `Scenographics` of the `StructBio` package to generate a spindle. The spindle should be colored green if it corresponds to a dehydron; otherwise, it is colored gray.

ii. Use the `HiLiter` class imported from `StructBio.Utilities.hiLite` to get small- and middle-sized spheres amid the wire representation seen in Figures 3.8 and 3.9. Entries for the primary list in the instantiation of `HiLiter` will be the atom objects corresponding to the carbonaceous atoms, and entries for the secondary list will be the two atom objects corresponding to the alpha carbons in the residues linked by the hydrogen bond.

iii. Define the microenvironment by using two transparent spheres that are ellipsoids with all semiaxes set to the `sphereRadius` value. Use the `Ellipsoid` class imported from `StructBio.Scenographics.ellipsoid` to generate the spheres along with the `displaySurface` method to put a colored transparent surface over the ellipsoid frame. Use color and transparency values that you consider to be most appropriate for your needs.

iv. Note: when the loop progresses from one hydrogen bond to the next, all the visual constructs associated with the previous hydrogen bond should be eliminated (use the `chimera.openModels.close` function).

Note that this script provides a significant number of visual constructs; but by using the classes provided by the `StructBio` package, they can be displayed with about two pages of script (about 40 lines).

12. Visual constructs for dehydrons (graphical user interface [GUI] script)

    Modify the script for Exercise 11 so that it works with a GUI. The following requirements should extend the capabilities of the script in Exercise 11:

    - The user should enter the PDB identifier via a text box.
    - The user should be able to select a single chain or the entire protein for further analysis.
    - In the case of nuclear magnetic resonance files, the user should be allowed to specify a model number.
    - There should be two options for the centers of the desolvation spheres:

  - The alpha carbon atoms of the residues holding the donor and acceptor atoms.
  - The donor and acceptor atoms.

- Before initiating function calls that find dehydrons, the user should be allowed to use default or text box–specified values for `distSlop`, `angleSlop`, `sphereRadius`, `wrapThreshold`, and `lineWidth`.
- Instead of using a `for` loop to cycle through all the hydrogen bonds, the visual constructs should be generated for a hydrogen bond that is selected in the Chimera display window. Note that Chimera does allow one to select a hydrogen bond. Designating a hydrogen bond by selecting an end atom (donor or acceptor) is not advised because in some situations a donor or acceptor may be involved with two hydrogen bonds and so there is some ambiguity in the choice of hydrogen bond.
- The GUI should allow the display representation to be either wire or ribbon. Since the path of the ribbon does not follow the backbone exactly (for the sake of smoothness, it is more like an approximation of the backbone), the spindle end points will have to be changed when the ribbon representation is used. Under these conditions, the coordinates of the spindle end would be taken from the `labelCoord` attributes of the donor and acceptor.
- The GUI should display data related to the hydrogen bond, such as the number of carbonaceous atoms, the length of the bond, and whether or not it is a dehydron.

13. Coloring H-bond plates according to wrapper count

    Recall the H-bond plates application described in Exercise 15 of Chapter 2. Modify the script contained in the `Generate_hBond_plates` class so that the plates are colored with respect to the number of carbonaceous atoms that are wrapping the hydrogen bond within each plate. Recall that this class is within the `StructBio.HBonds.generate_hBond_plates` package. You should allow the class to work with hydrogen bonds between all standard residues, not just the residues that are within helices and beta strands. The goal is to get structure visualizations like that shown in Figure 3.21, which illustrates STI-571 inhibition of a c-Kit tyrosine kinase [MD04].

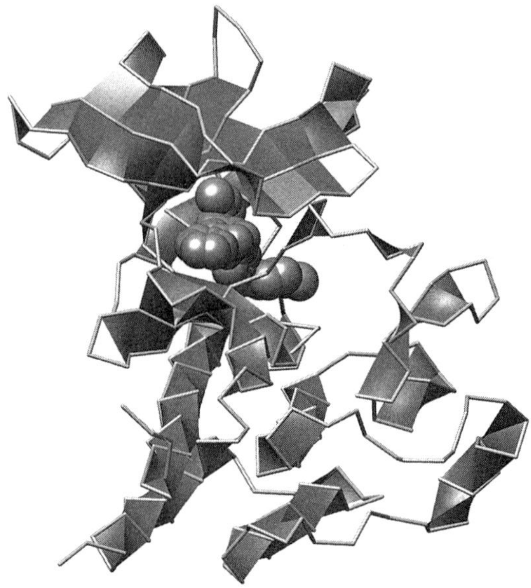

FIGURE 3.21 **(See color insert.)** Hydrogen bond plates colored with respect to wrapper atom count (1T46).

The following is a suggestion for the code snippet to compute the coloration:

```
# Normalize wrap count to a value in range [-1, 1].
normalizedWC = 0.1*(min(max(wC, wThrshld - 10),
                        wThrshld + 10) - wThrshld)
greenEntry = .5 - .5*normalizedWC
redBlueEntry = .5 + .5*normalizedWC
plateColor = (redBlueEntry, greenEntry, redBlueEntry)
```

where `wC` is the number of carbonaceous atoms around the hydrogen bond, and `wThrshld` sets the threshold for under wrapping (usually set to 19). With `wThrshld` set to 19, colors range from green to purple as the wrapper count goes from 9 to 29. The color gray indicates a count of 19.

Although the hydrogen plates of Exercise 15 in Chapter 2 may seem somewhat frivolous, the wrapper coloration gives them a biochemical relevance that is informative for a variety of discussions:

- The surfaces defined by the wrapper count–colored H-bond plates provide information about the local hydrophobic environment created by the hydrophobic collapse characterizing the protein folding process. Note that most of the H-bond plates in the interior of the protein have a purple color and the amphiphilicity of a helical secondary structure is immediately apparent.

- The chain traces bordering a plate and the knowledge that the plate is stabilized by the strength of the hydrogen bond across it serve to indicate where the protein has its most significant structural integrity.
- In some cases, the coloration of plates defining a binding pocket will help to elucidate the interaction of the protein and its substrate. For example, consider Figure 3.21. Wrapper atom counts did not include atoms in the STI-571 ligand, and so we see a wrapper deficiency for some of the hydrogen bonds near the binding pocket. For another example, see the note about HIV protease in Exercise 14.

14. Relating dehydrons to biological function

    Use the dehydron software built in Exercise 12 to investigate the occurrence of dehydrons in various proteins. For example, consider HIV protease and its peptide substrate. Figure 3.22 shows the apo form of HIV protease. Note the deficiently wrapped hydrogen bonds at the ends of the flap structures as indicated by the green coloration of the H-bond plates. When the protein interacts with its peptide substrate, hydrogen bonds are formed between residues in the flaps and residues in the substrate. This explains the extra plates that can

FIGURE 3.22 **(See color insert.)** HIV protease without substrate (1HIV with ligand removed).

be observed in Figure 3.23 immediately above and below the substrate. This figure also shows that the substrate has taken a position within the binding pocket that allows it to donate wrapper atoms to the desolvation spheres of the hydrogen bonds in the flaps, and so their plates are now colored purple.

Here are some more suggestions:

- Acetylcholine binding site: the apo form of the acetylcholine binding protein (PDB identifier: 2Y7Y) has a flap region that hovers over the binding site. It has three dehydrons defined by the atom pairs (CYS 189, TYR 186), (GLN 184, TYR 143), and (TYR 143, GLN 184). What happens to each of these dehydrons when the protein binds to cocaine (PDB identifier: 2PGZ) and to the toxin curare (PDB identifier: 3PMZ)?
- Protein dimers: formation of protein dimers is often accompanied by a loss of dehydrons (see [FS03]). Use the scripts built for Exercise 13 to illustrate this.
- Golden rule: as noted in Section 3.7.2, some proteins with an extraordinarily high number of dehydrons use disulfide bridges to achieve more structural stability [FB10]. This can be seen for proteins such as hirudin (1HIC, Figure 3.24) and ribonuclease A (1RHB, Figure 3.25). The antifreeze proteins found in some insects such as *Tenebrio molitor*, a darkling beetle that is the adult form of the mealworm larva, are more remarkable. This protein (1EZG) is

FIGURE 3.23 **(See color insert.)** HIV protease with substrate (1F7A).

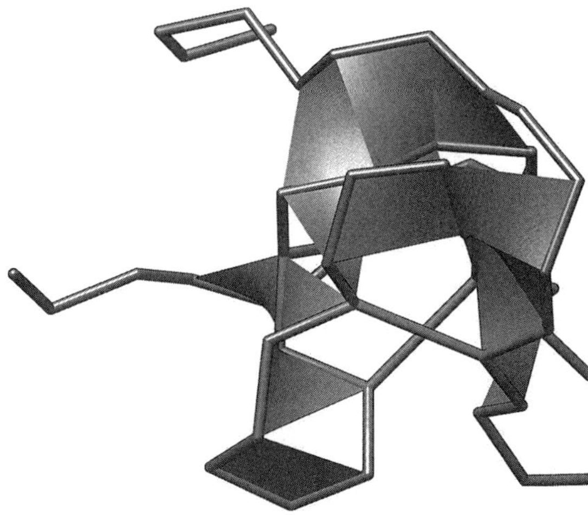

FIGURE 3.24 **(See color insert.)** Hirudin (1HIC).

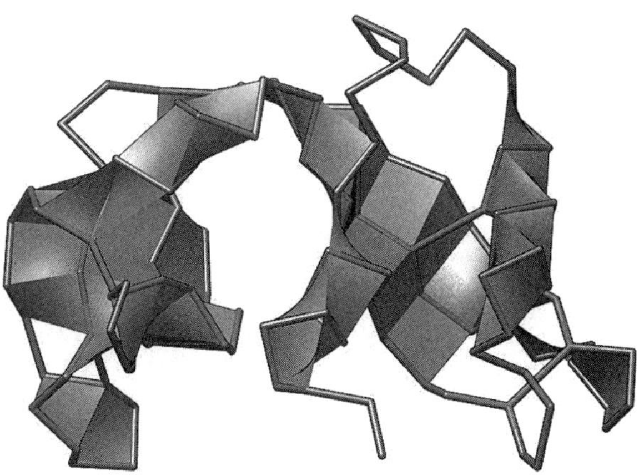

FIGURE 3.25 **(See color insert.)** Ribonuclease A (1RHB).

shown in Figure 3.26 using wrapper count–colored H-bond plates with disulfide bridges emphasized with a gold color. Recall that the color green is used to indicate a hydrogen bond that is deficient in hydrophobic protection. Another antifreeze protein (2PNE) can be seen in Figure 3.27. This protein can be found in the snow flea, which is actually a springtail insect that forages for algae on the surface of snow (http://www.youtube.com/watch?v=VjLKzogOj8Q). This protein has been studied by researchers [LG07] in Canada (a country not unfamiliar with snow).

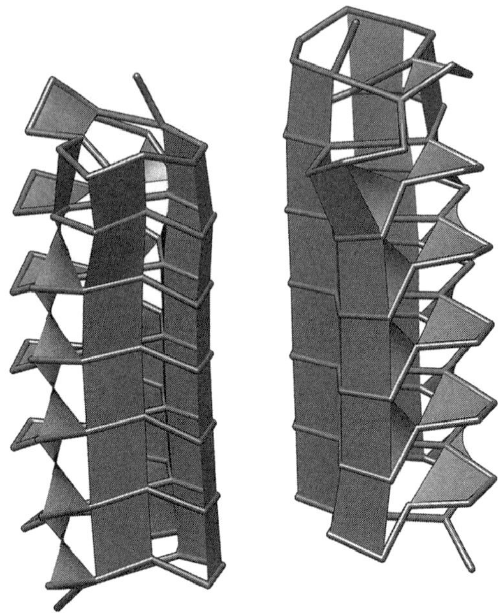

FIGURE 3.26 **(See color insert.)** *Tenebrio molitor* antifreeze protein (1EZG).

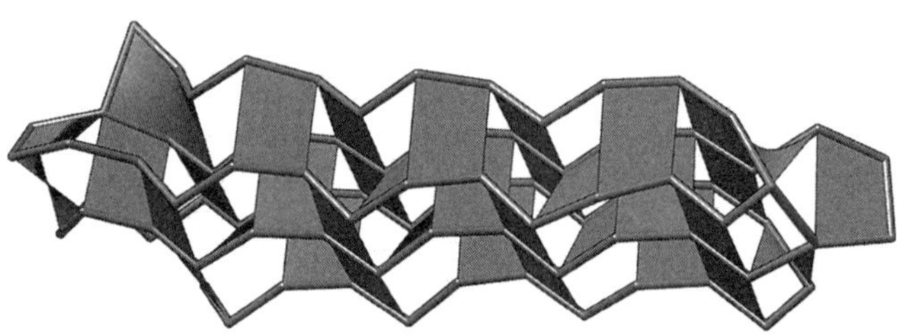

FIGURE 3.27 **(See color insert.)** Snow flea antifreeze protein (2PNE).

REFERENCES

[AM88] R. A. Abagyan and V. N. Maiorov. A simple quantitative representation of polypeptide chain folds: Comparison of protein tertiary structures. *Journal of Biomolecular Structure and Dynamics*, **5** (1988), 1267–1279.

[An73] C. B. Anfinsen. Principles that govern the folding of protein chains. *Science*, **181** (1973), 223–230.

[Aq86] J. Åqvist. A simple way to calculate the axes of an α-helix. *Computers & Chemistry*, **10** (1986), 97–99.

[CD09] D. M. Chenoweth and P. B. Dervan. Allosteric modulation of DNA by small molecules. *Proceedings of the National Academy of Sciences of the United States of America*, **106** (2009), 13175–13179.

[Ch05] D. Chandler. Interfaces and the driving force of hydrophobic assembly. *Nature*, **437** (2005), 640–647.

[EW82a] D. Eisenberg, R. M. Weiss, and T. C. Terwilliger. The helical hydrophobic moment: A measure of the amphiphilicity of a helix. *Nature*, **299** (1982), 371–374.

[EW82b] D. Eisenberg, R. M. Weiss, T. C. Terwilliger, and W. Wilcox. Hydrophobic moments and protein structure. *Faraday Symposia of the Chemical Society*, **17** (1982), 109–120.

[FB10] A. Fernández and R. S. Berry. Golden rule for buttressing vulnerable soluble proteins. *Journal of Proteome Research (ACS)*, **9** (2010), 2643–2648.

[FC08] A. Fernández and A. Crespo. Protein wrapping: A molecular marker for association, aggregation and drug design. *Chemical Society Reviews*, **37** (2008), 2373–2382.

[FE10] A. Fernández. *Transformative Concepts for Drug Design: Target Wrapping.* Springer-Verlag, Berlin, Germany, 2010.

[FK03] A. Fernández, J. Kardos, and J. Goto. Protein folding: Could hydrophobic collapse be coupled with hydrogen-bond formation? *Federation of European Biochemical Societies Letters*, **536** (2003), 187–192.

[FR00] J. Foote and A. Raman. A relation between the principal axes of inertia and ligand binding. *Proceedings of the National Academy of Sciences of the United States of America*, **97** (2000), 978–983.

[FR04] A. Fernández, K. Rogale, R. Scott, and H. A. Scheraga. Inhibitor design by wrapping packing defects in HIV-1 proteins. *Proceedings of the National Academy of Sciences of the United States of America*, **101** (2004), 11640–11645.

[FS02] A. Fernández, T. R. Sosnick, and A. Colubri. Dynamics of hydrogen-bond desolvation in folding proteins. *Journal of Molecular Biology*, **321** (2002), 659–675.

[FS03] A. Fernández and R. Scott. Dehydron: A structurally encoded signal for protein interaction. *Biophysical Journal*, **85** (2003), 1914–1928.

[FS04] A. Fernández, R. Scott, and R. S. Berry. The nonconserved wrapping of conserved protein folds reveals a trend toward increasing connectivity in proteomic networks. *Proceedings of the National Academy of Sciences of the United States of America*, **101** (2004), 2823–2827.

[FSS03] A. Fernández, R. Scott, and H. A. Scheraga. Amino acid residues at protein-protein interfaces: Why is propensity so different from relative abundance? *The Journal of Physical Chemistry B*, **107** (2003), 9929–9932.

[JN07] A.-H. Jin, H. Brandstaetter, S. T Nevin, C. C. Tan, R. J. Clark, D. J. Adams, P. F. Alewood, D. J. Craik, and N. L. Daly. Structure of α-conotoxin BuIA: Influences of disulfide connectivity on structural dynamics. *BioMed Central Structural Biology*, **7** (2007). Open Access: http://www.biomedcentral.com/1472-6807/7/28.

[JP03] A. Jewett, V. S. Pande, and K. W. Plaxco. Cooperativity, smooth energy landscapes and the origins of topology-dependent protein folding rates. *Journal of Molecular Biology*, **326** (2003), 247–253.

[LG07] F.-H. Lin, L. A. Graham, R. L. Campbell, and P. L. Davies. Structural modeling of snow flea antifreeze protein. *Biophysical Journal*, **92** (2007), 1717–1723.

[MD04] C. D. Mol, D. R. Dougan, T. R. Schneider, R. J. Skene, M. L. Kraus, D. N. Scheibe, G. P. Snell, H. Zou, B. C. Sang, and K. P. Wilson. Structural basis for the autoinhibition and STI-571 inhibition of c-Kit tyrosine kinase. *The Journal of Biological Chemistry*, **279** (2004), 31655–31663.

[OB92] C. A. Orengo, N. P. Brown, and W. R. Taylor. Fast structure alignment for protein databank searching. *Proteins: Structure, Function, and Genetics*, **14** (1992), 139–167.

[SF12] A. L. Saucedo, D. Flores-Solis, R. C. Rodríguez de la Vega, B. Ramírez-Cordero, R. Hernández-López, P. Cano-Sánchez, R. N. Navarro, et al. New tricks of an old pattern: Structural versatility of scorpion toxins with common cysteine spacing. *The Journal* of *Biological Chemistry*, **287** (2012), 12321–12330.

[WS80] K. D. Watenpaugh, L. C. Sieker, and L. H. Jensen. Crystallographic refinement of rubredoxin at 1.2 Å resolution. *Journal of Molecular Biology*, **138** (1980), 615–633.

[YS99] M. M. Young, A. G. Skillman, and I. D. Kuntz. A rapid method for exploring the protein structure universe. *Proteins*, **34** (1999), 317–332.

[ZK00] C. Zhang, S.-H. Kim. The anatomy of protein β-sheet topology. *Journal of Molecular Biology*, **299** (2000), 1075–1089.

[ZL04] C. Zhang, S. Liu, H. Zhou, and Y. Zhou. The dependence of all-atom statistical potentials on structural training database. *Biophysical Journal*, **86** (2004), 3349–3358.

CHAPTER 4

Algorithms Dealing with Angles

4.1 MOTIVATION

In addition to distance calculations, computations that derive bond angles and dihedral angles are fundamental to the evaluation of molecular structure. Usually, these evaluations are done for a static molecule. In a more dynamic setting, we may attempt to evaluate these quantities as they change with time due to flexibility of the molecule. For now we assume that the model is static and that we know the positions of atoms in three-dimensional (3D) space. In particular, we will be dealing with proteins for which the coordinates of all atoms are known with a fairly reasonable accuracy. Typically, we will get these coordinates from a file that is downloaded from the Protein Data Bank (PDB).

4.2 BOND ANGLE

Recall from linear algebra that an inner product of normalized vectors u and v can be viewed as the cosine of the angle between these vectors. This is expressed using the formula:

$$\cos\theta = \langle u, v \rangle / (\|u\|\|v\|). \tag{4.1}$$

Consider two atoms a and c (with coordinate vectors a and c) both bonded to a third atom b with coordinates given by vector b as in Figure 4.1.

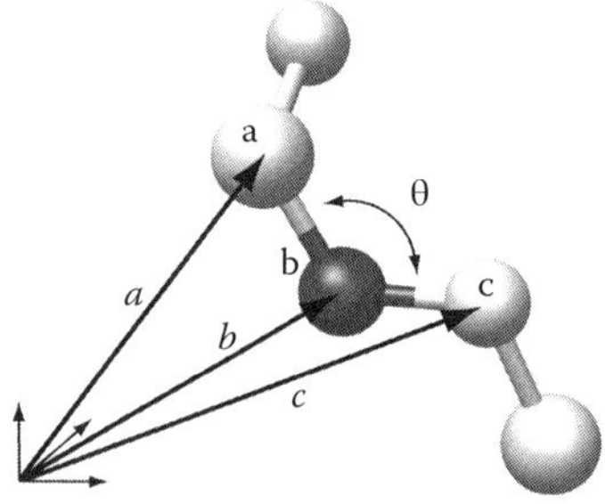

FIGURE 4.1 Bond angle defined via vectors.

If we set $u = a - b$ and $v = c - b$, then the cosine of the bond angle will be given by

$$\cos\theta = \frac{(a-b)^{T}(c-b)}{\|a-b\|\|c-b\|}. \tag{4.2}$$

Once this quantity is calculated, the required bond angle can be easily derived by taking the inverse cosine of the quantity on right-hand side of the last equation.

4.2.1 Calculating Bond Angles in a Script

If you need to calculate a very small number of bond angles, then it is advisable to use the Chimera menu invocation: **Tools... Structure Analysis... Angles/Torsions**, which will launch a dialog window that allows you to select atoms followed by a request for the angle calculation. In this section we are assuming that you need several, perhaps thousands, of these calculations and so a script is the only reasonable approach. The following code follows the formula given in Equation 4.2 and serves to calculate the bond angle formed when `atomB` is connected to both `atomA` and `atomC`:

```
def bondAngle(atomA, atomB, atomC):
    baVec = atomA.coord() - atomB.coord()
    bcVec = atomC.coord() - atomB.coord()
    baVec.normalize(); bcVec.normalize()
    angleRadians = arccos(baVec * bcVec)
    return angleRadians * 180.0/pi
```

The return value is the bond angle measured in degrees. An obvious modification of the code will allow you to get the bond angle in radians.

As one might expect, Chimera has a built-in function to compute angles. The following two lines will generate the same results:

```
angle = bondAngle(atomA, atomB, atomC)
angle = chimera.angle(atomA.coord(), atomB.coord(). atomC.coord()
```

4.3 DIHEDRAL ANGLES

4.3.1 Defining Dihedral Angles

Consider atoms in the protein backbone: N_1 $C\alpha_1$ C_1 N_2 $C\alpha_2$ C_2 N_3 $C\alpha_3$ C_3 N_4 .... As we move from residue to residue, the corresponding bond lengths do not change much.

For example: N_i- $C\alpha_i$ has approximately the same bond length as N_j- $C\alpha_j$ for arbitrary *i* and *j*. Similarly, corresponding bond angles tend to be the same (see Exercise 1 at the end of this chapter). You cannot make this same statement for dihedral angles. Dihedral angles are due to a "swivel" action around a single bond and are primarily responsible for defining the position of the backbone atoms in 3D space. Be sure to understand that changing a dihedral angle will not affect bond angles between successive atoms in the backbone. In fact, we could keep all bond angles constant while moving N_i relative to N_{i+1} by using a swiveling rotation around the $C\alpha_i$- C_i bond.

While a bond angle is determined by a sequence of three bonded atoms, a dihedral angle is defined by a sequence of four consecutive bonded atoms. In general, the first three atoms and the last three atoms of this sequence define two planes that intersect in a line that is coincident with the bond between the middle two atoms. For example, Figure 4.2 shows how the four consecutive atoms C_{i-1} N_i $C\alpha_i$ C_i of the backbone define two planes that intersect in the line that is coincident with the N_i- $C\alpha_i$ bond. The angle between these two planes is referred to as the ϕ (phi) angle associated with residue *i*. In a similar fashion, Figure 4.3 shows how the four consecutive atoms N_i $C\alpha_i$ C_i N_{i+1} of the backbone define two planes that intersect in the line that is coincident with the $C\alpha_i$- C_i bond. The angle between these two planes is referred to as the ψ (psi) angle associated with residue *i*. Since the backbone atoms go through consecutive repetitions of the atoms N, Cα, and C, we have discussed only two of the three possible starting positions for a sequence of four atoms that define a dihedral angle. In fact, there is another dihedral angle to be considered, namely the dihedral

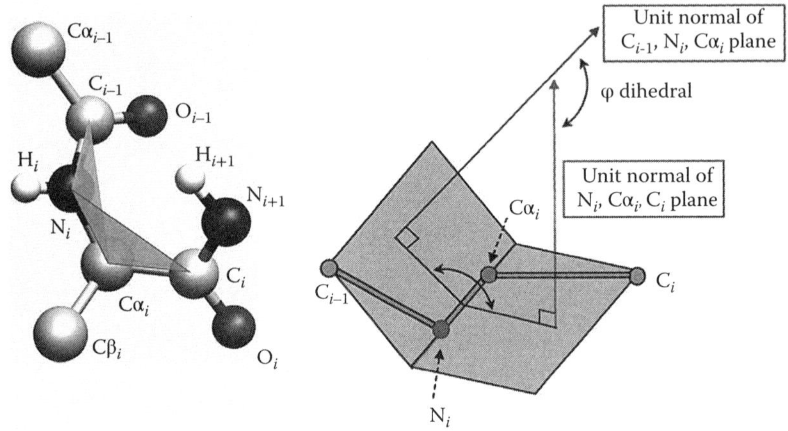

FIGURE 4.2 (**See color insert.**) Phi dihedral angle.

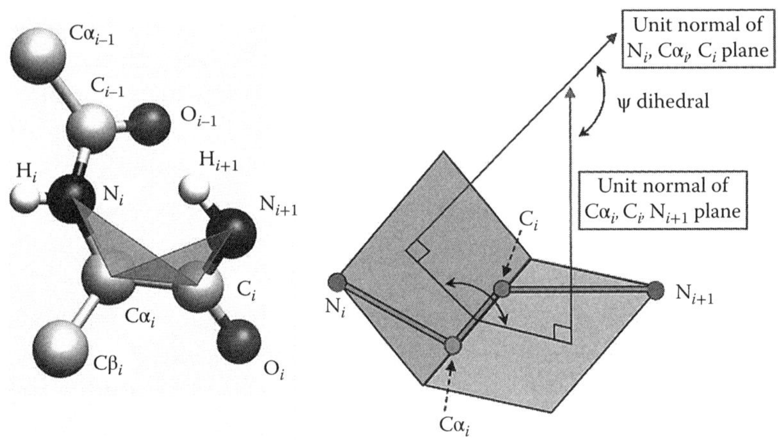

FIGURE 4.3 (**See color insert.**) Psi dihedral angle.

angle that is defined by the four-atom sequence $C\alpha_i$ C_i N_{i+1} $C\alpha_{i+1}$. This is called the ω (omega) angle and it has the carbon and nitrogen of the peptide bond as the middle two atoms. The peptide bond has a partial double-bond character due to resonance effects. Because of this and the steric interactions between the two side chains attached to the alpha carbon atoms at either end of the sequence, the ω dihedral angle typically shows a *trans* configuration for this four-atom sequence. In other words, $\omega = 180°$. As a consequence, $C\alpha_i$, C_i, N_{i+1}, $C\alpha_{i+1}$, the O atom attached to the carbonyl C_i atom and the H atom on the nitrogen are all very close to being coplanar. This planar nature of the peptide bond usually puts the side chains in positions that are as far apart as possible.

4.3.2 Mathematics for the Computation of a Dihedral Angle

The next few sections give a complete presentation of the theory needed for the computation of a dihedral angle. It is included for the sake of completeness and can be skipped if you simply need the script that does the calculation (Section 4.3.7).

Our definition of a dihedral angle as the angle between two planes has to be made more precise. Since an angle is determined by the intersection of two lines, we will need to derive two such lines from the planes defined by the sequence of four atoms. For example, we could have each line lie within a plane and have both lines perpendicular to the line of intersection of the two planes. The analytic derivation required by this approach is somewhat lengthy and is less preferable to a strategy that generates the needed lines by computing the normals of the two planes. These normals are illustrated in Figures 4.2 and 4.3. A short deduction using Euclidean geometry will show that the dihedral angle between the two planes is equal to the angle between the two normals.

Since the normal to a plane defined by two intersecting lines (in our case, two consecutive bonds) is easily computed using a cross product formula, we can use this formula as a convenient "algebraic subroutine."

4.3.3 Computation of a Normal

Two lines that intersect in 3D space define a plane. To construct a normal to this plane, we calculate a *cross product*. Suppose we have two vectors u and v. The cross product is a vector that is perpendicular to both u and v and it has a magnitude that is equal to the area of the parallelogram spanned by the vectors. It can be shown that this is

$$u \times v = \det \begin{bmatrix} i & j & k \\ u_1 & u_2 & u_3 \\ v_1 & v_2 & v_3 \end{bmatrix} = (u_2v_3 - u_3v_2)i + (u_3v_1 - u_1v_3)j + (u_1v_2 - u_2v_1)k \tag{4.3}$$

The result of this computation is a vector that is expressed as a linear combination of the three basis vectors i, j, and k. Alternatively, we could express the right-hand side of the previous equation using a column vector notation

$$u \times v = \begin{bmatrix} u_2 v_3 - u_3 v_2 \\ u_3 v_1 - u_1 v_3 \\ u_1 v_2 - u_2 v_1 \end{bmatrix} \tag{4.4}$$

To see why these formulae are correct, consider the following argument: If the magnitude of the normal is the area of the parallelogram defined by u and v, then its magnitude is given by the base times height formula $\|u\|h = \|u\|\|v\|\sin\theta$, where h is the height of the parallelogram and θ is the angle between u and v. So, we can write

$$\begin{aligned} \|u \times v\|^2 &= \|u\|^2 \|v\|^2 \sin^2\theta = \|u\|^2 \|v\|^2 (1 - \cos^2\theta) \\ &= \|u\|^2 \|v\|^2 - (\|u\|\|v\|\cos\theta)^2 \\ &= \|u\|^2 \|v\|^2 - (u^{\mathrm{T}} v)^2 \end{aligned} \tag{4.5}$$

Continuing

$$\|u \times v\|^2 = \|u\|^2 \|v\|^2 - (u^{\mathrm{T}} v)^2 = \left(\sum_{k=1}^{3} u_k^2\right)\left(\sum_{k=1}^{3} v_k^2\right) - \left(\sum_{k=1}^{3} u_k v_k\right)^2 \tag{4.6}$$

We can now apply Lagrange's Identity to see that the last expression is actually

$$\sum_{i=1}^{2} \sum_{j=i+1}^{3} \left(u_i v_j - u_j v_i\right)^2 \tag{4.7}$$

To prove this, you can simply expand both expressions (a somewhat tedious procedure).*

Now note that

$$\left\| \det \begin{bmatrix} i & j & k \\ u_1 & u_2 & u_3 \\ v_1 & v_2 & v_3 \end{bmatrix} \right\|^2 = \left\| (u_2 v_3 - u_3 v_2) i + (u_3 v_1 - u_1 v_3) j + (u_1 v_2 - u_2 v_1) k \right\|^2 \tag{4.8}$$

and we see that this is the same as the double sum just given.

* A proof of the general case can be found at en.wikipedia.org/wiki/Lagrange's_identity.

So far we have shown that working with this "area of parallelogram" definition, we get a vector that has a magnitude given by

$$\left|\det\begin{bmatrix} i & j & k \\ u_1 & u_2 & u_3 \\ v_1 & v_2 & v_3 \end{bmatrix}\right| \tag{4.9}$$

Is the vector defined by this determinant perpendicular to both u and v? Yes. If you calculate the inner product of this vector with u or v, you get 0. For example

$$\begin{bmatrix} u_2v_3 - u_3v_2 \\ u_3v_1 - u_1v_3 \\ u_1v_2 - u_2v_1 \end{bmatrix}^T \begin{bmatrix} u_1 \\ u_2 \\ u_3 \end{bmatrix} \equiv 0 \tag{4.10}$$

So, we have shown that the determinant-based calculation for the cross product has given the correct magnitude and we have the "perpendicular to both u and v" requirement. The resulting vector could be illustrated by either of the two diagrams contained in Figure 4.4. We now have to decide which of these is correct since they both show a vector perpendicular to both u and v. For $u \times v$, we use the "right-hand rule" that is also seen in the usual 3D Euclidean i, j, k coordinate system: If the fingers of your right hand curl around the k unit normal with fingers going in the direction from i to j, then the thumb points in the positive direction assigned to k. Similarly, if the fingers of the right hand curl around the $u \times v$ normal with fingers going in the direction from u to v, then the thumb points in the positive direction assigned to $u \times v$. So, the first diagram is for $u \times v$, while the next diagram is for $v \times u$ (which is the same as $-(u \times v)$).

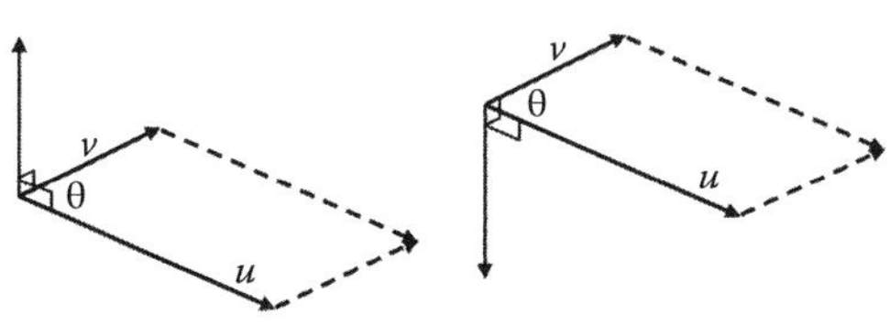

FIGURE 4.4 Right-hand rule.

One last vector algebra issue is that the cross product evaluated using the determinant does not necessarily have a unit length. Recall that our dihedral angle calculation requires vectors that are unit normals. In such a situation, we will have to normalize the cross product vectors. Given $u \times v$, we calculate

$$\frac{u \times v}{\left\| u \times v \right\|} \tag{4.11}$$

4.3.4 Calculating the Phi Dihedral Angle

Suppose we want to calculate the ϕ dihedral angle as shown in Figure 4.2. For simplicity, we will just use the atom names to label vectors that contain the coordinates of these atoms. We let $u = \mathrm{N}_i - \mathrm{C}_{i-1}$ and $v = \mathrm{C}\alpha_i - \mathrm{N}_i$. Then

$$u \times v = (u_2 v_3 - u_3 v_2) i + (u_3 v_1 - u_1 v_3) j + (u_1 v_2 - u_2 v_1) k \tag{4.12}$$

Dividing this by its norm will give the unit normal (call it $n(\mathrm{C}_{i-1}, \mathrm{N}_i, \mathrm{C}\alpha_i)$).

Similarly, to calculate the other required normal we let $v = \mathrm{C}\alpha_i - \mathrm{N}_i$ and $w = \mathrm{C}_i - \mathrm{C}\alpha_i$. Then calculate

$$v \times w = \det \begin{bmatrix} i & j & k \\ v_1 & v_2 & v_3 \\ w_1 & w_2 & w_3 \end{bmatrix} = (v_2 w_3 - v_3 w_2) i + (v_3 w_1 - v_1 w_3) j + (v_1 w_2 - v_2 w_1) k \tag{4.13}$$

Dividing this by its norm will give the second unit normal (call it $n(\mathrm{N}_i, \mathrm{C}\alpha_i, \mathrm{C}_i)$).

Finally with both unit normals computed, we can get the value of the ϕ dihedral for this alpha carbon using the inverse cosine function:

$$\phi = \tau(\mathrm{C}_{i-1}, \mathrm{N}_i, \mathrm{C}\alpha_i, \mathrm{C}_i) = \arccos\left(n(\mathrm{C}_{i-1}, \mathrm{N}_i, \mathrm{C}\alpha_i) \bullet n(\mathrm{N}_i, \mathrm{C}\alpha_i, \mathrm{C}_i) \right) \tag{4.14}$$

4.3.5 Sign of the Dihedral Angle

By convention, a dihedral angle is assumed to be in the range $[-\pi, \pi]$. Since the calculation of the `arccos` function may lead to an angle in the range $[0,\pi]$, we typically have to adjust the sign. To derive the sign of a dihedral angle, we look along the bond lying in the intersection of the two planes (be sure to look in the increasing *i* direction, as shown in Figure 4.5).

Consider the unit normal of the plane defined by the first three atoms. Compute its inner product with the vector going from the third atom to the last atom. If this inner product is positive (as in the diagram), then the sign of the dihedral angle is positive, otherwise it is negative. In the case of ϕ this computation is

$$n(C_{i-1}, N_i, C\alpha_i) \bullet (C_i - C\alpha_i) \tag{4.15}$$

For positive dihedral angles, the projection of the unit normal is in the same direction as the last bond. For negative dihedral angles, the projection of the unit normal is in the direction opposite to the last bond. In both cases, we are looking down along the central bond in the increasing *i* direction. The views for positive and negative dihedral angles are given in Figure 4.6.

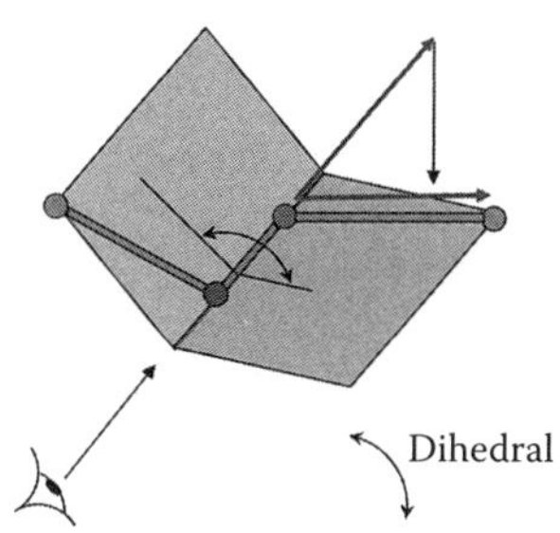

FIGURE 4.5 Sign of the dihedral angle.

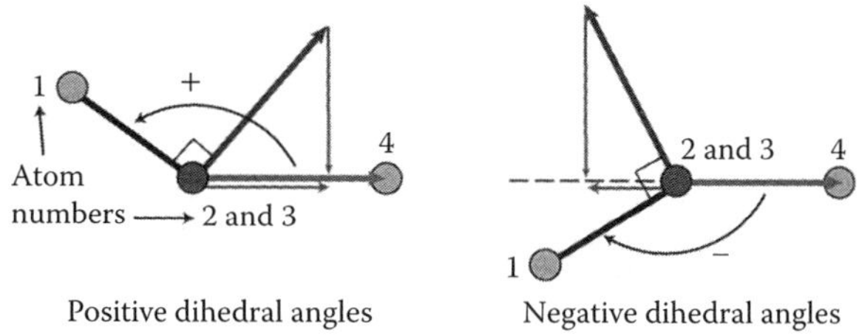

FIGURE 4.6 Sign of the dihedral angle.

4.3.6 Calculating the Psi Dihedral Angle

To calculate the ψ dihedral angle, we follow the same analysis but this time compute

$$\psi = \tau(\mathrm{N}_i, \mathrm{C}\alpha_i, \mathrm{C}_i, \mathrm{N}_{i+1}) = \arccos\left(n(\mathrm{N}_i, \mathrm{C}\alpha_i, \mathrm{C}_i) \bullet n(\mathrm{C}\alpha_i, \mathrm{C}_i, \mathrm{N}_{i+1})\right) \tag{4.16}$$

In other words, all the previous computations work to get ψ if you do the following substitutions:

$$\begin{aligned} \mathrm{C}_{i-1} &\leftarrow \mathrm{N}_i \\ \mathrm{N}_i &\leftarrow \mathrm{C}\alpha_i \\ \mathrm{C}\alpha_i &\leftarrow \mathrm{C}_i \\ \mathrm{C}_i &\leftarrow \mathrm{N}_{i+1} \end{aligned} \tag{4.17}$$

4.3.7 Calculating Dihedral Angles in a Script

Usually, explicit computation of dihedral angles can be avoided by simply extracting the appropriate attribute from the residue object. For example, consider the following user interactions with the **Python Shell** window:

```
>>> r = chimera.openModels.open("1CRN", type="PDB")[0].residues[4]
#1, chain A: crambin

>>> r.type
u'PRO'
>>> r.phi
-76.1815060916024
>>> r.psi
-18.979386491402792
>>> r.chi1
32.04248431338714
>>> r.chi2
-41.38196278264455
>>> r.chi3
>>>
```

The first line defines `r` to be the residue with index 4 within the protein 1CRN. The next lines indicate that it is a proline. After this, we have four statements that are used to retrieve the dihedral angles designated by the names: `phi`, `psi`, `chi1`, and `chi2`.

4.4 RAMACHANDRAN PLOTS

While the dihedral angles in a polypeptide can change with the conformation of the chain, they cannot freely adopt just any arbitrary values. It has been known since the early 1960s that steric collisions will prohibit certain combinations of the phi and psi angles on either side of a given alpha carbon atom. An observed combination of phi/psi angles is often represented by a single point in a two-dimensional plot that has a horizontal axis for phi ranging over values $[-\pi, \pi]$ (i.e., −180° to +180°) and a vertical axis for psi ranging over the same values. A collection of many such points can be used to record the observed phi/psi combinations corresponding to a set of alpha carbon atoms. This type of scatter plot is called a Ramachandran plot since Ramachandran and his colleagues (see [RR63]) were the first people to study the allowable phi/psi angles and the dependency of their values on the type of residue associated with the alpha carbon.

Figure 4.7 is derived from a Ramachandran scatter plot of 121,870 phi/psi combinations taken from 463 protein structures [MM92]. The

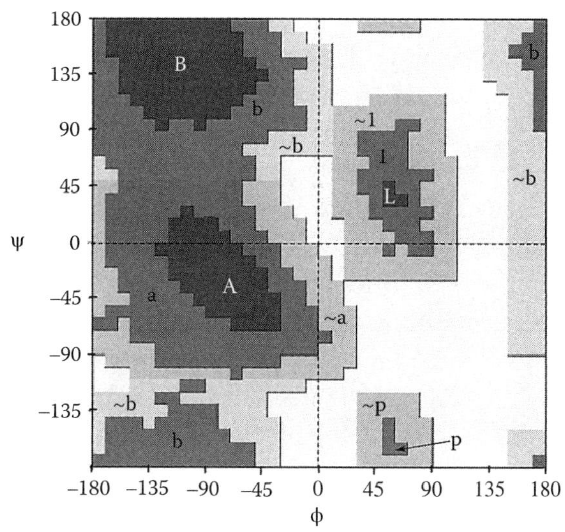

FIGURE 4.7 **(See color insert.)** Ramachandran plot for 121,870 residues (463 protein structures).

A - Core alpha	L - Core left-handed alpha
a - Allowed alpha	l - Allowed left-handed alpha
~a - Generous alpha	~l - Generous left-handed alpha
B - Core beta	p - Allowed epsilon
b - Allowed beta	~p - Generous epsilon
~b - Generous beta	

following procedure was used to generate this pixilated figure: The area of the original scatter plot was subdivided into square "pixels," each having a side length of 10°. In this way, the entire plot is covered with $36^2 = 1296$ pixels. A pixel was given a dark color and characterized as "core" if it covered at least 100 points in the scatter plot. A pixel was given a color of medium intensity and characterized as "allowed" if it was not a core pixel but covered at least eight points in the scatter plot. Pixels in the "generous" region are obtained by extending an allowed region with a border area of width 20°. The generous region is given a color that is lighter than an allowed region. Any pixel that is not core, allowed, or generous is given the lightest color or simply left as white. So, in summary, the darker areas correspond to regions of the plot where the density of points is highest and other colors represent regions of lower density. White regions or those with the lightest color are used to denote areas of the plot where there is the least likelihood of observing a point, for example, $\phi = 90°$ and $\psi = -90°$. The plot of Figure 4.7 was created using PROCHECK,* a program that checks the stereochemical quality of protein structures (see [LM93] and [MM92]).

As indicated in the list following Figure 4.7, various regions of the plot are labeled with single-letter designations that indicate some secondary structure properties such as alpha helices, beta strands, and epsilon turns. Alpha helices are typically right handed although left-handed helices do occur but with considerably less frequency. The epsilon region of the Ramachandran plot is for phi/psi combinations that are usually associated with a turn that is just prior to a helix or beta strand. A good example of this occurs in the protein 1JAL for residue VAL287, which occurs on a tight turn between two antiparallel beta strands. More information about two-residue turns can be found in [MP94].

Since an angle rotation of $180 + n$ degrees is the same as an angle rotation of $n - 180$ degrees, any Ramachandran plot has a "wraparound" continuity. We could consider the uppermost row of pixels to be adjacent to the lowest row of pixels and the leftmost column of pixels is essentially adjacent to the rightmost column. Inspection of Figure 4.7 will verify this.

As noted earlier, PROCHECK checks the stereochemical quality of a protein structure. An example of such processing is given by Figure 4.8, where the Ramachandran scatter plot for 1A4Y is superimposed on the

* http://www.ebi.ac.uk/thornton-srv/software/PROCHECK/.

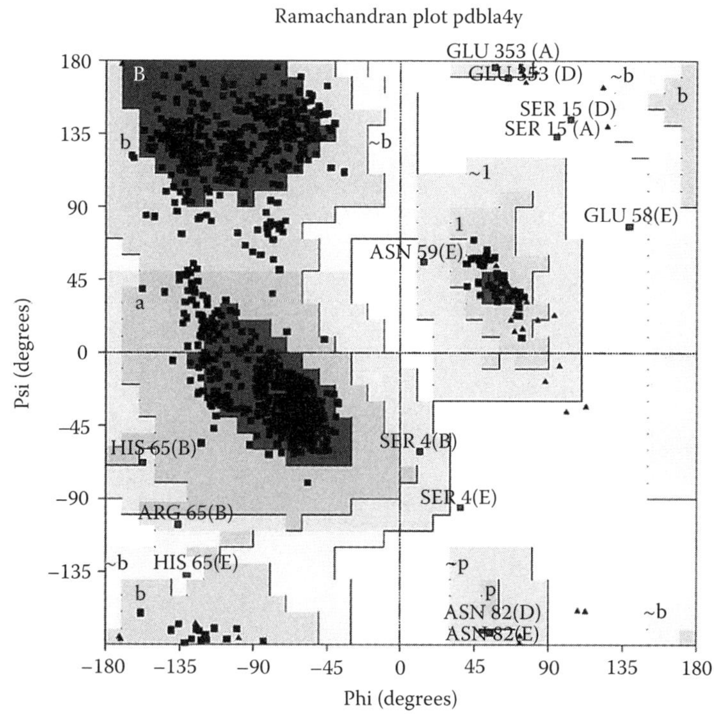

FIGURE 4.8 Ramachandran plot for residues of protein 1A4Y.

background distribution discussed earlier and illustrated by Figure 4.7. Note that most of the points reside in the core alpha and core beta regions. After the protein's backbone conformation has been computed, PROCHECK software can be used to detect the existence of any anomalous dihedral angles. For example, Figure 4.8 indicates that residues SER 15 (chains A and D), GLU 58 (chain E), and SER 4 (chain E) all have phi/psi angles that produce points in the white region. A point in the white region may indicate an error in the computation of the conformation or it may simply be a valid entry that rarely occurs.

To some extent, Ramachandran plots can also show how the dihedral angles on either side of an alpha carbon atom are affected by the type of residue associated with that alpha carbon. For example, the glycine residue essentially lacks a side chain and so it does not have the steric constraints of the other amino acids. Figure 4.9 shows the scatter plot for phi/psi angles that are limited to glycine residues. One can see that the distribution is somewhat different from the general case illustrated in Figure 4.7. Although there are fewer constraints on dihedral angles for glycine, there are still some distribution patterns that can be observed and Figure 4.9

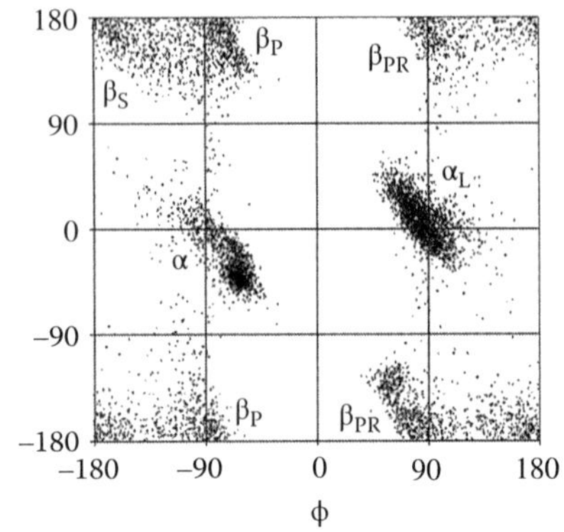

FIGURE 4.9 Ramachandran plot for glycine residues.

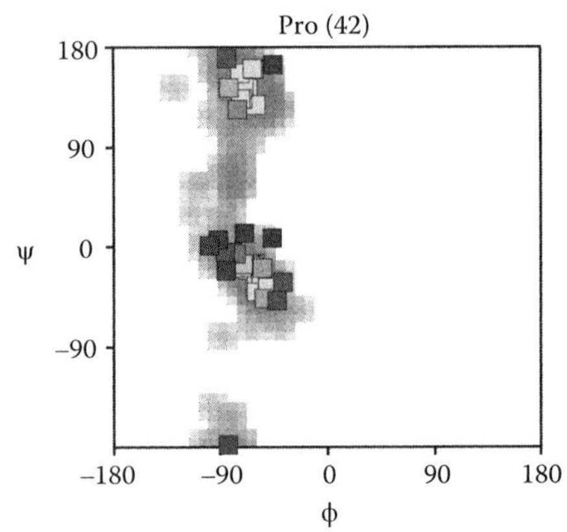

FIGURE 4.10 Ramachandran plot for proline residues.

shows five clusters within the scatter plot. An in-depth analysis of this plot can be found in [HB05].

While glycine provides an example of dihedral angles that are subject to few steric constraints, we find that the phi/psi angles for proline are severely constrained. The PROCHECK version of the Ramachandran plot for the 42 proline residues of 1A4Y is given in Figure 4.10. We see that phi is constrained to values that are within a narrow band typically ranging from −60° to −90° depending on the secondary structure of its local environment. Inspection of Figure 4.11 shows why this is expected. The phi dihedral angle just prior to the proline alpha carbon is severely restricted because the chain arising from the alpha carbon loops back to bond with the backbone nitrogen atom bonded to that alpha carbon. The psi angle

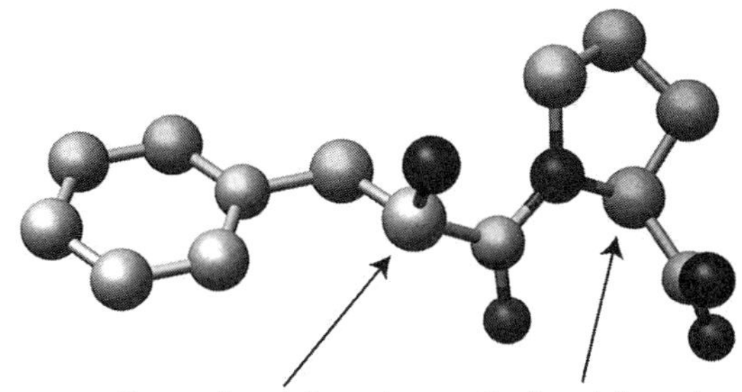

FIGURE 4.11 Phe 70 a pre-proline residue in 1AG9.

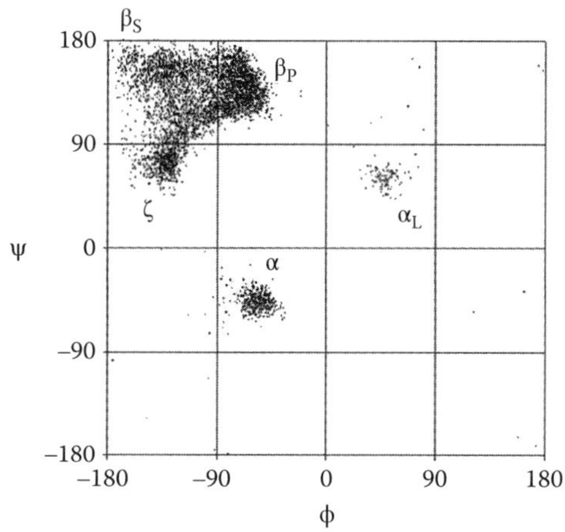

FIGURE 4.12 Ramachandran plot for pre-proline residues.

after the alpha carbon has much less restriction as can be seen in the plot of Figure 4.10.

Another interesting example of a constraint is seen in "pre-proline" residues. In this case, the side chain of the residue prior to proline bends away from the proline loop to avoid a steric clash. This is clearly illustrated in Figure 4.11, where the phenylalanine ring of residue 70 in 1AG9 veers away from the proline loop in the next residue. Figure 4.12 (taken from [HB05]) gives the Ramachandran plots and associated analysis for pre-proline residues.

4.4.1 Scripts for Generating Ramachandran Plots

Generating a Ramachandran plot can be done using the `matplotlib` module. This will require some familiarity with scatter plots. You should review the scatter plot section of Appendix D.

Script 4.1: Generating a Ramachandran Plot for a Protein

Script D.3 can be adapted to a variety of applications that need a scatter plot output. The next script shows how we can generate a Ramachandran plot for a given protein:

```
import matplotlib.pyplot as plt

def ramaPlot(res_L, saveFilePath = ""):
    plt.ion()
    fig = plt.figure(figsize = (8,8))

    ax = fig.add_subplot(111, aspect = 'equal')
    ax.set_title("Ramachandran Plot " + saveFilePath, fontsize = 14)
    ax.set_xlabel("PHI (degrees)", fontsize = 12)
    ax.set_ylabel("PSI (degrees)", fontsize = 12)
    ax.set_xlim([-180, 180])
    ax.set_ylim([-180, 180])
    ax.set_xticks([-180, -135, -90, -45, 0, 45, 90, 135, 180])
    ax.set_yticks([-180, -135, -90, -45, 0, 45, 90, 135, 180])
    ax.grid(True, linestyle = '-', color = '0.75')
    ax.hlines(0, -10, 10, colors='k', linestyles='dashed')
    ax.vlines(0, -10, 10, colors='k', linestyles='dashed')

    # Initialization:
    # List indicies: 0 for Helix, 1 for Strand, and 2 for Coil.
    colors_T = ('k', 'g', 'b') # Marker colors for Helix, Strand, Coil.
    shapes_T = ('o', 's', 'd') # Marker shapes for Helix, Strand, Coil.
    phi_L = [[], [], []]  # Phi lists for Helix, Strand, and Coil
    psi_L = [[], [], []]  # Psi lists for Helix, Strand, and Coil

    # Generate the phi, psi coordinate lists:
    for r in res_L:
        # Skip if residue has no phi, psi attributes:
        if r.phi is None or r.psi is None: continue
        if r.isHelix:
            phi_L[0].append(r.phi)
            psi_L[0].append(r.psi)
        if r.isStrand:
            phi_L[1].append(r.phi)
            psi_L[1].append(r.psi)
        if not (r.isHelix or r.isStrand):
            phi_L[2].append(r.phi)
            psi_L[2].append(r.psi)

    # Drop in the markers:
    for i in range(3):
        ax.scatter(phi_L[i], psi_L[i], s = 20,
                   color = colors_T[i], marker = shapes_T[i])

    # Save file if the saveFilePath parameter is not the empty string.
    if saveFilePath != "":
        fig.savefig(saveFilePath + "_RamachandranPlot", dpi = fig.dpi)
```

```
    raw_input("Hit Enter to dismiss the plot.")
    plt.close()


#=================================================================
pdbIDchars = raw_input("Type in 4 character PDB ID or PDB file name: \n")
prot = chimera.openModels.open(pdbIDchars, type="PDB")[0]

ramaPlot(prot.residues, pdbIDchars)
```

Note that this script gives some extra information by using different markers for phi, psi pairs situated in different secondary structures. Markers are black circles for helix residues, green squares for strand residues, and blue diamonds for coil residues. Plotting is set up in a function that can be called with parameters that specify a residue list and an optional string that can be used to generate the plot title and provide a file name for image saving. The script, available from the book's Web site (file name: `Script 4_01 Ramachandran plot for a protein.py`). Executing this script on 1YJH (the "bedbug spit" protein) shows the capabilities of the scatter plot because the protein has a good mix of helices, strands, and coils (see Figure 4.13).

4.4.2 Background Densities for Ramachandran Plots

While using a scatter plot to display the phi, psi pairings in a set of residues, it is helpful to see the plot displayed above a background image that shows the expected distribution of points. This strategy is illustrated by Figure 4.8. The colored background used by PROCHECK is a rendering of the statistical densities reported in the work done by Morris et al. (see page 350 of [MM92] to view the original version of this background distribution).

Software for this book includes a script that can generate this background (`Script 4_02 Generating the Ramachandran background distribution.py`). The script draws the background by using a scatter plot that uses square markers with a large size. The coloration of the markers is determined by a rather large list of lists forming a 36 by 36 array of color codes (`baseColorCodes` in the script). The idea is to generate the background which is then written to an image file. The image file can be cropped to eliminate titles and axes numbers and subsequently used as an image that can be loaded into other scatter plots. This image loading eliminates all the computations necessary to establish the background. Another advantage is that the background image will be

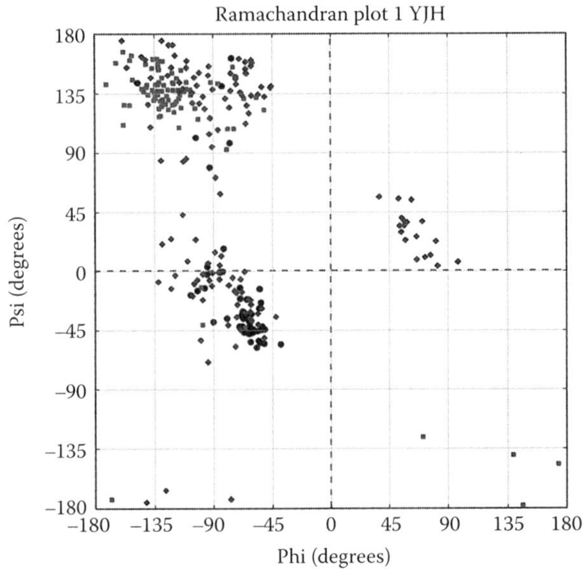

FIGURE 4.13 (**See color insert.**) Ramachandran plot for 1YJH.

automatically resized if the user changes the plot window size. The directory containing the script also contains the cropped version in the file: `croppedRamachandranBackGround.png`.

The next script (`Script 4_03 Ramachandran Plots with Background.py`) is the same as Script 4_01 but with the three extra statements that bring in the background as an image. If you run this script on the protein with ID 1A4Y, you should get a plot that is very similar to that shown in Figure 4.8. Running the script on chain A of the protein 1PTM produces the result in Figure 4.14. You should note the presence of six black circle markers within the core left-handed alpha region near (phi, psi) = (50, 40). This is unusual and it is due to the presence of a left-handed helix in 1PTM. More information about such anomalies can be found in [NK05].

4.4.3 3D Ramachandran Plots

The next script (see `Script 4_04 3D RamachandranPlot.py`) is offered as a novel way to give your Ramachandran plots some extra flair. Instead of using a scatter plot to track the frequency of the phi, psi pairs, Script 4.04 uses narrow box towers to represent the frequency distributions. Each of the 36 by 36 cells of the plot has a height that is

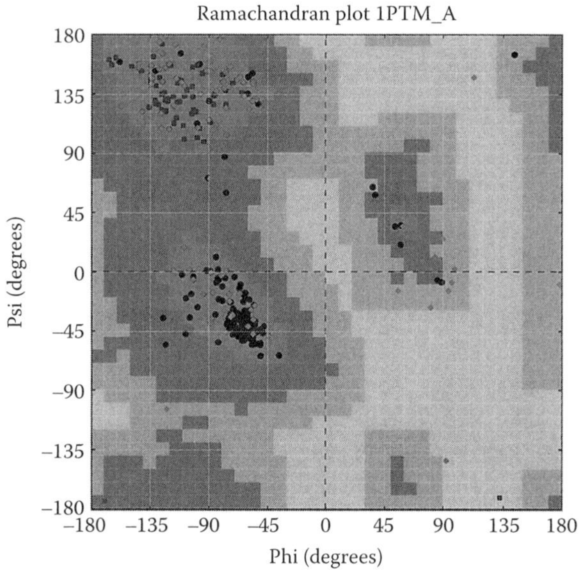

FIGURE 4.14 Ramachandran plot for 1PTM_A.

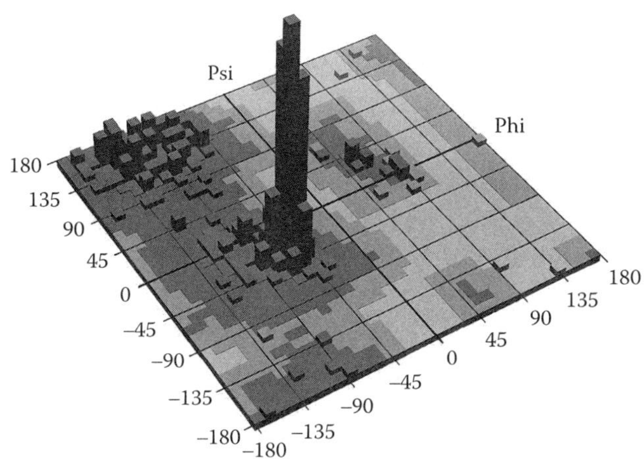

FIGURE 4.15 (**See color insert.**) Three-dimensional Ramachandran plot for 1PTM_A.

proportional to the number of phi, psi points that would fall into that cell if a scatter plot was used. Figure 4.15 shows the plot for the A chain of 1PTM. Rotating and zooming of the plot will give you the best perspective for the scene.

4.5 LEAST SQUARES PLANE

4.5.1 Motivation

In biochemistry, we have seen several structures in which the atoms are almost coplanar, for example, aromatic rings and heme rings. The last example is seen in the myoglobin protein: 1MBO (see Figure 4.13). In this section, we wish to calculate the equation of the "best" plane that goes among the "almost coplanar" atoms. In this case, the best plane will be the plane that minimizes the sum of the squares of the distances between the almost coplanar atoms and the plane. We call this the *least squares plane.*

Suppose we wish to find the equation of the least squares plane for the heme ring of 1MBO. We know that three points can define a plane, but which three should we pick? Since making a choice of some particular set of three atoms would introduce some bias, a reasonable approach would be to use more than three atoms, the idea being to find a plane that goes among the atoms of the heme ring and minimizes the sum of the squared distances measured from the atoms to that plane (Figure 4.16).

We will see that a mathematical representation of this least squares plane has several uses. For example, we may define a line in 3D space (such as the line coincident with a bond) and then we can measure the angle that this line makes with the plane.

4.5.2 Least Squares Plane: Mathematical Analysis

While the analysis of this problem follows a development that is similar to that for the inertial axis, there are some mathematical niceties that merit an individual treatment.

We will consider the almost coplanar atoms to make up a *target* list. In the discussion to follow, we will designate the target list atoms as $a^{(i)}$ $i = 1, 2, \ldots, n$. The ith atom in the list, $a^{(i)}$ will have coordinate vector

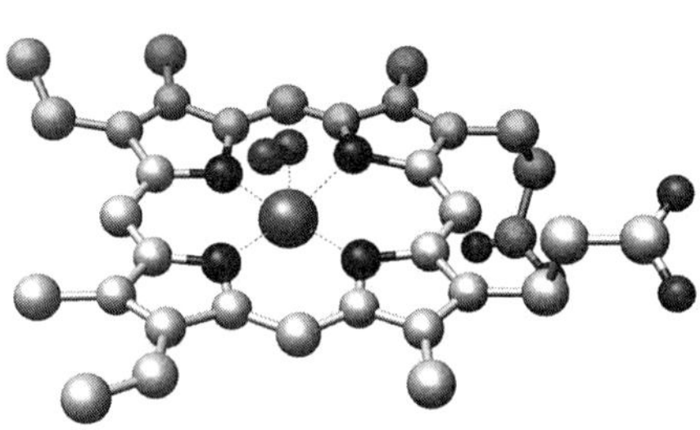

FIGURE 4.16 Hemoglobin ring with "almost coplanar" atoms.

$\left(a_x^{(i)}, a_y^{(i)}, a_z^{(i)}\right)^{\mathrm{T}}$. The goal of this section is to go through a sequence of steps that will derive the formula for the least squares plane. The derivation will be somewhat similar to the derivation of the inertial axis seen in the last section. We wish to minimize

$$S = \sum_{i=1}^{n} \|d_i\|^2 \tag{4.18}$$

where $\|d_i\|$ represents the perpendicular distance between atom $a^{(i)}$ and the least squares plane.

Note that, using vector notation, we can derive the equation of a plane as follows: Let R be a vector going from the origin to the plane meeting it in perpendicular fashion, that is, R is normal to the plane (Figure 4.17). Consider P to be any point in the plane. Then the vector $P - R$ can be considered to lie in the plane and hence perpendicular to R, as shown in the next figure (Figure 4.18). Since $P - R$ is perpendicular to R, we have the inner product of $P - R$ and R equating to zero. This is the equation of the plane since it characterizes all points P in the plane:

$$(P - R)^{\mathrm{T}} R = 0 \tag{4.19}$$

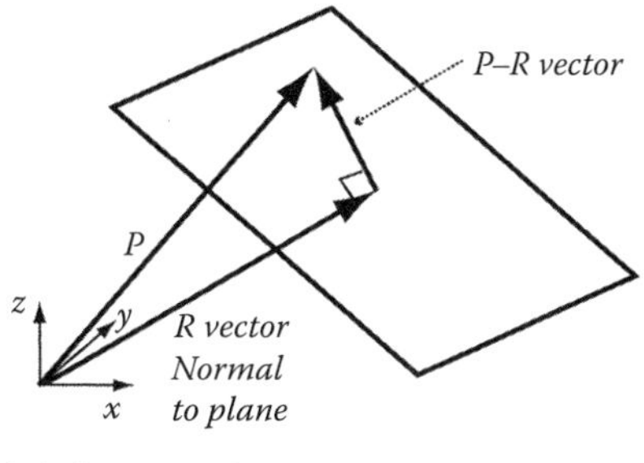

FIGURE 4.17 Normal defining a plane.

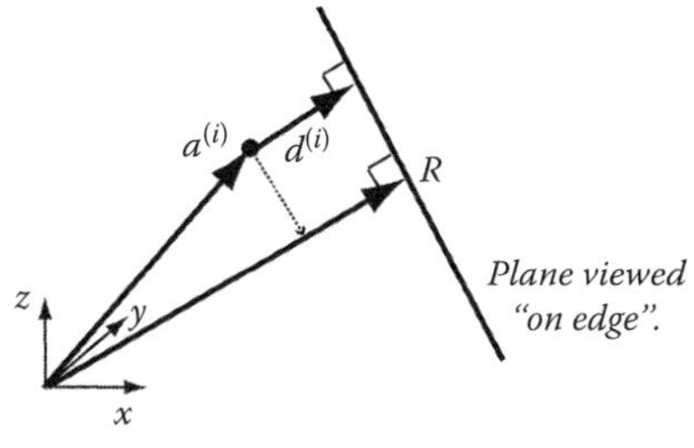

FIGURE 4.18 Plane viewed on edge.

This is a realistic expectation: We only need a single vector R to determine a plane. The direction of R will determine the orientation of the plane, while the length of R will establish the perpendicular distance from the origin to the plane.

A naive computational strategy could now proceed as follows: Given $a^{(i)}$, we can derive $\|d_i\|^2$ in terms of $a^{(i)}$ and the components of R. We can use Equation 4.18 to express S in terms of these components and then we would seek to determine the components of R that minimize S. If we assume that $R = \left(r_x, r_y, r_z\right)^{\mathrm{T}}$, this minimization can be attempted by calculating partial derivatives of S with respect to the components of R and then setting these partial derivatives to zero. Unfortunately, the equations arising from the partial derivatives are quite complicated and difficult to solve. You might want to verify this just to see how bad the situation can become. It so happens that a better representation of R is $R = mN$, where $N = \left(n_x, n_y, n_z\right)^{\mathrm{T}}$ is a unit vector that essentially specifies the direction of R (much like w did for the inertial axis in Section 3.6) and m is a simple scalar that specifies the length of R. This representation of R is somewhat counterintuitive because, instead of only three variables $\left(r_x, r_y, r_z\right)$, we now have to solve for four variables: n_x, n_y, n_z, and m. However, the number of degrees of freedom in the solution is still only 3 because N is a unit vector and so we have $N^{\mathrm{T}}N = 1$, which is a *constraint* acting on N. The presence of a constraint turns this into a Lagrange multiplier problem that has a reasonably straightforward solution.

The reader should note that the naive strategy demands that the components of R essentially define *both* the direction of R and its length. Setting $R = mN$ allows us to "disentangle" length information from direction information leading to a more effective mathematical analysis.

We now derive the Lagrange formulation. Suppose $a^{(i)}$ represents the coordinates of an atom that is **not** on the plane defined by $R = mN$. We can derive the formula for $\|d^{(i)}\|$ in terms of m, N, and $a^{(i)}$ by observing that the length of R is m, which is the sum of $\|d^{(i)}\|$ and the projection of $a^{(i)}$ onto the unit normal N, which is parallel to R. This is illustrated in Figure 4.14. Consequently,

$$\|d^{(i)}\| = m - N^{\mathrm{T}}a^{(i)} \tag{4.20}$$

We wish to minimize

$$S = \sum_{i=1}^{n} \|d^{(i)}\|^2 = \sum_{i=1}^{n} \left(m - N^{\mathrm{T}}a^{(i)}\right)^2 \tag{4.21}$$

subject to the constraint $N^{\mathrm{T}}N = 1$.

So, the Lagrange formulation is as follows: Find m, N, and λ to minimize $L(m,N,\lambda)$, where

$$L(m,N,\lambda)=\sum_{i=1}^{n}\left(m-N^{\mathrm{T}}a^{(i)}\right)^{2}-\lambda\left(N^{\mathrm{T}}N-1\right) \tag{4.22}$$

Taking partial derivatives of the Lagrangian, we get

$$\begin{aligned}
&\frac{\partial L}{\partial m}=0\Rightarrow\sum_{i=1}^{n}\left(m-N^{\mathrm{T}}a^{(i)}\right)=0\\
&\frac{\partial L}{\partial N}=0\Rightarrow\sum_{i=1}^{n}\left(m-N^{\mathrm{T}}a^{(i)}\right)a^{(i)}+\lambda N=0\\
&\frac{\partial L}{\partial \lambda}=0\Rightarrow N^{\mathrm{T}}N=1
\end{aligned} \tag{4.23}$$

Solving the first equation for m gives us

$$m=N^{\mathrm{T}}\left(\frac{1}{n}\sum_{i=1}^{n}a^{(i)}\right) \tag{4.24}$$

Now suppose that the coordinates of all the $a^{(i)}$ $i=1,2,\ldots,n$ are not the coordinates given by the PDB file but instead the PDB coordinates translated so that they are in a frame of reference with the origin at the centroid of the atoms in the target list. That is, suppose $a^{(i)}=p^{(i)}-C$ was computed prior to the beginning of our discussion. Here, $p^{(i)}$ $i=1,2,\ldots,n$ are the PDB coordinates of the target list atoms and C is the centroid of all these atoms. In other words

$$C=\frac{1}{n}\sum_{i=1}^{n}p^{(i)} \tag{4.25}$$

It is easily seen that this gives us $\sum_{i=1}^{n}a^{(i)}=0$ because

$$\sum_{i=1}^{n}a^{(i)}=\sum_{i=1}^{n}\left(p^{(i)}-C\right)=\sum_{i=1}^{n}p^{(i)}-\sum_{i=1}^{n}C=nC-nC=0 \tag{4.26}$$

The value of m is computed to be

$$m=N^{\mathrm{T}}\left(\frac{1}{n}\sum_{i=1}^{n}a^{(i)}\right)=\frac{N^{\mathrm{T}}}{n}\left(\sum_{i=1}^{n}a^{(i)}\right)=\frac{N^{\mathrm{T}}}{n}(0)=0 \tag{4.27}$$

Since $m = 0$, we can now consider our frame of reference to have its origin situated among the atoms of the target list and the least squares plane will go through the origin. The unit vector N will be a direction vector that determines the orientation of the plane.

With $m = 0$, the first equation in the set of Equation 4.23 will be consistent with Equation 4.26 and second equation can be restated as

$$\begin{aligned}\frac{\partial L}{\partial N} = 0 &\Rightarrow \sum_{i=1}^{n}\left(m - N^{\mathrm{T}}a^{(i)}\right)a^{(i)} + \lambda N = 0\\ m = 0 &\Rightarrow \sum_{i=1}^{n}\left(N^{\mathrm{T}}a^{(i)}\right)a^{(i)} = \lambda N\end{aligned} \tag{4.28}$$

Since the terms in the parentheses form a simple inner product of two vectors, we can reverse the order of N and $a^{(i)}$ to get

$$\sum_{i=1}^{n}\left(a^{(i)\mathrm{T}}N\right)a^{(i)} = \lambda N \tag{4.29}$$

If we set

$$A = \begin{bmatrix} a^{(1)} & a^{(2)} & \cdots & a^{(n)} \end{bmatrix} \tag{4.30}$$

then A^{T} has rows that are the transposes of the $a^{(i)}$ so the n by 1 matrix $A^{\mathrm{T}}N$ has the form:

$$A^{\mathrm{T}}N = \begin{bmatrix} a^{(1)\mathrm{T}}N \\ a^{(2)\mathrm{T}}N \\ \vdots \\ a^{(n)\mathrm{T}}N \end{bmatrix} \tag{4.31}$$

and the n by 1 matrix $AA^{\mathrm{T}}N$ can be calculated as

$$A(A^{\mathrm{T}}N) = \begin{bmatrix} a^{(1)} & a^{(2)} & \cdots & a^{(n)} \end{bmatrix}\begin{bmatrix} a^{(1)\mathrm{T}}N \\ a^{(2)\mathrm{T}}N \\ \vdots \\ a^{(n)\mathrm{T}}N \end{bmatrix} = \begin{bmatrix} \sum_{i=1}^{n}(a^{(i)\mathrm{T}}N)a_1^{(i)} \\ \sum_{i=1}^{n}(a^{(i)\mathrm{T}}N)a_2^{(i)} \\ \vdots \\ \sum_{i=1}^{n}(a^{(i)\mathrm{T}}N)a_m^{(i)} \end{bmatrix} = \sum_{i=1}^{n}\begin{bmatrix} (a^{(i)\mathrm{T}}N)a_1^{(i)} \\ (a^{(i)\mathrm{T}}N)a_2^{(i)} \\ \vdots \\ (a^{(i)\mathrm{T}}N)a_m^{(i)} \end{bmatrix}.$$

However, this column vector can be rewritten as $AA^{\mathrm{T}}N = \sum_{i=1}^{n}\left(a^{(i)\mathrm{T}}N\right)a^{(i)}$.

Consequently, Equation 4.29 can be rewritten as an eigenvector equation:

$$AA^{\mathrm{T}}N = \lambda N \tag{4.32}$$

Now referring to Equation 4.31 and Equation 4.21 with $m = 0$, consider the inner product:

$$\begin{aligned}\left(A^{\mathrm{T}}N\right)^{\mathrm{T}}\left(A^{\mathrm{T}}N\right) &= \begin{bmatrix} a^{(1)\mathrm{T}}N & a^{(2)\mathrm{T}}N & \cdots & a^{(n)\mathrm{T}}N \end{bmatrix}\begin{bmatrix} a^{(1)\mathrm{T}}N \\ a^{(2)\mathrm{T}}N \\ \vdots \\ a^{(n)\mathrm{T}}N \end{bmatrix} \\ &= \sum_{i=1}^{n}\left(N^{\mathrm{T}}a^{(i)}\right)^{2} = S \end{aligned} \tag{4.33}$$

But the right-hand side of this can be simplified by appealing to Equation 4.32:

$$\left(A^{\mathrm{T}}N\right)^{\mathrm{T}}A^{\mathrm{T}}N = N^{\mathrm{T}}AA^{\mathrm{T}}N = N^{\mathrm{T}}\lambda N = \lambda \tag{4.34}$$

and so we see that the value of S is simply the eigenvalue. Consequently, when solving Equation 4.32 we will use the eigenvector N corresponding to the minimum eigenvalue if we wish to minimize S.

4.5.3 A Script for Computing the Least Squares Plane

We now have all the mathematical analysis complete. So how do we use it to build software that implements a least squares plane? Before going any further, let us review what has been accomplished. The input for the least squares plane is simply a collection of Chimera atom objects. The mathematical analysis tells us that we can use the coordinates of all these atoms to build the A matrix, which leads to the matrix Equation 4.32. After computing the eigenvalues, we end up with two arrays (each a three-component vector) that constitute a solution of the problem, namely: the centroid C of the atoms and the eigenvector N that determines the direction of a line that is normal to the plane. As discussed earlier, this point in 3D space and the normal vector are enough to *implicitly* specify a plane. They will be stored as attributes for a class that will allow us to instantiate objects that "**behave like**" least squares planes. To understand that these attributes, C and N, are sufficient for our needs, let us consider the methods that such a class would support:

- `projectionOnPlane()`: Given a point P not on the plane, determine Q, the projection of P onto the plane. This will be the intersection of the plane with a line from P that is normal to the plane.
- `distanceToPlane()`: Given a point P not on the plane, calculate the distance between P and its projection Q on the plane.
- `angleToPlane()`: Given two points P and Q, calculate the angle made by the line through P and Q with the plane.

We now show how each of these calculations would use the vectors C and N.

`projectionOnPlane()`: Consider Figure 4.19. We are given the arrays N, C, and P, representing the vectors CN, OC, and OP, respectively. Both OC and OP are defined relative to the frame of reference established by the PDB file, while CN is considered to have an origin that is the centroid C. We need to derive the coordinates of Q (the projection of P onto the plane) represented in Figure 4.19 by the vector OQ (again in the PDB frame of reference). Since we know both C and P we can compute $P - C$ to get an array that is represented in the figure by the vector CP (as an equation: $CP = OP - OC$). The vector CP essentially represents the point P in a frame of reference that has C as its origin. Now we are able to compute the dot product of CP with CN (the normal to the plane) to get P_N, which is the projection of P onto that normal. If we subtract the vector CP_N from OP, we get the vector OQ, which represents the required projection (relative to the PDB frame of reference). In other words: $PQ = -CP_N$ and adding this to OP (as in Figure 4.19) gives us the needed projection.

`distanceToPlane()`: Again, consider Figure 4.19. We are given the coordinates of P and we wish to find the distance from P to the plane. This will simply be the length of vector PQ, so this function will call the `projectionOnPlane()` function to get Q and then calculate the norm of PQ.

`angleToPlane()`: Consider Figure 4.20. We are given the arrays P_A and P_B representing the vectors OP_A and OP_B, respectively. If the line segment from P_B to P_A is extended until it intersects the plane at point V, then our goal is to find the angle θ between this line and the line from Q_B to Q_A, which also goes through point V in the plane. In the figure, it is intended

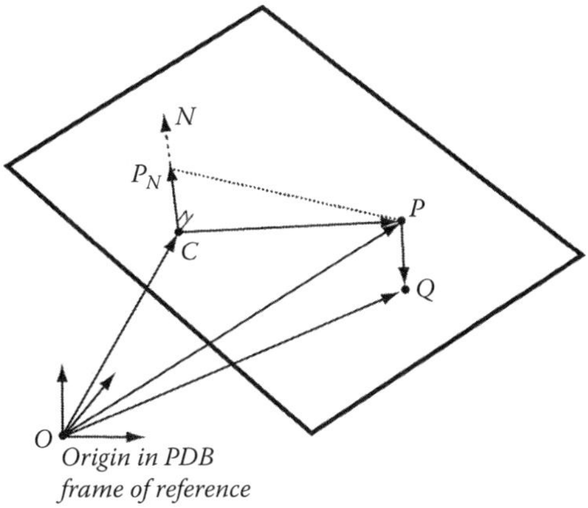

FIGURE 4.19 Calculating a projection onto the plane.

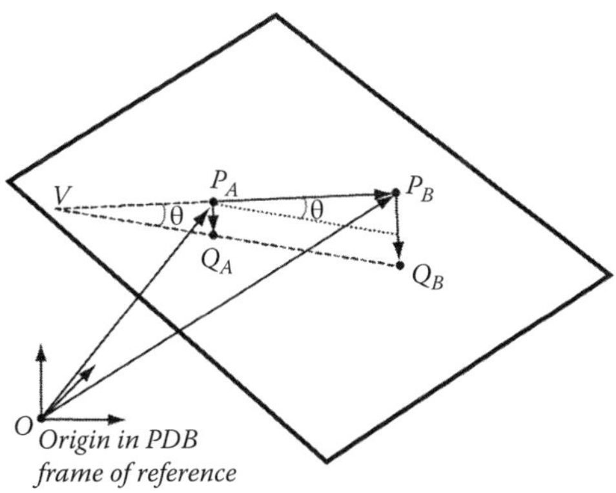

FIGURE 4.20 Calculating the angle between a line segment and a plane.

that Q_A and Q_B are the projections onto the plane of P_A and P_B, respectively. We can obtain θ by first evaluating the cosine of θ, which is the ratio $\|VQ_B\|/\|VP_B\|$. Note that the dotted line above Q_AQ_B has the same length as Q_AQ_B. We can use a similarity of triangles deduction to show that

$$\cos(\theta) = \frac{\|Q_AQ_B\|}{\|P_AP_B\|} \tag{4.35}$$

In summary, this function will use the `projectionOnPlane()` function to calculate both Q_A and Q_B and then calculate $\arccos\left(\|Q_AQ_B\|/\|P_AP_B\|\right)$ to get θ.

We now have enough mathematical background to implement the `LeastSquaresPlane` class. However, before doing so, it should be noted that there are applications that need the generation of a plane object that is simply defined by means of a point and plane normal. For example, in Exercise 8, we need a plane that is to be perpendicular to the inertial axis of a membrane protein. These considerations give us a wonderful opportunity for the illustration of how Python can elegantly support both strategies for the specification of a plane. We use the Python object-oriented inheritance mechanism. We will implement a base class called `Planes` with an `__init__` method that accepts a point *C* and a normal *N*. It will offer the methods just described. We then implement a derived class called `LeastSquaresPlane` with an `__init__` method that accepts a list of Chimera atom objects. This class will be derived from the `Planes` class and so it will also have the methods just described.

Both classes are in the file called planes (available from the book's website) using

```
from StructBio.LeastSquares.planes import Planes
from StructBio.LeastSquares.planes import LeastSquaresPlane
```

Here is the script:

```
__import chimera, numpy
from numpy import array, linalg, zeros, dot, arccos

#==================================================================
# Planes CLASS

class Planes(object):
    def __init__(self, c, n):
        # The next two arrays (3-component vectors) define
        # a plane in the PDB frame of reference.
        self.center = c
        self.normal = n/linalg.norm(n)

    # --------------------------------------------------------------
    # Given an atom object or an array object this function
    # returns the coordinates.
    def getCoords(self, atomORarray):
        if type(atomORarray) is chimera.Atom:
                coords = array(atomORarray.coord())
        if type(atomORarray) is numpy.ndarray:
                coords = atomORarray
        if not(type(atomORarray) is chimera.Atom or
               type(atomORarray) is numpy.ndarray):
           print "Bad argument type in getCoords function call."
           return None
        return coords
```

```
    # ------------------------------------------------------------------
    # Given an atom object or an array this function gets the
    # coordinates (or directly uses the array) to define the
    # coordinates of a point P (coordsP) in the PDB frame of reference.
    # It then projects these coordinates onto the plane by:
    #   1) Subtract the center c array to get the "local" coordinates
    #      localP (in a frame of reference that has c as the origin.
    #   2) Calculate the projection of localP on the normal vector
    #      (self.normal).
    #   3) Subtract this result from coordsP to get the projection
    #      on the plane relative to the PDB frame of reference.
    def projectionOnPlane(self, atomORarray):
        # Compute coordinates relative to an origin at the centroid.
        coordsP = self.getCoords(atomORarray)
        localP = coordsP - self.center  # Step 1
        projOnNormal = self.normal*dot(self.normal, localP) # Step 2
        return coordsP - projOnNormal # Step 3

    # ------------------------------------------------------------------
    # Given an atom object or an array object this function computes the
    # perpendicular distance from that point to the least squares plane.
    def distanceToPlane(self, atomORarray):
        return linalg.norm(self.getCoords(atomORarray) -
                           self.projectionOnPlane(atomORarray))

    # ------------------------------------------------------------------
    # Given two atom objects or array objects this function computes
    # the angle between the line joining these two points and the
    # projection of this line in the LS plane.
    def angleToPlane(self, a_atomORarray, b_atomORarray):
        # Compute the coordinates:
        a_coords = self.getCoords(a_atomORarray)
        b_coords = self.getCoords(b_atomORarray)
        distBTWNcoords = linalg.norm(a_coords - b_coords)

        a_coordsProj = self.projectionOnPlane(a_atomORarray)
        b_coordsProj = self.projectionOnPlane(b_atomORarray)
        distBTWNprojections = linalg.norm(a_coordsProj - b_coordsProj)

        return arccos(distBTWNprojections/distBTWNcoords)*180/numpy.pi

#======================================================================
# Least Squares Plane CLASS

class LeastSquaresPlane(Planes):
    def __init__(self, atoms_L):
        self.atoms_L = atoms_L
        self.centroid = self.centroid()
        # Vectors providing local frame of reference:
        self.localAxes_L = []
        # The normal to the plane will be localAxes_L[0]:
        self.normalToPlane = None
```

```
        # List of atom coordinates relative to centroid:
        coords _ L = []
        for a in self.atoms _ L:
            coords _ L.append(array(a.coord()) - self.centroid)

        # Use coords _ L to build matrixA:
        matrixA = array(coords _ L)
        eigvals, eigvecs = linalg.eig(dot(matrixA.transpose(),
                                            matrixA))
        # Sort the eigenvalues to get the needed eigenvector:
        ind = eigvals.argsort()
        # Important: We have to extract eigenvectors
        # from the transpose of eigvecs!  We consider the
        # first eigenvector to be the normal to the plane
        # and (in the coordinate system set up at centroid),
        # it will be the z-axis. The other two eigenvectors
        # will be passed back to give us the unit vectors
        # defining the x-axis and y-axis.
        # They are kept in self.localAxes_L (just in case
        # they are needed for some other application).
        for i in range(3):
            self.localAxes_L.append(eigvecs.transpose()[ind[i]])
        self.normalToPlane = self.localAxes _ L[0]

        Planes.__init__(self, self.centroid, self.normalToPlane)

    # ---------------------------------------------------------------
    # Function to compute the centroid of the atoms in the list.
    def centroid(self):
        sum = zeros(3, float)
        for a in self.atoms _ L:
            sum += a.coord()
        return sum/float(len(self.atoms _ L))
```

Reviewing the code for the initialization function, we see that the list of atom objects is established in `self.atoms _ L` and the centroid is computed. When the eigenvalue problem is solved, we will get a new frame of reference with origin established at the centroid. The axes for the frame of reference will be kept in `self.localAxes _ L`. For ease of reference we will put the first axis (index 0) into the variable named `self.normalToPlane`. We consider this normal to the plane to be the z axis and so the plane itself will have a coordinate system defined by the x and y axes stored in `self.localAxes _ L[1]` and `self.localAxes _ L[2]`. These are kept as class attributes because they may be needed by some other applications that require an orthonormal basis set with origin at the centroid.

The temporary variable `coords_L` stores a list of atomic coordinates each one being the PDB coordinates shifted so that the origin of their new frame of reference is at the centroid. These coordinates are assembled into an array referenced as `matrixA`. This gives us the matrix A needed in Equation 4.32. The script then forms the product AA^T and solves the eigenvalue problem. We extract eigenvectors from the transpose of `eigvecs` and store them in `self.localAxes_L` as promised earlier.

The methods `projectionOnPlane`, `distanceToPlane`, and `angleToPlane` follow the implementation details described earlier. Note that, for the sake of convenience, these methods will take arguments that are either simple arrays or Chimera atom objects. The function `getCoords` checks the Python type of these objects and returns the PDB coordinates for either case. Note that an object instantiated from this class will not produce any visual representation in the Chimera window, but the object can be used in conjunction with additional code to draw planar objects such as disks.

4.6 EXERCISES

The following exercises will deal with angle calculations and will include problems that incorporate the various classes covered in this chapter.

1. Computing CYS–Fe–CYS angles in 1BRF

   A frequent task in structural biology is the measurements of bond angles determined by two atoms bonded to a third atom. This involves finding the atoms, getting their coordinates, and then applying Equation 4.2 to compute the bond angle. This "warm-up" exercise involves writing the function that can do this computation. To test the function, your script should fetch the PDB file 1BRF (rubredoxin from *Pyrococcus furiosus*) and then compute the following bond angles:

```
(CYS 5.A SG) - (FE 54.A FE) - (CYS 8.A SG)
(CYS 5.A SG) - (FE 54.A FE) - (CYS38.A SG)
(CYS 5.A SG) - (FE 54.A FE) - (CYS41.A SG)
(CYS 8.A SG) - (FE 54.A FE) - (CYS38.A SG)
(CYS 8.A SG) - (FE 54.A FE) - (CYS41.A SG)
(CYS38.A SG) - (FE 54.A FE) - (CYS41.A SG)
```

Entries within a set of parentheses identify an atom. The first entry is the residue type. The second entry is the residue position and residue chain identifier separated with a period. The third entry is the atom name. In your script, it will be sufficient to use the second and third entries to uniquely identify an atom. To get the atoms, recall the script used in Exercise 2 of Chapter 2.

The relationship between the iron atom and its neighboring CYS residues will be very similar to that shown in Figure 3.1. Compare your results with those published in [BR98] (page 490).

2. Consistency of bond angles in the protein backbone

Collecting statistics related to proteins gives us data that are useful for various applications. An example would be checking the results of structure predictions from nuclear magnetic resonance experiments to see if there are any anomalies in bond angles. In particular, it has been observed that repeated bond angle measurements taken from a protein backbone tend to have a small standard deviation. In this exercise, we gather evidence to test this assertion. Using the Chimera menu invocation **Tools... Structure Analysis... Angles/Torsions**, we can determine the following bond angles (see Figure 4.21):

SER 6 C_α – SER 6 C – LEU 7 N 116.086
SER 6 C – LEU 7 N – LEU 7 C_α 121.881

Are these observations typical for C_α–C–N and C–N–C_α bond angles? Write a script that gets several hundred observations of each type of bond angle and use this data to calculate both the mean and the standard deviation of the bond angles in the protein backbone. You should also report the minimum and maximum lengths for each type of angle. Note any significant differences in these observations if you restrict the amino acids to be within a helix, strand, or loop.

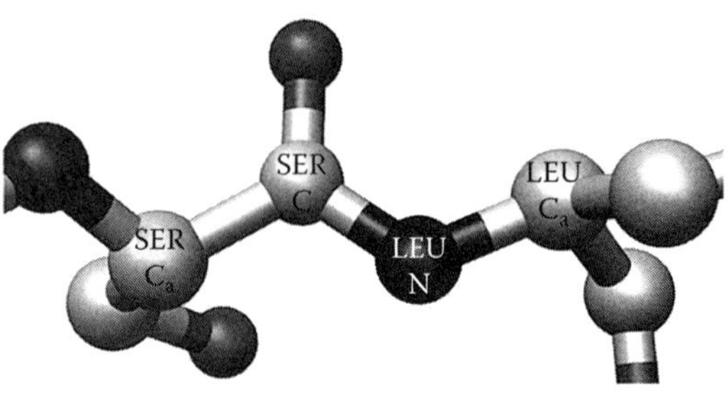

FIGURE 4.21 Angles in the peptide backbone.

Here is the output that would be expected when the script processes the protein 1A4Y:

```
N-CA-C angles:
Secondary        Mean      Standard       Max        Min       Sample
Structure:                 Deviation                           Size
Helix:          111.51       2.72       120.35     103.97       448
Strand:         109.79       2.97       115.39     100.08       176
Loop:           111.59       3.62       121.32      98.81       541

CA-C-N angles:
Secondary        Mean      Standard       Max        Min       Sample
Structure:                 Deviation                           Size
Helix:          116.38       0.83       119.85     113.81       448
Strand:         116.12       0.83       118.53     112.44       176
Loop:           116.16       5.03       123.42      15.93       541

C-N-CA angles:
Secondary        Mean      Standard       Max        Min       Sample
Structure:                 Deviation                           Size
Helix:          121.44       1.07       124.43     118.43       448
Strand:         121.45       1.06       124.70     118.19       176
Loop:           121.36       4.47       127.62      54.22       541
```

3. Computing the dihedral angle in a disulfide bridge

   Modify the script for Exercise 3 in Chapter 3 so that it will also compute the dihedral angle formed by the four atoms (CB–SG–SG–CB) in a disulfide bridge (see the example in Figure 3.10). Computation of a dihedral angle should be done using a function called `dihedralAngle` that accepts four atom objects representing the four consecutive atoms that form the dihedral angle and returns the dihedral angle measured in degrees. In a separate program you can compare your results with those that can be obtained by using the `chimera.dihedral()` function.

4. Consistency of the phi, psi dihedral angles

   a. Use the `dihedralAngle` function from the previous exercise to compute the phi, psi dihedral angles for alpha carbon C_α [i] when given the coordinates of C[i – 1], N[i], C_α [i], C[i], and N[i + 1].

      As done in the previous exercises, get 1000 observations of each type of dihedral angle and use this data to calculate both the average and the standard deviation. You should also report the minimum and maximum lengths for each type of angle. The observed data spread should be much wider than that computed for bond angles. What are the significant differences in these observations if you restrict the amino acids to be within a helix? Repeat these calculations for beta sheets and loops.

b. Repeat the previous question using the phi, psi attributes of a residue object to get dihedral angle values. Note that some residues will not have phi and psi attributes, so your code will have to take this into account when doing further computations involving phi and psi.

5. Ramachandran plots for a specific residue

Write a script that reads a text file with the following format:

First line is informative text for the user and is ignored by the script.

Second line identifies a residue type in the first three columns.

Third and succeeding lines contain PDB IDs in the first four columns.

Sample text file:

```
Proteins used for generating a Ramachandran plot:
GLY
1CRN
1TIM
1AON
```

The script should read all the PDB files in the list, and for each protein it should get the phi and psi dihedral angles for all residues that have the type specified by the second line. It should then generate a Ramachandran plot for these residues. The script should provide information about the number of residues encountered. For example:

```
Number of GLY residues found: 828
Number of residues within helices: 215
Number of residues within strands: 58
Number of residues within coils: 555
```

Note that the plot should show a distribution that is very similar to that seen in Figure 4.9. Implementation suggestions: Modify Script 4.03 so that it accepts `phi_L` and `psi_L` lists as function arguments instead of computing them from a list of residues. Use a modification of Script 2.03 to fetch the PDB files (note the use of `chimera.openModels.close()` to eliminate protein models after they have been processed).

6. Checking chirality at the alpha carbon

   As described in Chapter 1, the CA atom and its four neighbors form a chiral center for all residues except glycine. In most biological systems, this chirality is described as being in the L-configuration (see Figure 1.7), but it is possible to find proteins that have residues with a D-configuration. Write a function called `chiralityExaminer` that checks the chirality of a chiral center within a residue. The input for the function will be the CA atom object of a residue and the output will be a single letter code: "L" for L-configuration, "D" for D-configuration, and "N" for not applicable (returned when a chiral center is not present). Test the function on a mainline program that reads a specified PDB file and then prints a list of all residues (identified by type, position, and chain ID) that produce a `chiralExaminer` result that is either "D" or "N." Test your program using a mainline that fetches a protein file (PDB ID provided by user input) and then prints a list with entries: residue position, residue type, and chiral code. Test the program using PDB ID "1AN1," a tryptase inhibitor from leech [DP97]. The chapter discusses the residue at position 115 in chain E.

7. Introducing the Solids class to display a disk

   This exercise will get you acquainted with the `addDisk` function of the `Solids` class. This will be used in Exercise 8. The `addDisk` function can be used to display a set of disks that can act as visual stand-in for various biological structures (e.g., membrane layers). The function is in the file named `solids.py`. Use the following import:

   ```
   from StructBio.Scenographics.solids import Solids
   ```

   The class can be instantiated with an empty argument list, for example:

   ```
   diskScene = Solids()
   ```

   If you so desire, a character string acting as a model name can be passed as an instantiation parameter. After this, we can bring disks into the scene by adding them to the instance, for example:

   ```
   diskScene.addDisk(center, normal, thickness, radius, rgb)
   ```

The significance of the parameters should be fairly clear. The position of the disk is specified by an array that defines a 3D point called the `center`. The orientation of the disk is specified by a three-component array that defines a vector acting as a normal for the plane of the disk. The argument `rgb` is a three-component color vector. Finally, a statement such as

```
diskScene.display()
```

will cause the disks to be shown on the screen. To test the class you should generate a few lines of code that will display 12 disks arranged in a dodecahedral configuration. Your displayed scene should look something like Figure 4.22. Many viruses have an icosahedral coat, but its dual structure, a dodecahedron, has come up in various research works dealing with viruses (see [FS03] and [TJ03]).

To write the script, read the Wikipedia article at: http://en.wikipedia.org/wiki/Icosahedron. You will see that the 12 vertices of an icosahedron are really simple to calculate if you consider them to be the vertices of three rectangular Borromean rings. These rings have their own interesting mathematical and cultural significance (see http:/en.wikipedia.org/wiki/Borromean_rings).

The 12 vertices can be written as three sets of four vertices specified by

$$(0,\pm1,\pm\varphi),\ \ (\pm1,\pm\varphi,0),\ \ (\pm\varphi,0,\pm1)$$

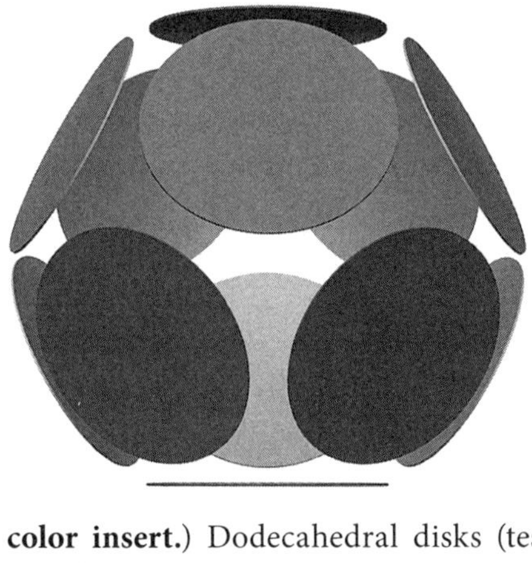

FIGURE 4.22 **(See color insert.)** Dodecahedral disks (testing the `addDisk` function in the `Solids` class).

where $\varphi = \left(1+\sqrt{5}\right)/2$ is the golden ratio (more cultural and mathematical excursions can be discovered at: http://en.wikipedia.org/wiki/Golden_ratio). The four vertices specified within a set of parentheses are derived by taking all four possible combinations of the plus and minus signs. When working with the `addDisk` method, each of the 12 triplets can be used to define both the disk center and the normal to the disk. The line going from the origin to the disk center will be normal to that disk. You should experiment with a radius value that will get a reasonable set of disks. Colors can be obtained from the Chimera Web site: http://www.cgl.ucsf.edu/chimera/docs/UsersGuide/colortables.html.

You can convert a hexadecimal TK color code, stored in a character string, to the three component tuple needed by the disk code by importing `struct` and then using

```
tuple(array(struct.unpack('BBB', cStr.decode('hex')))/255.0)
```

8. Movement of helices in mechanosensitive channels

   Bacterial mechanosensitive channels open in response to membrane tension so that small ions and molecules can leave the cell. An elementary description of this functionality can be found in [MA92]. Researchers have managed to do x-ray analysis of a mechanosensitive channel from *Escherichia coli* (see references [SB07] and [WB08]).

   You can get PDB files for this channel in its open conformation (PDB ID = 2VV5) and in its closed conformation (PDB ID = 2OAU). The protein has seven chains each with six helices. Researchers have determined that of these 42 helices, 21 are within the membrane of *E. coli*. When the channel changes its state from closed to open, these helices move in a manner that is reminiscent of the mechanical movement of a camera iris.

   The script for this exercise has two objectives:

   a. Fetch the protein file (done for both 2VV5 and 2OAU) and add various visual components to the display: A disk represents the cytoplasmic side of the cell membrane and a tube, perpendicular to the disk, represents the primary axis of the channel. The disk should be positioned so that the helices within the membrane are situated above the disk as in

Figure 4.23. Finally, we want each helix in chain C to have a tube that defines the position and orientation of its inertial axis. We only need this to be done for the first three helices but it is easier to display all of them. Unlike chain A, chain C starts with three distinct helices that will suit our purpose.

b. Generate an object instantiated from the `Plane` class that represents a plane parallel to the disk. Working with the inertial axis of each of the three helices at the beginning of chain C, calculate the angle that the inertial axis makes with the plane. The script should calculate these three angles for both proteins and then print out all six angles followed by a list of the

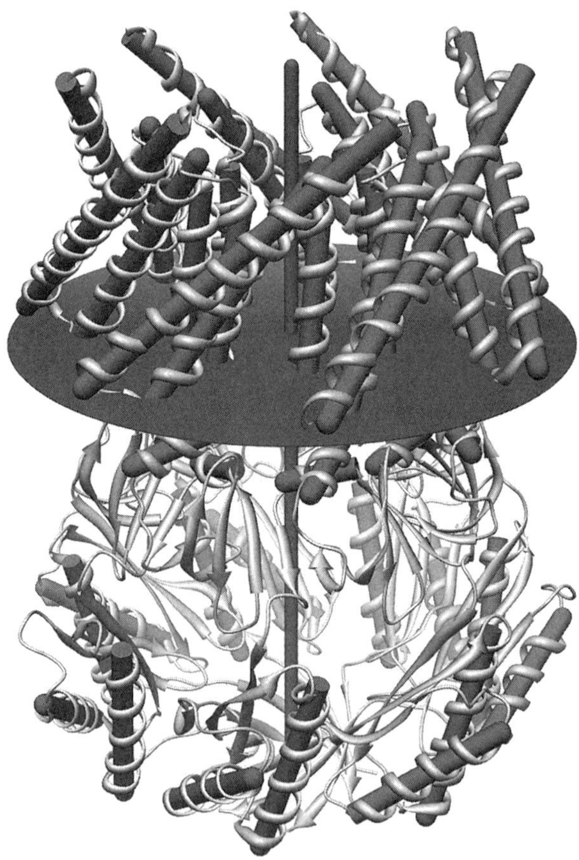

FIGURE 4.23 MscS in open conformation (PDB ID 2VV5).

differences in the angles when the protein goes from the closed conformation to the open conformation.

Hints and suggestions:

- Use the `TubesForSSEs` class to generate the helix tubes.
- Use the `InertialAxis` class to generate the primary axis of symmetry that goes down the center of the channel.
- With the primary axis in place, you can use projections onto the axis to get the top and bottom points for a tube that will be positioned along the primary axis. Another projection (use a residue atom that would be close to the bottom of the membrane) will get a position for the center of the disk.
- Use the `Solids` class to generate a disk perpendicular to the primary axis.
- Use the `Planes` class to get an implicitly defined plane that is parallel to the disk.
- Note that an instance of the `TubesForSSEs` class has the attribute `axes` and this has the attribute `inertialAxes _ L`. By using indices 0, 1, and 2, we can get the first three inertial axes objects and their attributes: `startPoint` and `endPoint` that can define two points to define the position of a helix axis. These points can be used as arguments for the `angleToPlane` method that is provided within your `Planes` class object.

9. Least squares plane for a heme ring

Part (a):

If you use Chimera to view the heme ring of the myoglobin protein 1MBO "on edge," you will see that most of the atoms are coplanar except for nitrogen "NA," which seems to be pulled out of the plane, probably due to forces exerted by the chain ending in "O1A" and "O2A." So, to define the position and orientation of a plane that is closest to the atoms of the porphyrin ring, we will make the somewhat arbitrary decision to go with 12 carbon atoms in the ring, as specified in the following Python "hit list":

```
hitList = ["CHA","C1A","C4A","CHB","C1B","C4B","CHC","C1C","C4
C","CHD","C1D","C4D"]
```

In this exercise, you should use the `LeastSquaresPlane` class (in the `planes.py` module) and `Surfaces` classes to put a 12-sided polygon into the ring. Figure 4.24 shows the intended result.

Part (b):
Researchers have determined that the bond within the oxygen molecule is tilted so that the two oxygen atoms and the iron atom are *not* collinear. This deviation from collinearity is essential for the eventual removal of oxygen from the hemoglobin or myoglobin protein. Let us suppose that we want to measure the angle made by the bond in the O_2 molecule with the plane of the heme ring as illustrated by Figure 4.25.

The angle α may be determined using the equation

$$\sin\alpha = (g - h)/b$$

where b is the bond length, h is the distance of O1 to the plane, and g is the distance of O2 to the plane. O1 and O2 are the names of

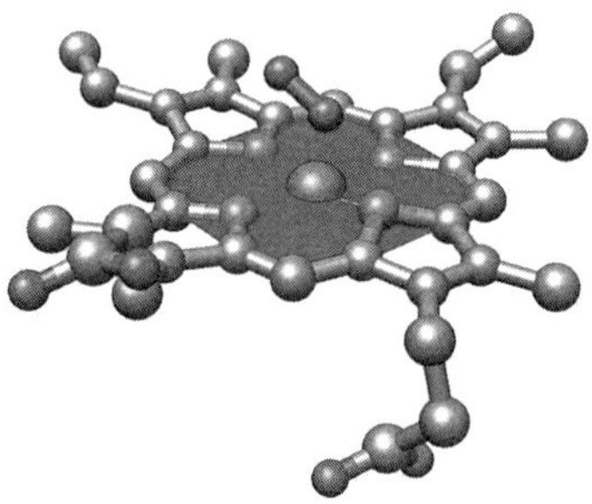

FIGURE 4.24 Polygonal plane within a hemoglobin ring.

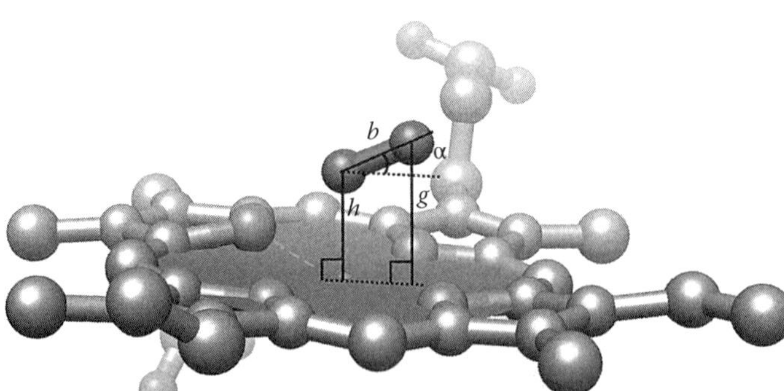

FIGURE 4.25 Measurements needed to calculate the angle.

the oxygen atoms in the oxygen ligand molecule. You can also use a cosine formula if you first compute the distance between the points representing the intersections of these perpendiculars with the plane. In your script you could use the `angleToPlane` method after generating the least squares plane object.

This part of the exercise involves the following steps:

- Modify Part (a) so that the script will calculate the angle between the least squares plane and the O_2 molecule (measured in degrees).
- Heme rings are found in both hemoglobin and myoglobin. The iron in the heme ring can also interact with carbon monoxide and cyanide. In each of the latter two cases, the bond in the molecule also makes an angle with the heme plane, but it tends to be closer to 90°. Use the `angleToPlane()` method of the `LeastSquaresPlane` class to fill in the table that follows:

Source	**Molecule**	**PDB ID**	**ligand name**	**Angle**
Corynebacterium diphtheria	Heme oxygenase	1V8X	OXY	
Human	Oxyhemoglobin	1HHO	OXY	
Sperm Whale	Oxymyoglobin	1MBO	OXY	
Human	Carbonmonoxyhemoglobin	1BBB	CMO	
Mycobacterium tuberculosis	Hemoglobin with cyanide	1RTE	CYN	

The protein 1V8X has a functionality that is different from the other proteins in the table. Do some searching on the Internet to find out what that difference is.

10. CO orientation above the heme plane

Read the work done by Kachalova, Popov, and Bartunik [KP99]. In Figure 1 of their work, they discuss the orientation of the carbon monoxide ligand relative to a plane that passes through the four nitrogen atoms in the heme ring. Carefully read the work to get the definitions of four angles: α, τ, ϕ_{FeC}, and ϕ_{CO} that define the orientation of the CO ligand above the heme plane. Use the `angleToPlane()` method of the `LeastSquaresPlane` class to evaluate α and τ. Generate code to evaluate ϕ_{FeC} and ϕ_{CO}. Compare your results with that of [KP99].

11. Helix movement relative to the heme plane

    Read the work done by Kachalova, Popov, and Bartunik [KP99]. In Figure 2 of their work, they illustrate the change in heme geometries that are seen in myoglobin bound with and without carbon monoxide. This amounts to a change in the position of two helices with respect to the heme plane. Write a script to do computations that evaluate the angle changes in these helices. You will need two PDB files for myoglobin (with and without CO, 1BZR and 1BZP, respectively). Compute the inertial axes of the helices that are involved. For each protein and for each of the helices (both E and F), compute the angle of the helix axis relative to the plane of the heme group. The plane should be determined by the four nitrogen atoms of the heme ring. Show how these helix angles change when myoglobin goes from deoxyMb to MbCO.

    You should note that the acceptance of the CO ligand causes the Fe atom to be driven down slightly below the heme plane. The script should also calculate the amount of the shift in the position of the Fe atom.

12. A modeless dialog to do Ramachandran plots

    The material in Appendix A covers the implementation of modeless dialogs. Use the canvas widget as a container for the display of a Ramachandran plot. An `Entry` widget working with a `Button` widget should be used to fetch the protein file. The function invoked by the Save button should be used to store the plot as an image. Consider Figure 4.26 as a guide for the graphical user interface (GUI) design.

13. A modeless dialog to do Ramachandran plots for residues

    Use the canvas widget as a container for the display of a Ramachandran plot for specified residues. A `Button` widget should be used to fetch a file containing a list of several proteins. The plot should be based on several protein files to get a good statistical coverage of the (phi, psi) angle pairs. To avoid overrepresentation of data coming from identical chains, the file should list PDB identifiers and for each a letter should indicate the single chain that is to be used. After all the protein files have been read and the dihedral angle data extracted, the dialog should allow the user to specify the residue that is to be characterized by the plot. Recall from Figures 4.11 and 4.12 that the residue prior to a proline residue will typically have its dihedral angles restricted by the proline side chain. The dialog deals with all 20 residues and such a pre-proline residue.

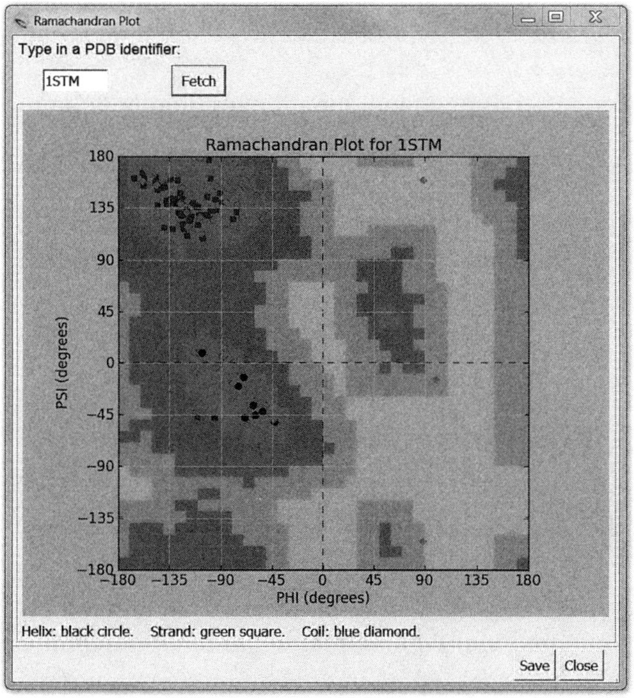

FIGURE 4.26 A GUI for Ramachandran plots applied to proteins.

The function invoked by the Save button should be used to store the plot as an image. Consider Figure 4.27 as a guide for the GUI design.

14. A modeless dialog to change torsion angles of side chains

    As described in Section 2.7.5 of Chapter 2, it is possible to get the dihedral (or torsion) angles of a residue by accessing the attributes `chi1`, `chi2`, and so on. It is also possible to assign values to these attributes and the conformation of the side chain will change accordingly. The objective of this exercise is to implement a GUI that will allow a user to change the torsional angles for a selected residue. Selection of a residue can be done through a **Select** menu invocation or by using a sequence dialog. It is also convenient to simply select an atom in the display and then use the statements

```
from chimera import selection
atm_L = selection.currentAtoms()
res = atm_L[0].residue
```

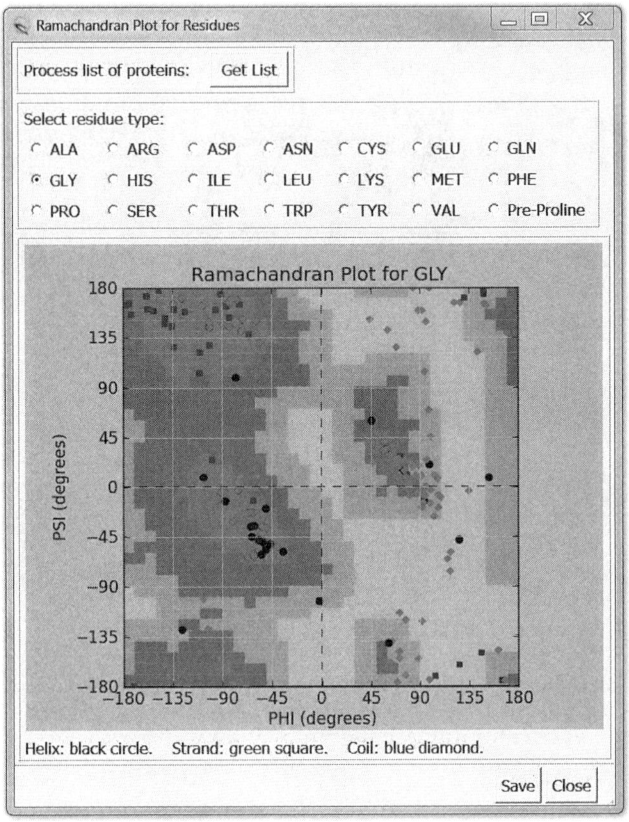

FIGURE 4.27 (**See color insert.**) A GUI for Ramachandran plots applied to residues.

The `selection.currentAtoms()` function will return a list of selected atoms. Ideally, the list contains only one atom but if the list contains more than one, we can simply work with the first atom and get the residue containing that atom. The selection of the residue should be done in response to a click of an Accept button, and a Label box can be used to identify the residue as shown in Figure 4.28.

Once the residue is selected, the script should set the positions of slider widgets in the dialog to indicate the values of the torsion angles. Sliders should be disabled if they correspond to torsion angles that are not present in the residue. A Label widget can be used to indicate the values of the torsion angles. Moving the slider controls should produce conformational changes of the selected

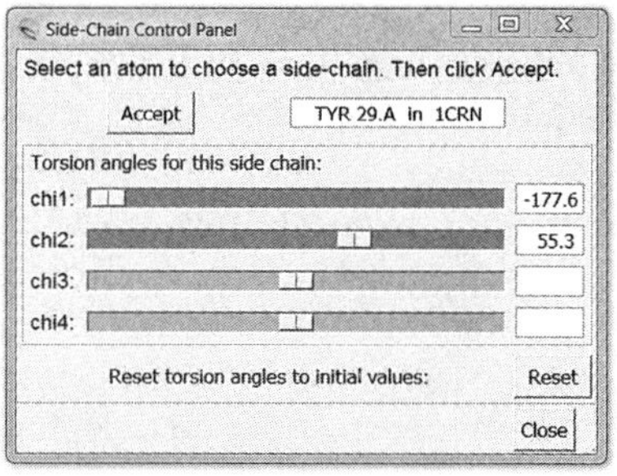

FIGURE 4.28 A GUI for the side chain control panel.

residue. Finally, a Reset button should allow the user to recover the torsional angles that were present in the residue when the Accept button was clicked.

Note that Chimera already has a dialog to modify torsional angles (Tools… Structure Editing… Adjust Torsions). Nonetheless, doing this exercise is a worthwhile experience for dialog implementation. In particular, the components of the GUI must work together to manage the state settings of the various widgets during the user interactions. Consequently, the complexity of the logic is more significant than that seen in Exercise 13 (for example, widgets are enabled and disabled depending on activities related to the Accept button). Furthermore, we will be using this dialog as a starting point for a more capable dialog dealing with rotamers (see Chapter 7).

REFERENCES

[BR98] R. Bau, D. C. Rees, D. M. Kurtz, R. A. Scott, H. Huang, M. W. W. Adams, and M. K. Eidsness. Crystal structure of rubredoxin from *Pyrococcus furiosus* at 0.95 Å resolution, and the structures of N-terminal methionine and formylmethionine variants of Pf Rd. Contributions of N-terminal interactions to thermostability. *Journal of Biological Inorganic Chemistry*, **3** (1998), 484–493.

[DP97] S. Di Marco and J. P. Priestle. Structure of the complex of leech-derived tryptase inhibitor (LDTI) with trypsin and modeling of the LDTI-tryptase system. *Structure*, **5** (1997), 1465–1474.

[FS03] P. Fender, G. Schoehn, J. Foucaud-Gamen, E. Gout, A. Garcel, E. Drouet, and J. Chroboczek. Adenovirus dodecahedron allows large multimeric protein transduction in human cells. *Journal of Virology*, **77** (2003), 4960–4964.

[HB05] B. K. Ho and R. Brasseur. The Ramachandran plots of glycine and pre-proline. *BMC Structural Biology*, **5** (2005), http://www.biomedcentral.com/1472-6807/5/14.

[KP99] G. S. Kachalova, A. N. Popov, and H. D. Bartunik. A steric mechanism for inhibition of CO binding to heme proteins. *Science*, **284** (1999), 473–476.

[LM93] R. A. Laskowski, M. W. MacArthur, D. S. Moss, and J. M. Thornton. PROCHECK: a program to check the stereochemical quality of protein structures. *Journal of Applied Crystallography*, **26** (1993), 283–291.

[MA92] B. Martinac. Structural plasticity in MS channels. *Nature Structural & Molecular Biology*, **12** (2005), 104–105.

[MM92] A. L. Morris, M. W. MacArthur, E. G. Hutchinson, and J. M. Thornton. Stereochemical quality of protein structure coordinates. *Proteins*, **12** (1992), 345–364.

[MP94] C. Matos, G. A. Petsko, and M. Karplus. Analysis of two-residue turns in proteins. *Journal of Molecular Biology*, **238** (1994), 733–747.

[NK05] M. Novotny and G. J. Kleywegt. A survey of left-handed helices in protein structures. *Journal of Molecular Biology*, **347** (2005), 231–241.

[RR63] G. N. Ramachandran, C. Ramakrishnan, and V. Sasisekharan. Stereochemistry of polypeptide chain configurations. *Journal of Molecular Biology*, **7** (1963), 95–99.

[SB07] S. Steinbacher, R. B. Bass, P. Strop, and D. C. Rees. Structures of the prokaryotic mechanosensitive channels MscL and MscS. *Current Topics in Membranes* in *Mechanosensitive Ion Channels, Part A* (O. P. Hamill, ed), **58** (2007), 1–24.

[TJ03] L. Tang, K. N. Johnson, L. A. Ball, T. Lin, M. Yeager, and J. E. Johnson. The structure of Pariacoto virus reveals a dodecahedral cage of duplex RNA. *Natural Structural Biology*, **8** (2003), 77–83.

[WB08] W. Wang, S. S. Black, M. D. Edwards, S. Miller, E. L. Morrison, W. Bartlett, C. Dong, J. H. Naismith, and I. R. Booth. The structure of an open form of an *E. coli* mechanosensitive channel at 3.45 A resolution. *Science*, **321** (2008), 1179–1183.

CHAPTER 5

Structure Overlap and Alignment

5.1 MOTIVATION

Structure comparison deals with various algorithms that analyze two or more proteins in an effort to describe structural similarities. Why should we want to do this? The primary goal is to understand functionality, but this is a very complicated endeavor. In Chapter 1, we presented various behavioral aspects related to a protein's structure. On reviewing these points, we see that protein structure must provide for the following "life cycle functionalities":

- Uniqueness: The peptide chains must fold into a particular low-energy conformation that has the desired biological utility. The protein should not have other competitive low-energy conformations that are characterized by biological dysfunction.
- Stability: Once the protein is in its folded state, it must retain that folded state even though some changes in conformation may occur due to flexibility requirements.
- Interaction: The final conformation must have an interaction capability, for example, a ligand binding site or a surface for protein–protein interactions. This capability may also include the flexibility noted in the previous list item.

- Degradation: In many cases, the structure should be degradable so that the protein is conveniently disassembled, its constituent amino acids used in the construction of other proteins.

Unfortunately, each of these functional behaviors is generally difficult to relate to structure. In some cases, the mechanism of a binding site may be thoroughly studied in a drug design project and the interactions of the drug with the binding site are known with high precision. The conformational changes of a protein when it interacts with a ligand are one of the motivations for structural comparisons.

However, in most situations the connection between function and structure is much more difficult to assess and we must rely on observation in lieu of theory. In other words, when a new protein is being studied we typically compare its structure with that of similar proteins in an attempt to draw inferences about its behavior based on the known behavior of the other proteins in the same category.

But how do we organize this categorization? Since the final structure must satisfy all of the aforementioned requirements, evolutionary processes will have produced structures that attempt to optimize fitness criteria with multiple objectives. Since a particular protein strives to satisfy multiple objectives, it is very difficult to determine the structural properties that are most important to its functionality. For example, do we focus on the configuration of atom types in a binding site or do we stress the organization of secondary structural elements that lead to the stability of a hydrophobic core? The question has no definitive answer.

As an example, consider the overlap of 1MBN and 1JEB, both illustrated in Chapter 1, Figure 1.26. If we focus on this pair of proteins, it seems that the region of highest sequence conservation corresponds to the 16 residues at the very beginning of the protein. This is a helix that does not seem to be directly related to the heme binding site. The region does contain the sixth amino acid that leads to sickle-cell anemia if it mutates to valine in hemoglobin, but this is only a single residue. One may speculate that there seems to be some additional critical functionality that is being preserved by this helix. To investigate this issue more completely, it would be necessary to consider other proteins in the globin family.

In summary, we do structural comparisons for various reasons: we may wish to study the conformational change that is seen in an interaction between a protein and another molecule. We may also study the geometry of binding sites and seek to understand the chemical mechanism of that interaction. The other issues (uniqueness, stability, and degradation) are somewhat more difficult to assess. We understand that a hydrophobic core is necessary for stability, but understanding the attainment of all three objectives and its relation to the consecutive arrangement of amino acids in a peptide chain is a major challenge. In other words, we do not have algorithms that can examine a primary sequence and then accurately predict whether the peptide would fold to a unique and stable conformation that could be enzymatically degraded at a later time. To understand or predict a protein's behavior, we must consider its membership in a family of proteins with similar structure. Therefore, we are concerned with the structural differences between a given protein and other proteins that have an evolutionary relationship with it. In particular, what has been structurally conserved?

We may also study the *structural plasticity* of a family. This topic deals with the degree of structural variability seen in a family. It seems that some families show more acceptance of such diversity than do others. Algorithms that deal with multiple structure alignments are important for these investigations. Experiments such as those reported in [VB96] and [SB06] have demonstrated the effect that amino acid insertions have had on stability and the tendency to maintain local helix formation by propagating the effects of the insertion into a bend or loop. Another interesting discussion about structural plasticity and its relation to evolution may be found in [MO78].

Finally, it should be mentioned that proteins can assume similar folds even though there is no obvious evolutionary relationship among them. These *analogs* may be the result of a limited number of configurations available when packing helices and sheets to form a hydrophobic core (see [OJ03]).

Research in structural comparison seems to reflect the wide variety of concerns and viewpoints on the topic. To get some perspectives on this, the reader may consult [BO96] written by Brown, Orengo, and Taylor. This paper suggests a systematic nomenclature that helps to organize the diverse topics contained in approximately 50 or so research papers written on protein structure comparison methodologies prior to 1996. Since then, considerably more papers have been written with a corresponding increase in the range of comparison methods.

5.2 INTRODUCTION

5.2.1 Specifying the Problem

To understand the wide range of topics related to structural comparison, we list some of the attributes that characterize a study of the problem:

- Level of detail: A comparison study is done at a particular resolution or level of structural detail. Comparison algorithms may deal with the positions of various types of structure:
    - Individual atoms
    - The locations of residues (specified by coordinates of the alpha carbon atoms or beta carbon atoms or the center of mass of the side chains)
    - The position and orientation of secondary structural elements (e.g., treating helices as cylinders)
    - The similarity of folds at the tertiary structure level
- Physicochemical properties: The quantification of similarity may only involve the relative positions of structural entities in three-dimensional (3D) space, or it may also include additional descriptive attributes that are important for some application. For example, an atom may be categorized with a physicochemical label that describes it as being a hydrogen bond donor, being a hydrogen bond acceptor, having a positive charge or a negative charge, or being hydrophobic. Hydrophobicity or hydrophilicity may also be assigned to entire residues. Depending on the objectives of the comparison, a property may be associated with a point in space or it may be spread out over some portion of a surface constructed around the underlying atoms.
- Extent of comparison: The comparison may be global, dealing with the entire protein, or more local, being restricted in some fashion. Various types of restriction may be applied to limit the comparison to a protein domain, a contiguous subsequence of amino acids, or the subset of atoms in a binding pocket.
- Number of proteins: A comparison study may involve one, two, or several proteins. In the case of a single protein, the study typically compares the change in atom positions when the protein changes its conformation. Good examples of this would be the

conformational change associated with an allosteric movement or the shift of side chains in a binding pocket during ligand docking. When two proteins are involved, the objectives of the study usually depend on how different the proteins are in sequence composition. A simple case would be the study of conformational change due to the substitution of one or two residues. A more complicated comparison would deal with two proteins that have different lengths and perhaps very little sequence identity when a sequence alignment is performed. In the latter case, it is typical that the proteins have a homologous relationship and the study is designed to assess the extent of structural similarity that has been conserved in an evolutionary descent. Studies involving many proteins are often designed to evaluate similarities in proteins that are closely related as established by sequence alignment. Possible objectives include an analysis of all proteins in the group with the intention of characterizing atom types in a binding site (looking for similarities) or an analysis of evolutionary divergence (this time looking for differences). Conversely, studies may establish similarities in proteins due to structural conformations only independent of sequence similarity. If the sequence similarity is then found to be very low, it is reasonable evidence that structural similarity is due to evolutionary convergence.

These points should convince the reader that there are many approaches to structural comparison. In fact, an extended treatment of the various methodologies that are used could form the subject matter of an entire book. To set a limit on this extensive material, we continue with the discussion of only a few comparison algorithms.

5.3 TECHNIQUES FOR STRUCTURAL COMPARISON

Suppose we are given two proteins that we will designate as P and Q. Algorithms for structural comparison of P and Q generally fall into two broad categories:

1. Superposition algorithms: These algorithms try to maximize some type of overlap of two proteins in 3D space. The overlap may deal with all the residues of the protein or some subset of the residues. If subsets of residues are being considered, then there must be some specification of atom pairs that will participate in the overlap.

2. Algorithms that compare geometric relationships within a protein: We start by defining a set of relationships between elements in the protein *P*. For example, we might calculate for each alpha carbon in *P* a set of distances to all the other alpha carbons in *P*. Each distance is seen as a relationship between the two atoms. Then, we do the same calculations for *Q*. The comparison algorithm then compares the relationships within *P* with those within *Q*. There are many variations on this theme: instead of atoms, we can use fragments or secondary structure components such as helices or strands. Instead of all such elements, concentrate on only those that are nearby or within the same domain. The strategies for comparing the relationships can also vary from one algorithm to the next.

5.4 SCORING SIMILARITIES AND OPTIMIZING SCORES

In all comparison algorithms, we try to evaluate some successful demonstration of similarity. All comparison algorithms must clearly specify the following:

1. Score evaluation: Given two proteins *P* and *Q*, a comparison algorithm may reveal several different possibilities, each demonstrating the similarity between *P* and *Q*. For example, there may be several ways to overlap two proteins in 3D space. Consequently, it is necessary to have some measure of similarity. This evaluation of similarity is usually called a *score*.

2. Score optimization: Typically, there will be a strategy to optimize this score. For example, the algorithm will derive a best structural overlap, or compute some type of correspondence for the atoms of *P* and *Q*, so that the score will be optimized.

When we discuss the various comparison algorithms, both of these points are covered.

5.5 SUPERPOSITION ALGORITHMS

5.5.1 Overview

A frequent concern is whether two proteins have the same or a very similar structure. An assessment of this can be done by attempting to superimpose the two proteins in 3D space. The proteins may have the same residues, or they may be fairly similar (e.g., homologs).

There are various situations when such a comparison would be calculated:

- The proteins may have the same sequence but differ in conformation. The difference might be due to a backbone flexing of the protein or may be due to a change in conformation when a ligand is present in a binding site.
- The proteins may have almost the same sequence. For example, a mutation has caused an amino acid to change. What is the effect on conformation?
- The proteins may have more extensive differences in their sequences, but it is possible that they nonetheless show a lot of similarity in conformation. The question is as follows: How similar are these conformations?
- The proteins may have considerable differences in their sequences, but it is possible that they share similar structure in various regions. We would want to compare these similar regions. Recall that structure is more conserved than sequence.

In some situations, a transformation of coordinates is necessary to compare different conformations of the same protein because the coordinates are derived from protein data bank (PDB) files that assume different orientations of the protein. For example, 1IZI, 1MSN, and 1EBW all deal with the same protein, that is, HIV protease. A casual comparison of the coordinates of the first two atoms in each file gives the following:

```
1IZI:
ATOM  1  N   PRO A  1           -12.600  38.218  3.719
ATOM  2  CA  PRO A  1           -12.444  38.367  2.244

1MSN:
ATOM  1  N   PRO A  1             0.421  40.709  18.682
ATOM  2  CA  PRO A  1            -0.422  39.511  18.905

1EBW:
ATOM  1  N   PRO A  1            29.101  40.309  5.484
ATOM  2  CA  PRO A  1            30.105  39.343  4.986
```

So, it is clear that the PDB does not put proteins in any standard orientation. This was also obvious if you viewed both conformations of the MscS protein (2OAU and 2VV5) in Exercise 8 of Chapter 4.

5.5.2 Characterizing the Superposition Algorithm

In this algorithm, we consider the structural comparison of two proteins that we designate as P and Q. We will not consider any physicochemical properties, just atomic coordinates. So, at this level of detail it is expected that the input to the algorithm will be 3D coordinates of the atoms involved in the comparison. The algorithm will require us to specify a set called an *equivalence*. This is a set of matching atom pairs that are to be brought into 3D superposition. Each pair will designate an atom in P and its corresponding atom in Q. The algorithm will then find the translation and rotation operations that minimize the sum of squares of the distances between the atoms in the pairs.

In the description of the algorithm, we assume that this equivalence is generated using some technique that is appropriate to the application that uses the structure overlap algorithm. Various approaches can be listed:

- If the proteins have the same sequence, then the equivalence can be generated simply by matching up the corresponding atoms. Then, we can extract a subset of these pairs that is appropriate to the needs of the study. For example, we might deal solely with the alpha carbons in the backbone or some part of it.
- If the proteins have a different sequence, then a sequence alignment between P and Q can identify matching pairs of residues and then the corresponding alpha carbon atoms can be used to define the equivalence.

In some cases, the sequence similarity is so vague that it is quite difficult to generate the equivalence set by doing a sequence alignment. In these cases, a different algorithm is required, one that attempts to do a structural comparison using coordinate information with little or no help from a sequence alignment.

Before going any further, it is necessary to settle on some terminology. The terms overlap and superposition are used interchangeably and we assume that the computation that performs the superposition has access to a set of atom pairings that are used to guide an optimal overlap. For protein structures (this will not apply to protein sequences), we use the term structural alignment when dealing with an algorithm that strives to find common structure in two proteins with little or no reliance on the protein sequences involved.

5.5.3 Formal Problem Description

Although different choices could be made for the atoms in an equivalence set, we assume that the superposition algorithm deals with alpha carbon atoms. We are given two sequences of alpha carbon 3D coordinates:

$$P=\left\{p^{(i)}\right\}_{i=1}^{|P|} \qquad Q=\left\{q^{(i)}\right\}_{i=1}^{|Q|} \tag{5.1}$$

where $|P|$ and $|Q|$ are the number of residues in protein P and protein Q, respectively.

We start with a formal description of the equivalence set followed by the notion of an alignment. Some definitions are as follows:

Definition 5.1: An equivalence is a set of pairs

$$\left\{\left(p^{(\alpha_1)}, q^{(\beta_1)}\right),\left(p^{(\alpha_2)}, q^{(\beta_2)}\right),\ldots,\left(p^{(\alpha_N)}, q^{(\beta_N)}\right)\right\} \tag{5.2}$$

indicating the correspondence between the amino acids in P and Q.

Definition 5.2: An alignment M for P and Q is an equivalence such that $\alpha_1 < \alpha_2 < \cdots < \alpha_N$ and $\beta_1 < \beta_2 < \cdots < \beta_N$.

We extract the alpha carbons from each list in the alignment as follows:

$$\begin{aligned} M(P)&=\left(p^{(\alpha_1)}, p^{(\alpha_2)},\ldots, p^{(\alpha_N)}\right) \\ M(Q)&=\left(q^{(\beta_1)}, q^{(\beta_2)},\ldots, q^{(\beta_N)}\right) \end{aligned} \tag{5.3}$$

The structural overlap of the two proteins P and Q will involve modifying the coordinates of atoms to do the following:

1. All the atoms of each protein are moved (translated) so that the centroids of $M(P)$ and $M(Q)$ coincide at the origin.
2. An optimal rotation of protein P is then done to get the maximal amount of overlap between $M(P)$ and $M(Q)$.

What do we mean by maximal overlap? Our measure of success is to minimize the sum of squares of norms that measure the distance between matching alpha carbons. This is minimizing in the least-squares sense:

$$E(M(P_{\text{transformed}}), M(Q)) = \frac{1}{2}\sum_{i=1}^{N}\left\| Rp^{(\alpha_i)} + T - q^{(\beta_i)} \right\|^2 \quad (5.4)$$

Our objective is to find the rotation matrix R and the translation vector T that will minimize E. We consider E to be an evaluation of the similarity score and the purpose of the algorithm is to optimize this score by calculating R and T so that E is minimized. Section 5.5.4 gives a full treatment of the mathematical analysis. If you simply want the final algorithm, you can fast-forward to Section 5.5.5.

5.5.4 Computations to Achieve Maximal Overlap

We start by defining the centroids of the alpha carbons used in the superimposition.

Let

$$p^{(c)} = \frac{1}{N}\sum_{i=1}^{N} p^{(\alpha_i)} \qquad q^{(c)} = \frac{1}{N}\sum_{i=1}^{N} q^{(\beta_i)} \quad (5.5)$$

Then let

$$x^{(i)} = p^{(\alpha_i)} - p^{(c)} \qquad y^{(i)} = q^{(\beta_i)} - q^{(c)} \quad (5.6)$$

We now consider $x^{(i)}$ and $y^{(i)}$ $i = 1, 2, \ldots, N$ to be the coordinates of matching alpha carbons in proteins P and Q, respectively. For each case, we have essentially translated the entire protein so that its centroid is at the origin.*

As seen earlier in Chapter 4, these new coordinates have a property that will be useful when we have to simplify expressions, namely:

$$\sum_{i=1}^{N} x^{(i)} = 0 \quad \sum_{i=1}^{N} y^{(i)} = 0 \quad (5.7)$$

* Be sure to remember that we should calculate the center of mass; if the atoms in this calculation do not have the same atomic weight.

So, working with our new coordinate system, we see that we want to find R and T to minimize E:

$$E = \frac{1}{2}\sum_{i=1}^{N}\left\|Rx^{(i)} + T - y^{(i)}\right\|^2 \tag{5.8}$$

Expanding this, we get the following:

$$\begin{aligned} E &= \frac{1}{2}\sum_{i=1}^{N}\left(Rx^{(i)} + T - y^{(i)}\right)^{\mathrm{T}}\left(Rx^{(i)} + T - y^{(i)}\right) \\ &= \frac{1}{2}\sum_{i=1}^{N}\left(\left(Rx^{(i)} - y^{(i)}\right)^{\mathrm{T}} + T^{\mathrm{T}}\right)\left(\left(Rx^{(i)} - y^{(i)}\right) + T\right) \\ &= \frac{1}{2}\sum_{i=1}^{N}\left[\left\|Rx^{(i)} - y^{(i)}\right\|^2 + \left(Rx^{(i)} - y^{(i)}\right)^{\mathrm{T}} T + T^{\mathrm{T}}\left(Rx^{(i)} - y^{(i)}\right) + T^{\mathrm{T}}T\right] \\ &= \frac{1}{2}\sum_{i=1}^{N}\left[\left\|Rx^{(i)} - y^{(i)}\right\|^2 + T^{\mathrm{T}}T\right] \end{aligned} \tag{5.9}$$

Note that in the second-last line, the coefficients of T sum to 0 because the centroid of each protein is centered at the origin. So, we end up with

$$E = \frac{1}{2}\sum_{i=1}^{N}\left[\left\|Rx^{(i)} - y^{(i)}\right\|^2 + \left\|T\right\|^2\right] \tag{5.10}$$

If we wish to get a minimum E, it is clear that we want $T = 0$ since this will zero out the $\|T\|^2$. Note that we could also get this result by computing $\partial E/\partial T$ and setting this to 0 will also make $T = 0$. So, finally, we see that we want to find the rotation matrix R that will minimize

$$E = \frac{1}{2}\sum_{i=1}^{N}\left\|Rx^{(i)} - y^{(i)}\right\|^2 \tag{5.11}$$

We want to find the matrix R that will minimize E, but it must be subject to the rotation constraint $R^{\mathrm{T}}R = I$. Lagrange multipliers can take care of this. But before we derive a Lagrangian, there is still another simplification that can be made. Since $R^{\mathrm{T}}R = I$, we can write

$$\begin{aligned} E &= \frac{1}{2}\sum_{i=1}^{N}\left\|Rx^{(i)} - y^{(i)}\right\|^2 = \frac{1}{2}\sum_{i=1}^{N}\left(Rx^{(i)} - y^{(i)}\right)^{\mathrm{T}}\left(Rx^{(i)} - y^{(i)}\right) \\ &= \frac{1}{2}\sum_{i=1}^{N}\left(x^{(i)T}R^{T} - y^{(i)\mathrm{T}}\right)\left(Rx^{(i)} - y^{(i)}\right) \\ &= \frac{1}{2}\sum_{i=1}^{N}\left(x^{(i)\mathrm{T}}R^{\mathrm{T}}Rx^{(i)} - x^{(i)\mathrm{T}}R^{\mathrm{T}}y^{(i)} - y^{(i)\mathrm{T}}Rx^{(i)} + y^{(i)\mathrm{T}}y^{(i)}\right) \\ &= \frac{1}{2}\sum_{i=1}^{N}\left(\left\|x^{(i)}\right\|^2 - x^{(i)\mathrm{T}}R^{\mathrm{T}}y^{(i)} - y^{(i)\mathrm{T}}Rx^{(i)} + \left\|y^{(i)}\right\|^2\right) \\ &= \frac{1}{2}\sum_{i=1}^{N}\left(\left\|x^{(i)}\right\|^2 + \left\|y^{(i)}\right\|^2\right) - \sum_{i=1}^{N}y^{(i)\mathrm{T}}Rx^{(i)} \end{aligned} \tag{5.12}$$

Notice that the first sum is independent of R. Consequently, we can minimize E by maximizing H where

$$H = \sum_{\gamma=1}^{N} y^{(\gamma)\mathrm{T}}Rx^{(\gamma)} \tag{5.13}$$

R is the following 3×3 matrix:

$$R = \begin{bmatrix} r_1^1 & r_1^2 & r_1^3 \\ r_2^1 & r_2^2 & r_2^3 \\ r_3^1 & r_3^2 & r_3^3 \end{bmatrix} \tag{5.14}$$

subject to $R^{\mathrm{T}}R = I$, a constraint that we will rewrite as follows:

$$\sum_{\gamma=1}^{3} r_\gamma^\alpha r_\gamma^\beta = \delta_\alpha^\beta = \begin{cases} 0 & \text{if } \alpha \neq \beta \\ 1 & \text{if } \alpha = \beta \end{cases} \tag{5.15}$$

Since α= 1, 2, and 3 and β= 1, 2, and 3, there are nine of these constraints.

The Lagrangian will be $G = H - F$, where

$$H = \sum_{\gamma=1}^{N} y^{(\gamma)\mathrm{T}}Rx^{(\gamma)} \text{ and } F = \frac{1}{2}\sum_{\alpha=1}^{3}\sum_{\beta=1}^{3}\lambda_\beta^\alpha\left[\left(\sum_{\gamma=1}^{3} r_\gamma^\alpha r_\gamma^\beta\right) - \delta_\alpha^\beta\right] \tag{5.16}$$

When multiple constraints are set up in a Lagrangian, we use a linear combination of all the constraints. The λ_β^α values represent the nine

Lagrange multipliers. We have chosen to index them with α and β. This notation is a convenience that will be useful later when representing the equations in matrix form. Note that the constraint does not change when we interchange α and β. This symmetry implies that $\lambda_\beta^\alpha = \lambda_\alpha^\beta$.

How does H depend on the components of R? Noting that $Rx^{(\gamma)}$ is just a 3D vector,

$$Rx^{(\gamma)} = \begin{bmatrix} \sum_{\beta=1}^{3} r_1^\beta x_\beta^{(\gamma)} & \sum_{\beta=1}^{3} r_2^\beta x_\beta^{(\gamma)} & \sum_{\beta=1}^{3} r_3^\beta x_\beta^{(\gamma)} \end{bmatrix}^T \tag{5.17}$$

we can fully expand H to get the following equation:

$$H = \sum_{\gamma=1}^{N} \sum_{\alpha=1}^{3} y_\alpha^{(\gamma)} \sum_{\beta=1}^{3} r_\alpha^\beta x_\beta^{(\gamma)} \tag{5.18}$$

We need to take the partial derivatives of G with respect to all the nine components of the R matrix. First working with H,

$$\frac{\partial H}{\partial r_i^j} = \sum_{\gamma=1}^{N} \frac{\partial}{\partial r_i^j} \left(\sum_{\alpha=1}^{3} y_\alpha^{(\gamma)} \sum_{\beta=1}^{3} r_\alpha^\beta x_\beta^{(\gamma)} \right) = \sum_{\gamma=1}^{N} y_i^{(\gamma)} x_j^{(\gamma)} \tag{5.19}$$

The simplification to the final sum follows from the fact that when taking derivatives only those terms with $\beta = j$ and $\alpha = i$ will contribute.

To simplify our equations, we set

$$\sum_{\gamma=1}^{N} y_i^{(\gamma)} x_j^{(\gamma)} = c_i^j \qquad \Rightarrow \qquad C = \sum_{\gamma=1}^{N} y^{(\gamma)} x^{(\gamma)T} \tag{5.20}$$

There are nine of these c_i^j values, all derived from the input data, and we are storing them in a 3 by 3 matrix called C.

The partial derivatives of F are a bit more complicated since the R matrix entries appear in a quadratic fashion: $r_\gamma^\alpha r_\gamma^\beta$. We have to apply the product rule when taking derivatives.

$$\begin{aligned} \frac{\partial F}{\partial r_i^j} &= \frac{1}{2} \sum_{\alpha=1}^{3} \sum_{\beta=1}^{3} \lambda_\beta^\alpha \left[\left(\sum_{\gamma=1}^{3} \frac{\partial}{\partial r_i^j} r_\gamma^\alpha r_\gamma^\beta \right) - \delta_\alpha^\beta \right] \\ &= \frac{1}{2} \sum_{\beta=1}^{3} \lambda_\beta^j r_i^\beta + \frac{1}{2} \sum_{\alpha=1}^{3} \lambda_j^\alpha r_i^\alpha = \sum_{\beta=1}^{3} r_i^\beta \lambda_\beta^j \end{aligned} \tag{5.21}$$

In the last line, the first sum comes from contributions made when $\alpha = j$ and $\gamma = i$. The second sum is from contributions made when $\beta = j$ and $\gamma = i$. The final simplification utilizes the symmetry $\lambda_\beta^j = \lambda_j^\beta$.

Finally, since the Lagrangian is $G = H - F$,

$$\frac{\partial G}{\partial r_i^j} = 0 \;\Rightarrow\; \frac{\partial H}{\partial r_i^j} = \frac{\partial F}{\partial r_i^j} \;\Rightarrow\; \sum_{\gamma=1}^{N} y_i^{(\gamma)} x_j^{(\gamma)} = c_i^j = \sum_{\beta=1}^{3} r_i^\beta \lambda_\beta^j \quad \forall i, j \tag{5.22}$$

We can consider these variables to be entries in arrays R, λ, and C defined as follows:

$$R = \begin{bmatrix} r_1^1 & r_1^2 & r_1^3 \\ r_2^1 & r_2^2 & r_2^3 \\ r_3^1 & r_3^2 & r_3^3 \end{bmatrix} \quad \lambda = \begin{bmatrix} \lambda_1^1 & \lambda_1^2 & \lambda_1^3 \\ \lambda_2^1 & \lambda_2^2 & \lambda_2^3 \\ \lambda_3^1 & \lambda_3^2 & \lambda_3^3 \end{bmatrix} \quad C = \begin{bmatrix} c_1^1 & c_1^2 & c_1^3 \\ c_2^1 & c_2^2 & c_2^3 \\ c_3^1 & c_3^2 & c_3^3 \end{bmatrix} \tag{5.23}$$

By adopting this notation, we can rewrite this last equation in a very succinct fashion as follows:

$$C = R\lambda \tag{5.24}$$

We know C. How do we solve for λ and then R? We have used the equation $R^T R = I$ to do various simplifications before we created the Lagrangian, but this constraint has not yet been used as a constraint for the Lagrangian analysis itself. So, note

$$C^T C = \lambda^T R^T R \lambda = \lambda^T \lambda \tag{5.25}$$

Since $C^T C$ is a square symmetric matrix, we can do an eigen decomposition:

$$\lambda^T \lambda = C^T C = V S^2 V^T \tag{5.26}$$

If we can use this to find an appropriate λ, then we set $R = C\lambda^{-1}$ and we are done. What is meant by an appropriate λ? Selection of λ must be made with due attention to two issues that have not yet been addressed:

1. The rotation matrix must not introduce a reflection that changes chirality.
   a. Note that preservation of angles and lengths does not guarantee a final conformation that has an L-configuration (as discussed in Section 1.4.1 of Chapter 1).
2. Although we wanted to minimize E, there is nothing in the Lagrange strategy that guarantees this. The procedure could also lead to an R that maximizes E.
   b. The Lagrange strategy only gets critical rotations that produce extreme values of E.

To ensure a proper rotation, we insist that the determinant of the rotation matrix is +1, that is, det $(R) = 1$. To ensure that E is minimized by our choice of R, we must look more deeply into the quantity

$$H = \sum_{\gamma=1}^{N} y^{(\gamma)\mathrm{T}} R x^{(\gamma)} \tag{5.27}$$

to see how its value is determined by the choice of R. In particular, the construction of λ will involve the determination of signs of the square roots of the three entries on the diagonal matrix within VS^2V^{T}. A very elegant strategy for the computation of R starts with the singular value decomposition of C. The theory for a singular value decomposition (SVD) tells us that we can write as follows:

$$C = USV^{\mathrm{T}} \quad \Rightarrow \quad C^{\mathrm{T}}C = \lambda^{\mathrm{T}}\lambda = VS^2V^{\mathrm{T}} \tag{5.28}$$

where $S^2 = \mathrm{diag}(s_1^2, s_2^2, s_3^2)$. It is easy to show that $\lambda = V\mathrm{diag}(\sigma_1 s_1, \sigma_2 s_2, \sigma_3 s_3)V^{\mathrm{T}}$ with $\sigma_i = \pm 1 \quad i = 1,2,3$. Since $R\lambda = C$, we can write as follows:

$$\begin{aligned} R = C\lambda^{-1} &= USV^{\mathrm{T}}V\mathrm{diag}\left(\sigma_1 s_1^{-1}, \sigma_2 s_2^{-1}, \sigma_3 s_3^{-1}\right)V^{\mathrm{T}} \\ &= U\mathrm{diag}\left(\sigma_1, \sigma_2, \sigma_3\right)V^{\mathrm{T}} \end{aligned} \tag{5.29}$$

So

$$C = USV^{\mathrm{T}} \quad \Rightarrow \quad R = U\mathrm{diag}(\sigma_1, \sigma_2, \sigma_3)V^{\mathrm{T}} \tag{5.30}$$

Now we put this into our H quantity:

$$\begin{aligned} H &= \sum_{\gamma=1}^{N} y^{(\gamma)\mathrm{T}} R x^{(\gamma)} = \sum_{\gamma=1}^{N} y^{(\gamma)\mathrm{T}} U\mathrm{diag}(\sigma_1, \sigma_2, \sigma_3) V^{\mathrm{T}} x^{(\gamma)} \\ &= \sum_{\gamma=1}^{N}\sum_{k=1}^{3} \sigma_k y^{(\gamma)\mathrm{T}} u^{(k)} v^{(k)\mathrm{T}} x^{(\gamma)} = \sum_{\gamma=1}^{N}\sum_{k=1}^{3} \sigma_k \left(y^{(\gamma)} \bullet u^{(k)}\right)\left(v^{(k)} \bullet x^{(\gamma)}\right) \\ &= \sum_{\gamma=1}^{N}\sum_{k=1}^{3} \sigma_k \left(u^{(k)} \bullet y^{(\gamma)}\right)\left(x^{(\gamma)} \bullet v^{(k)}\right) = \sum_{\gamma=1}^{N}\sum_{k=1}^{3} \sigma_k u^{(k)\mathrm{T}} y^{(\gamma)} x^{(\gamma)\mathrm{T}} v^{(k)} \\ &= \sum_{k=1}^{3} \sigma_k u^{(k)\mathrm{T}} \left[\sum_{\gamma=1}^{N} y^{(\gamma)} x^{(\gamma)\mathrm{T}}\right] v^{(k)} = \sum_{k=1}^{3} \sigma_k u^{(k)\mathrm{T}} C v^{(k)} \end{aligned} \tag{5.31}$$

The big dots in Equation 5.31 indicate dot product computations. Note that this derivation makes use of a useful formula:

$$U\mathrm{diag}(\sigma_1, \sigma_2, \ldots, \sigma_l)V^{\mathrm{T}} = \sum_{k=\in 1}^{l} \sigma_k u^{(k)} v^{(k)\mathrm{T}} \tag{5.32}$$

The SVD of C tells us that $Cv^{(k)} = u^{(k)} s_k$, and so we get a very concise value for H:

$$H = \sum_{k=1}^{3} \sigma_k u^{(k)\mathrm{T}} C v^{(k)} = \sum_{k=1}^{3} \sigma_k u^{(k)\mathrm{T}} u^{(k)} s_k = \sum_{k=1}^{3} \sigma_k s_k \tag{5.33}$$

Recall that E was minimized when H was maximized, so the best E occurs when $\sigma_i = +1 \quad i = 1,2,3$. This gives us

$$R = U\mathrm{diag}(\sigma_1, \sigma_2, \sigma_3)V^{\mathrm{T}} = UV^{\mathrm{T}} \tag{5.34}$$

Now that we have the final elegantly simple result that $R = UV^{\mathrm{T}}$, are we finally done? Not quite yet. Remember that we stated that we must have $\det(R) = 1$. It is possible that the matrix C has a singular value decomposition that leads to $\det(UV^{\mathrm{T}}) = -1$. This is called an *improper rotation*, and it introduces a reflection. We can still get a proper rotation by defining R as follows:

$$R = U\mathrm{diag}(1, 1, -1)V^{\mathrm{T}} \tag{5.35}$$

Why does this work? The determinant of R has changed sign because the determinant of the diagonal matrix has the value −1. Now it is a proper rotation. However, the value of H is $s_1 + s_2 - s_3$ and so it is not as large as $s_1 + s_2 + s_3$. So, we have somewhat compromised E to get a proper rotation.

Note: to get the maximum value of H and hence the minimum value of E under these circumstances, we ensure that s_3 is the smallest of the three values. That is to say, we are assuming s_1, s_2, and s_3 are sorted to be in the descending order.

Figure 5.1 shows the results of a structural overlap of 1MBN and 1JEB. Both sets of coordinates were given to the Chimera molecular modeling system, which then displayed a visualization showing the overlap of secondary structures in rounded ribbon form. Only one of the proteins provided the heme structure.

5.5.5 Summary

Steps for the 3D overlap of proteins P and Q are as follows:

1. Determine the subsequences of alpha carbons to be used in the 3D overlap:

$$\begin{aligned} M(P) &= \left(p^{(\alpha_1)}, p^{(\alpha_2)}, \ldots, p^{(\alpha_N)}\right) \\ M(Q) &= \left(q^{(\beta_1)}, q^{(\beta_2)}, \ldots, q^{(\beta_N)}\right) \end{aligned} \tag{5.36}$$

2. Calculate the centroids $p^{(c)}$ and $q^{(c)}$ using Equation 5.5.

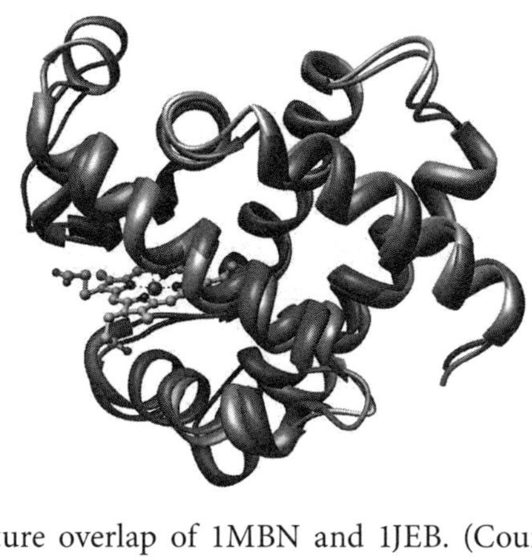

FIGURE 5.1 Structure overlap of 1MBN and 1JEB. (Courtesy of Shuo [Alex] Xiang, a former graduate student at the University of Waterloo, Ontario, Canada.)

3. Shift the proteins so that the centroids are at the origin (using Equation 5.6). We are then working with $x^{(i)}$ and $y^{(i)}$ coordinate sets.

4. Calculate the C matrix as defined by Equation 5.20, and compute its SVD. This gives $C = USV^{\mathrm{T}}$. If necessary, reorder the singular values so that $s_1 > s_2 > s_3$.

5. Compute the rotation matrix

$$R = UV^{\mathrm{T}} \tag{5.37}$$

6. Check to see if det $(R) = 1$. If this determinant is negative, then we must redefine the rotation matrix to be

$$R = U\mathrm{diag}(1,1,-1)V^{\mathrm{T}} \tag{5.38}$$

7. Apply the rotation matrix to the $x^{(i)}$ coordinates.

5.5.6 Measuring Overlap

5.5.6.1 Calculation of the Root Mean Square Deviation

After the rotation matrix is applied to the $x^{(i)}$ coordinates, we can compute the squared distance between each $x^{(i)}$ point and its corresponding $y^{(i)}$ point:

$$d^{(i)2} = \left\| Rx^{(i)} - y^{(i)} \right\|^2 \tag{5.39}$$

Having done this, we can compute the root mean square deviation (RMSD) for that set of corresponding points:

$$\mathrm{RMSD}(P,Q) = \sqrt{\frac{\sum_{i=1}^{N} d^{(i)2}}{N}} \tag{5.40}$$

One should be careful about the interpretation of this calculation. When P and Q represent different conformations of the same protein and N is the number of atoms of this protein, RMSD represents the conventional measure of the similarity of P and Q as measured by the degree of overlap. If the RMSD is 0 (or very close to 0), we regard P and Q as being identical. If the RMSD is between 1 and 3 Å, then we consider P and Q to be very similar. If the RMSD is over 3 Å, then we typically consider P and Q to have similarity that is low or not present.

If P and Q have the same lengths but some differences in primary sequence, then we cannot compute $d^{(i)2}$ for atoms in corresponding but different residues. This is true because it is impossible to get a correspondence between the atoms in these residues. In this case, one may compute the RMSD with respect to the alpha carbons only. This is still reasonable in that we would have an evaluation of the similarity of the backbone conformations of P and Q.

If P and Q have different lengths, then the RMSD calculation can still be applied to the atoms in the equivalence set. In this case, one must bear in mind that it is not measuring overlap across the entire extents of P and Q.

5.6 A SIMPLE SCRIPT TO DO STRUCTURAL SUPERIMPOSITION

Structural superimposition requires that we translate and rotate one structure so as to get maximal overlap with another structure, the goal being to minimize the distances between particular atom pairs. This is the simple scenario in which we know which atom pairs to use and these pairings do not change as the algorithm progresses.

Before discussing scripts that can do structural superimposition, it must be emphasized that Chimera has menu invocations that provide a sophisticated level of structural alignment capabilities. If you are working with only a few protein structures that require a conventional approach to structural alignment, then it is much easier to use a menu invocation and thus avoid the extra work of dealing with a script. So, why bother with a script? Here are some reasons:

- The script may be needed within a larger program that needs a structural superimposition capability.
- Your application might need to carry out several, perhaps thousands of, structural superimpositions.
- Your algorithm is using some alignment strategy that is different from that offered by Chimera.

Our script to accomplish structural superimposition has been implemented as a class called `Overlapper`. Instantiation of the class creates an `Overlapper` object that has data structures tracking atomic coordinates, centroids, and the rotation matrix involved in the superimposition.

When an `Overlapper` object is instantiated, it is given a Chimera molecule object designated as `mol` in the `__init__` function (e.g., a protein molecule object):

```
class Overlapper(object):
    def __init__(self, mol):
        self.molecule = mol
        self.atomCoords_D = {}
        # Overlap list for this molecule (molecule P to be moved):
        self.overlapCoordsP_L = []
        # Overlap list for target molecule (P moved to target Q):
        self.overlapCoordsQ_L = []
        self.newOverlapCoordsP_L = []
        self.centroidP = zeros(3) # Centroid of overlapCoordsP_L
        self.centroidQ = zeros(3) # Centroid of overlapCoordsQ_L
        self.rotationMatrix = zeros((3,3))

        for a in self.molecule.atoms:
            # Saving original coordinates:
            self.atomCoords_D[a] = a.coord()
```

The `moveToOverlap` method within the `Overlapper` class can be used to translate and rotate the molecule so that it achieves optimal overlap with another molecule when `moveToOverlap` is given a list of atom pairings. The `undoMove` method allows us to move `mol` back to its original position in the 3D space. By using both `moveToOverlap` and `undoMove` methods, we can cause `mol` to overlap with any number of target molecules one after another in a sequential fashion. The following is the code for the `moveToOverlap` and `undoMove` methods:

```
    def moveToOverlap(self, pairs_L):
        # Get lists of coordinates:
        # We assume the first entry of a pair refers to the self
          molecule and the second entry refers to
          the target protein.
        for pair in pairs_L:
            self.overlapCoordsP_L.append(getCoords(pair[0]))
            self.overlapCoordsQ_L.append(getCoords(pair[1]))

        self.centroidP = sum(self.overlapCoordsP_L)/len(pairs_L)
        self.centroidQ = sum(self.overlapCoordsQ_L)/len(pairs_L)

        # Compute shifted cverlap coords (centroids at origin):
        shiftedOverlapCoordsP_L = []
        shiftedOverlapCoordsQ_L = []
```

```
        for i in range(len(pairs_L)):
            shiftedOverlapCoordsP_L.append(self.overlapCoordsP_L[i] -
                                           self.centroidP)
            shiftedOverlapCoordsQ_L.append(self.overlapCoordsQ_L[i] -
                                           self.centroidQ)

        self.rotationMatrix = computeRotationMatrix \
                     (shiftedOverlapCoordsP_L,
                      shiftedOverlapCoordsQ_L)

        # Apply the transformation to all atoms in the molecule:
        for a in self.molecule.atoms:
            newCoords = dot(self.rotationMatrix,
                    array(a.coord())-self.centroidP)+self.centroidQ
            a.setCoord(Point(newCoords[0], newCoords[1], newCoords[2]))

    def undoMove(self):
        for a in self.molecule.atoms:
            a.setCoord(self.atomCoords_D[a])
```

The operation of `undoMove` is quite straightforward, but `moveToOverlap` requires some additional explanation. The function `getCoords` increases the usefulness of `moveToOverlap` by allowing the atom pairings to be expressed in any one of three different ways: an entry in a pair can be a Chimera atom object or its 3D coordinates expressed as a `Point` object or as a three-component array. Details of its implementation can be found in the `Overlapper` class (available from the book's website) accessed via

```
from StructBio. StructureOverlap.overlapper import Overlapper
```

Once all the atom pairings have been used to define the two lists `shiftedOverlapCoordsP _ L` and `shiftedOverlapCoordsQ _ L`, generation of the rotation matrix is accomplished by invoking the function `computeRotationMatrix`. Subsequent statements apply the rotation matrix to get the translations and rotation of `mol` to achieve the optimal overlap. Note that our mathematical derivation of the overlap algorithm involved a translation of both molecules so that their centroids were moved to the origin of the 3D space. In the implementation, this is not done because we do not want to shift the atoms in the target protein (just those in `mol`). Consequently, `mol` is shifted and rotated and then it is translated by a distance that is determined by the centroid of the target molecule.

Code for computing the rotation matrix is as follows:

```
def computeRotationMatrix(coordsP_L, coordsQ_L):
    coordsP_A = zeros((3,len(coordsP_L)), float)
    coordsQ_A = zeros((3,len(coordsQ_L)), float)
    for ix in range(len (coordsP_L)):
        coordsP_A[:, ix] = coordsP_L[ix]
        coordsQ_A[:, ix] = coordsQ_L[ix]

    c_A = dot(coordsQ_A, coordsP_A.transpose())
    u, s, vt = linalg.svd(c_A)

    # NOTE: we are assuming that s has
    # singular values in descending order.
    rotMatrix = dot(u,vt)
    if linalg.det(rotMatrix) < 0:
        rotMatrix = dot (u*array([1.0, 1.0, -1.0]), vt)

    return rotMatrix
```

This is a reasonably straightforward implementation of the steps 4, 5, 6, and 7 of the summary in Section 5.5.5. If the molecules to be overlapped have an obvious set of atom pairings, then it is easy to generate a list of these pairs to produce an optimal overlap (see Exercise 1 at the end of this chapter).

5.7 PROTEIN SEQUENCE ALIGNMENT

Up to this point, we have been assuming that structural overlap is preceded by some algorithm that determines the pairs of atoms (see Equations 5.3) that are to be structurally overlapped. If the two proteins have the same sequence, then this is a trivial exercise. If they are very different, then the pairing can be rather difficult. Although there are various purely structural approaches to this problem, a simple strategy is to start with a local alignment of the primary sequences. We do not cover this in minute detail because the information is readily available from other sources such as [DE98].

Our objective is to derive a short (hopefully simple) script that provides a sequence alignment class that can be readily embedded into other scripts that need this type of preprocessing prior to a structural overlap. In effect, we are striving to initiate a structural alignment by starting with a sequence alignment. Such an alignment is really an attempt to assess sequence similarity.

As an example of similarity, consider the following two amino acid sequences:

P = WATERMANMIKE and Q = MERMANETHEL

To show the similarity of these two sequences, we can write them as two rows with vertical bars between matching letters.

```
P =   WATERMANMIKE-
         |||||  :|
Q =   --MERMANETHEL
```

These two rows represent a *global alignment* of the two sequences. As just noted, we want a *local* alignment, but we first discuss global alignments because they are simpler to explain.

Definition: An optimal global alignment of two sequences P and Q is obtained by writing the two sequences in such a way as to maximize the occurrence of matching pairs of letters, one in P and the other in Q. Gaps (written as dashes) may be introduced into either sequence or at either end of a sequence in such a way as to have both strings of equal length and to facilitate the largest numbers of matches.

5.7.1 Partial Matches

Another consideration in the generation of an alignment is that amino acids may be accepted as a pair even though they do not match identically. This approach takes into account the biochemical properties of the amino acids. For example, D (aspartate) and E (glutamate) are both acidic amino acids and so they are good candidates for a partial match. Other partial matches would be biologically justified: hydrophobic amino acids (W, F, Y, L, I, V, M, A), and basic amino acids (K, R, H).

For long sequences, it may be possible to generate many different optimal alignments, each having the maximal number of exact matches. When partial matches contribute to the score of an optimal alignment, the analysis becomes more complicated because we must decide how much they contribute. For example, should the score increase if the alignment brings in three partial matches at the expense of foregoing one exact match? To answer such questions, researchers have constructed *substitution matrices* that essentially define the score contributions of both exact and partial matches.

Figure 5.2, adapted from http://www.ncbi.nlm.nih.gov/Class/FieldGuide/BLOSUM62.txt, gives the BLOSUM62 (see [HH92]) version of such a substitution matrix. There are other BLOSUM matrices with suffixes such as 45, 50, 80, and 90. The BLOSUM62 matrix is often used as the standard matrix when doing ungapped alignments, whereas BLOSUM50 is often used for gapped alignments. In general, BLOSUM80 works best for sequences that show less divergence, whereas BLOSUM45, at the other end of the application scale, is used for sequences that show more evolutionary divergence. The score contribution of an exact match can be found on the main diagonal. These entries have integer values ranging from 4 to 11 and take into account the relative frequency of occurrence of an amino acid. For example, the score contribution made by an exact match for W (tryptophan) is 11, which is the highest value since tryptophan has the lowest frequency of occurrence. In general, the entries of the substitution matrix are based on issues other than simple frequency of occurrence. More precisely, it describes the frequency of observed changes of characters in related protein segments or *blocks* that have been conserved in an evolutionary descent.

	A	R	N	D	C	Q	E	G	H	I	L	K	M	F	P	S	T	W	Y	V
A	4	−1	−2	−2	0	−1	−1	0	−2	−1	−1	−1	−1	−2	−1	1	0	−3	−2	0
R	−1	5	0	−2	−3	1	0	−2	0	−3	−2	2	−1	−3	−2	−1	−1	−3	−2	−3
N	−2	0	6	1	−3	0	0	0	1	−3	−3	0	−2	−3	−2	1	0	−4	−2	−3
D	−2	−2	1	6	−3	0	2	−1	−1	−3	−4	−1	−3	−3	−1	0	−1	−4	−3	−3
C	0	−3	−3	−3	9	−3	−4	−3	−3	−1	−1	−3	−1	−2	−3	−1	−1	−2	−2	−1
Q	−1	1	0	0	−3	5	2	−2	0	−3	−2	1	0	−3	−1	0	−1	−2	−1	−2
E	−1	0	0	2	−4	2	5	−2	0	−3	−3	1	−2	−3	−1	0	−1	−3	−2	−2
G	0	−2	0	−1	−3	−2	−2	6	−2	−4	−4	−2	−3	−3	−2	0	−2	−2	−3	−3
H	−2	0	1	−1	−3	0	0	−2	8	−3	−3	−1	−2	−1	−2	−1	−2	−2	2	−3
I	−1	−3	−3	−3	−1	−3	−3	−4	−3	4	2	−3	1	0	−3	−2	−1	−3	−1	3
L	−1	−2	−3	−4	−1	−2	−3	−4	−3	2	4	−2	2	0	−3	−2	−1	−2	−1	1
K	−1	2	0	−1	−3	1	1	−2	−1	−3	−2	5	−1	−3	−1	0	−1	−3	−2	−2
M	−1	−1	−2	−3	−1	0	−2	−3	−2	1	2	−1	5	0	−2	−1	−1	−1	−1	1
F	−2	−3	−3	−3	−2	−3	−3	−3	−1	0	0	−3	0	6	−4	−2	−2	1	3	−1
P	−1	−2	−2	−1	−3	−1	−1	−2	−2	−3	−3	−1	−2	−4	7	−1	−1	−4	−3	−2
S	1	−1	1	0	−1	0	0	0	−1	−2	−2	0	−1	−2	−1	4	1	−3	−2	−2
T	0	−1	0	−1	−1	−1	−1	−2	−2	−1	−1	−1	−1	−2	−1	1	5	−2	−2	0
W	−3	−3	−4	−4	−2	−2	−3	−2	−2	−3	−2	−3	−1	1	−4	−3	−2	11	2	−3
Y	−2	−2	−2	−3	−2	−1	−2	−3	2	−1	−1	−2	−1	3	−3	−2	−2	2	7	−1
V	0	−3	−3	−3	−1	−2	−2	−3	−3	3	1	−2	1	−1	−2	−2	0	−3	−1	4

FIGURE 5.2 The BLOSUM62 substitution matrix.

5.7.2 Gaps

Consider the following two alignments:

```
Alignment #1:

M  I  C  H  A  E  L  S  W  A  T  E  R  M  A  N
      |  |     |  |  :           |  |  |  |  |
-  E  T  H  -  E  L  M  -  -  -  E  R  M  A  N
0 -3 -1  8  0  5  4 -1  0  0  0  5  5  5  4  6

Alignment #2:

M  I  C  H  A  E  L  S  W  A  T  E  R  M  A  N
      |  |     |  |        :     |  |  |  |  |
-  E  T  H  -  E  L  -  -  M  -  E  R  M  A  N
0 -3 -1  8  0  5  4  0  0 -1  0  5  5  5  4  6
```

The numbers in each column of the alignment are the score contributions given by the BLOSUM62 matrix. The score for the entire alignment is simply the sum of these contributed scores. This means that alignment #1 has a score of 37 while alignment #2 also has a score of 37. However, most alignment programs will offer alignment #1 as the better alignment because alignment #2 initiates a gap four times, whereas the former does so only three times. To support this choice, it is possible for the alignment score to incorporate a term that leads to an extra penalty whenever a new run of gap symbols is started.

Let us consider why a gap appears. A gap is a sequence of one or more spaces drawn as dashes in the alignment display. Each space can represent an insertion or a deletion. For example, the space below the first A represents either of the following:

1. The insertion of an A into the first string, which would have been `MICHELSWATERMAN` prior to the insertion
2. The deletion of some letter that was originally between `H` and `E` of `ETHELMERMAN`

Very often, we do not have enough evidence to determine which scenario occurred and so either possibility is acceptable and the space is referred to as an indel (insertion or deletion). The key fact is that indels are assumed to be much less likely than substitutions from a biological perspective. Consequently, their initiation and extension lead to the addition of negative contributions (penalties) in the score computation. These penalties have been assessed using empirical observations gathered from several examples of homologous proteins that have been previously aligned.

5.7.3 Summary

We are now in a position to summarize the main points of the mathematical model for global sequence alignment.

Given two strings representing the residues in two proteins *P* and *Q*, an optimal global alignment seeks to demonstrate maximal similarity by writing one string beneath the other so that residues in columns either identically or partially match. The objective of attaining maximal similarity must address the following issues:

- Gaps (written as dashes) may be introduced into either sequence or at either end of a sequence in such a way as to have both strings of equal length and to facilitate the largest numbers of matches.
- Although gaps may be introduced, we attempt to do this in a parsimonious fashion and, so, in the formulation of the problem we reduce the score of an alignment by introducing a *gap penalty*.

5.7.4 Computation

The optimal global alignment problem can be solved using dynamic programming. This solution of the problem was done in 1970 by Needleman and Wunsch (see [NW70]). A modern computer scientist would see this algorithm as an application of dynamic programming, a strategy that was pioneered by Richard Bellman in 1954 (see [Be54]). A review of dynamic programming is provided in Appendix E.

In the design of a dynamic programming algorithm, the most important issue is the specification of a subproblem (see Appendix E). We know that a subproblem should be smaller in some sense and capable of contributing to the solution of a larger problem.

It is possible to define subproblems that have too many degrees of freedom in their specifications. For example, we might state that a subproblem should be the global alignment of two substrings extracted from *P* and *Q*, that is, $P[i..j]$ and $Q[k..m]$, where the notation $[i..j]$ represents the string of contiguous characters between and including character positions i and j. This subproblem would be parameterized by four integers, i, j, k, and m. This approach would involve evaluation of all cells in a four-dimensional array, and this would be an extremely costly evaluation for lengthy strings.

In fact, we need not go to such an extreme when specifying substrings. It is shown in Lemma 5.1 that the selected substrings can be *prefixes* of the given strings.

Definition: the ith prefix of a string X denoted by X_i is the substring of X consisting of the first i characters of X, that is, $X_i = x_1, x_2, \ldots, x_{i-1}, x_i = X_{i-1}, x_i$ Note the use of comma-separated lowercase letters to represent consecutive characters within the prefix. With this agreement about notation, we proceed with the specification in Section 5.7.4.1.

5.7.4.1 Subproblem Specification

The dynamic programming subproblem will be parameterized by two arguments i and j, and its solution will be the optimal global alignment of the two prefixes P_i and Q_j.

5.7.4.2 Scoring Alignments

We score an alignment by adding the scores derived from each column in the alignment. Suppose that after adding dashes we get an alignment that involves m_{ij} columns when we do a global alignment of P_i and Q_j. Then, the score for this alignment will be given by

$$NW(i,j) = \sum_{k=1}^{m_{ij}} s_c(k) \tag{5.41}$$

where $s_c(k)$ is the score for the kth column in the alignment. We have used the notation $NW\ (i, j)$ to emphasize that this score is for the global alignment algorithm due to Needleman and Wunsch. Based on the earlier discussion concerning gaps and partial matches in Section 5.7.2, we can define the column score using

$$s_c(k) = \begin{cases} B(p'_k, q'_k) & \text{in a column with a match} \\ -d & \text{in a column with a dash} \end{cases} \tag{5.42}$$

In this equation, the quantity d represents the penalty that is given for the occurrence of a gap. The gap penalty d is a positive quantity, so the presence of any gap in the alignment will cause this penalty to be subtracted from the total alignment score. We use the notations p'_k and q'_k to indicate the residues of P_i and Q_j that appear in the kth column when the global alignment is written out, complete with dashes. The function B (*row_residue, column_residue*) is used to extract an entry from the row and column of the substitution matrix that is being used for the alignment. In Section 5.7.1, we used the BLOSUM62 matrix as an example of such a substitution matrix.

5.7.4.3 Suitability of the Subproblem

There are undoubtedly many ways to define a subproblem, many of them ill-fitted to the needs of a dynamic programming algorithm. However, our specification of a subproblem is appropriate for the dynamic programming algorithm. This is demonstrated by proving that there is a relationship between the subproblem specified as the optimal global alignment of prefixes P_i and Q_j and other optimal global alignment subproblems that work on shorter prefix strings. This gives us a recursive evaluation leading to the computation of the full solution: the full problem depends on smaller subproblems, each of which depends on even smaller subproblems, and so on until eventually we have subproblems so small that they are trivial to solve. We need the following lemma:

Lemma 5.1

Let P_i and Q_j be prefixes of the strings P and Q, respectively. If we consider the optimal global alignment of P_i and Q_j, there are three possible cases when describing the appearance of the last column in the alignment:

1. Character p_i matches character q_j in the optimal global alignment, which can then be drawn as follows:

$$\begin{matrix} P_{i-1}, p_i \\ Q_{j-1}, q_j \end{matrix} \tag{5.43}$$

   Claim: in this case, the matching of characters seen prior to p_i and q_j is, in fact, the optimal global alignment of P_{i-1} and Q_{j-1}.

2. Character p_i does not participate in any match, that is, the optimal global alignment can be drawn as follows:

$$\begin{matrix} P_{i-1}, p_i \\ Q_j, - \end{matrix} \tag{5.44}$$

Claim: in this case, the matching of characters seen prior to p_i is the optimal global alignment of P_{i-1} and Q_j.

3. Character q_j does not participate in any match, that is, the optimal global alignment can be drawn as follows:

$$\begin{array}{l} P_i, \ - \\ Q_{j-1}, q_j \end{array} \tag{5.45}$$

Claim: in this case, the matching of characters seen prior to q_j is the optimal global alignment of P_i and Q_{j-1}.

Proof of claim for case (a):

Assume that the matching of P_{i-1} and Q_{j-1} as seen in case (a) is not the optimal global alignment of P_{i-1} and Q_{j-1}. Then, suppose that we replaced this alignment with the optimal global alignment of P_{i-1} and Q_{j-1}. This will raise the score associated with the columns prior to the last column. Then, by simply appending p_i to P_{i-1} and q_j to Q_{j-1}, we would get an alignment with a higher score than that seen in case (a). This is a contradiction because we have already assumed that case (a) represents an optimal global alignment.

Proof of claim for case (b):

Reasoning along similar lines: suppose that the matching of P_{i-1} and Q_j as seen in case (b) is not the optimal global alignment of P_{i-1} and Q_j. Then, suppose we replaced this alignment with the optimal global alignment of P_{i-1} and Q_j. This will raise the score associated with the columns prior to the last column. Then, by simply appending p_i to P_{i-1} and writing a dash below it we would get an alignment that is characterized as being case (b) but with a higher score. This is a contradiction because we have already assumed that case (b) represents an optimal global alignment.

Proof of claim for case (c):

The proof is a simple variation of the proof for case (b).

Computation of $NW(i, j)$: Lemma 5.1 is a formal discussion that brings out the relationships among subproblems and serves to justify the formulation of our subproblem as the optimal global alignment of P_i and Q_j. At the moment, the implications these cases provide for the dependencies among the scores associated with the subproblems are more important for us. Each case in Lemma 5.1 defines a possible relationship between $NW(i, j)$ and the score derived from the solution of some smaller alignment problem.

Case (a):

By matching the last characters in each string of residues, the score will be incremented by an amount that is accessed from the substitution matrix:

$$NW(i,j)=NW(i-1,j-1)+B(p_i,q_j) \tag{5.46}$$

Case (b):

By placing a dash in the bottom row of the alignment, we decrease the score in going from *NW* (*I* – 1, *j*) to *NW* (*i*, *j*):

$$NW(i,j)=NW(i-1,j)-d \tag{5.47}$$

Case (c):

By placing a dash in the top row of the alignment, we decrease the score in going from *NW* (*i*, *j* – 1) to *NW* (*i*, *j*):

$$NW(i,j)=NW(i,j-1)-d \tag{5.48}$$

Considering these possibilities, we see that there are three ways to define *NW* (*i*, *j*). Since we wish to maximize the alignment score, it makes sense to always choose the option that will maximize *NW* (*i*, *j*). Consequently, we use the following equation for the computation of $NW(i,j)$:

$$NW(i,j)=\max\begin{Bmatrix} NW(i-1,j-1)+B(p_i,q_j), \\ NW(i-1,j)-d, \\ NW(i,j-1)-d \end{Bmatrix} \tag{5.49}$$

We now have enough information about the mathematical model to go through the dynamic programming design steps suggested in Section E.4 of Appendix E. These steps are summarized as follows:

1. Identification of subproblems: The subproblem will be parameterized by two arguments *i* and *j*, and its solution will be the optimal global alignment of the two prefixes P_i and Q_j.

2. Specify evaluation of score values: We will use the notation $|P|$ and $|Q|$ to represent the number of residues in proteins P and Q, respectively. The $NW\ (i, j)$ score values will be retained in an array with $|P| + 1$ rows and $|Q| + 1$ columns. There will be an extra row "0" and an extra column "0" to store initial values required by the recursion. Considering the dependencies of $NW\ (i, j)$ on other score values in the array, we see that the score stored in cell (i, j) depends on the scores stored in cells $(i - 1, j - 1)$, $(i - 1, j)$, and $(i, j - 1)$. Consequently, it will be sufficient to evaluate the array entries in a row-by-row fashion. We will need to know the score of the full global alignment: this will be found in the bottom right corner of the array (the intersection of the last row and the last column).

3. Set values for the base case: the entries in row 0 and column 0 will hold scores that correspond to penalties that are introduced because of the insertion of initial dashes that might appear in the alignment before the characters of either the P string or the Q string. As we evaluate along row 0, the penalties increase so that $NW\ (0, j) = -jd$. Similarly, for column 0, we have $NW\ (i, 0) = -id$.

4. Recover the solution configuration: the traceback operation involves keeping track of the cell that produced the maximum value when $NW\ (i, j)$ was calculated using Equation 5.49. This can be done using a separate trace-back array having $|P|$ rows and $|Q|$ columns. The entry stored in cell (i, j) of this array would be a code indicating whether the maximum value was taken from $(i - 1, j - 1)$, $(i - 1, j)$, or $(i, j - 1)$. When describing a trace-back operation, one can superimpose the trace-back array over the score array with trace-back dependencies illustrated by means of arrows.

5.7.4.4 A Global Alignment Example

We will assume a gap penalty of $d = -1$. The results are displayed in Figure 5.3.

Working with the trace-back, one gets the following alignment:

```
 -  M  I  C  H  A  E  L  S  W  A  T  E  R  M  A  N
    |        |     |  |           |  |  |  |  |  |
 E  T  -  -  H  -  E  L  -  -  -  M  E  R  M  A  N
-1 -1 -1 -1  8 -1  5  4 -1 -1 -1 -1  5  5  5  4  6
```

				E		T		H		E		L		M		E		R		M		A		N
		0	←	**−1**	←	−2	←	−3	←	−4	←	−5	←	−6	←	−7	←	−8	←	−9	←	−10	←	−11
M	↑	−1	←	−2	↖	**−2**	←	−3	←	−4	↖	−2	↖	0	←	−1	←	−2	←	−3	←	−4	←	−5
I	↑	−2	←	−3	↑	**−3**	←	−4	←	−5	↖	−2	↑	−1	←	−2	←	−3	↖	−1	←	−2	←	−3
C	↑	−3	←	−4	↑	**−4**	←	−5	←	−6	↑	−3	↑	−2	←	−3	←	−4	↑	−2	↖	−1	←	−2
H	↑	−4	↖	−3	←	−4	↖	**4**	←	3	←	2	←	1	←	0	←	−1	←	−2	↑	−2	↖	0
A	↑	−5	↑	−4	↖	−3	↑	**3**	↖	3	←	2	←	1	←	0	←	−1	←	−2	↖	2	←	1
E	↑	−6	↖	0	←	−1	↑	2	↖	**8**	←	7	←	6	↖	6	←	5	←	4	←	3	←	2
L	↑	−7	↑	−1	↖	−1	↑	1	↑	7	↖	**12**	←	11	←	10	←	9	←	8	←	7	←	6
S	↑	−8	↑	−2	↖	0	↑	0	↑	6	↑	**11**	↖	11	↖	11	←	10	←	9	↖	9	←	8
W	↑	−9	↑	−3	↑	−1	↑	−1	↑	5	↑	**10**	↑	10	↑	10	←	9	↖	9	←	8	←	7
A	↑	−10	↑	−4	↑	−2	↑	−2	↑	4	↑	**9**	↑	9	↑	9	↖	9	←	8	↖	13	←	12
T	↑	−11	↑	−5	↖	1	←	0	↑	3	↑	8	↖	**8**	↑	8	↑	8	↖	8	↑	12	↖	13
E	↑	−12	↑	−6	↑	0	↖	1	↖	5	↑	7	↑	7	↖	**13**	←	12	←	11	↑	11	↑	12
R	↑	−13	↑	−7	↑	−1	↑	0	↑	4	↑	6	↑	6	↑	12	↖	**18**	←	17	←	16	←	15
M	↑	−14	↑	−8	↑	−2	↑	−1	↑	3	↖	6	↖	11	↑	11	↑	17	↖	**23**	←	22	←	21
A	↑	−15	↑	−9	↑	−3	↑	−2	↑	2	↑	5	↑	10	↑	10	↑	16	↑	22	↖	**27**	←	26
N	↑	−16	↑	−10	↑	−4	↖	−2	↑	1	↑	4	↑	9	↖	10	↑	15	↑	21	↑	26	↖	**33**

FIGURE 5.3 Global alignment example.

5.8 VARIATIONS IN THE GLOBAL ALIGNMENT ALGORITHM

For many problems solved by dynamic programming, the computed solutions are provably optimal. The pairwise alignment of two protein sequences is a much more complicated problem because of the many parameters that must be properly set before the dynamic program can proceed. This complexity is related to the mathematical model for the treatment of gaps and the handling of partial matches. As described in Section 5.7.1, substitution matrices such as BLOSUM62 are used to specify the score contribution of a partial match. Here are some typical approaches for the computation of penalties arising from gaps:

- Linear gap model: the presence of a dash introduces a penalty that is seen in the recursion as a decrease of the score by the value d_L. In the simplest model, the dependence is linear and the presence of a gap with g dashes leads to a decrease of gd_L.
- Affine gap model: the affine gap model uses a penalty formula given by $d_A + (g - 1)\, e$, where d_A is the gap-open penalty and e is the gap extension penalty. Typically, $e < d_A$, and this means that the insertion of a long sequence of dashes into the alignment penalized less than it would be if a linear model was used.

Typical values are $d_L = 8$ for the linear case and $d_A = 12$ and $e = 2$ for the affine case. Although these values get good results in practice, there is no rigorous theory or rule stipulating the optimal values for these parameters. Furthermore, the choice of substitution matrix may not be obvious. One can find guidelines for their use (see page 84, [ZB07]), but the choice is often decided by empirical observation and experience rather than a mathematically rigorous decision process.

5.9 PERCENT IDENTITY COMPARISON

Given the amount of choice in the construction of an alignment algorithm, how does one objectively choose the best alignment for a given pair of sequences? Ideally, we want some quantitative assessment that indicates the goodness of an alignment. The simplest approach is to count the number of identical matches in an alignment. Then, *percent identity* can be derived by dividing this count by the length of the aligned region and multiplying by 100. It should be noted that each column in the alignment of two random proteins would have a 1 out of 20 or 5% chance of producing an identical match. This calculation implicitly assumes that all amino acids occur with the same probability. It is well known that this is not the case. The most frequent amino acids glycine, alanine, and leucine account for roughly one-quarter of all residues, and any one of them typically appears four times more often than any one of the least frequent amino acids tryptophan, methionine, histidine, and cysteine. Doolittle [Do81] notes that when these observed frequencies are taken into account, the percent identity for random sequences goes up from 5% to 6%. Consequently, we would expect the alignment of homologous proteins to have a percent identity that is well above 6%, but then we must ask the following: What is the threshold of percent identity above which we will accept the alignment as producing evidence of homology?

In 1999, Burkhard Rost reported the results of an experiment that provides this threshold (see [Ro99]). He analyzed more than a million sequence alignments between protein pairs with known structures. Since the structures were known, he was able to determine whether a pair of proteins had similar or nonsimilar structure. This structure comparison was used to provide a definitive test of homology for the protein pairs. The following points describe some of the main results:

- When the percent identity is high, over 40% for long alignments, the sequence alignment will unambiguously distinguish between similar and nonsimilar structures.

- For alignments showing more than 30% sequence identity, about 90% of the pairs were homologous.
- For alignments showing less than 25% sequence identity, less than 10% of the pairs were homologous.

The region between 35% and 20% percent identity was designated as the *twilight zone*. Here, pair homology may exist but cannot be reliably assured without extra evidence. More than 95% of all pairs in the twilight zone had nonsimilar structure. Sequence percent identity below 20% has been called the *midnight zone* and typically requires an assessment of structure since the alignment is not reliable.*

5.10 LOCAL ALIGNMENT

It is possible that two proteins have evolved from a distant ancestor and the evidence of this is seen as a conserved domain. Suppose that for each protein the portion of the primary sequence belonging to this domain is small in comparison with the rest of the protein that corresponds to a sequence bearing no relation to the other protein. Under these circumstances, a global alignment of the proteins may obscure the similarity of the domain because the algorithm is forced to find an alignment that deals with the entire sequence of each protein. The insertion of gaps necessary to achieve this global alignment will cause a reduction of the score, and a calculation to evaluate the significance of the alignment score may be undeservedly low. In this situation, the discovery of homology is best handled by computing a local alignment rather than a global alignment. We want the best alignment between shorter regions of the given proteins. We should think of global alignment as being most applicable for the alignment of two proteins that have roughly the same length and are somewhat similar over their entire lengths.

The dynamic programming algorithm that produces such an alignment is the Smith–Waterman algorithm (see [SW81]). The dynamic program for Smith–Waterman uses the same two-dimensional array that we used for the Needleman–Wunsch global alignment algorithm. However, because we only want to align shorter regions of the given pair of proteins, the trace-back path will not necessarily start in the lower right corner of the array and will not necessarily end in the upper left corner of the array. More specifically, the trace-back path will start with the highest score and, in general, it will

* Recall from Chapter 1, Section 1.4.3.3 that the globin family has sequence identity at around 15%, but structural analysis reveals their homology.

traverse only part of the array, ending when we encounter a cell with score value 0. Moreover, we will not assign penalties when the given strings P and Q do not produce alignments at the ends of the strings. In other words, dashes at either end are inserted with no penalty. This means that even though the alignment algorithm may use a linear or affine gap model, the initial scores set up in the cells of the first row and column will be 0 and never negative. Furthermore, during the evaluation of scores in the array we can change any negative score (computed in the same manner as Needleman–Wunsch) by setting such a score to 0. Consequently, the recursion becomes as follows:

$$SW(i,j) = \max\begin{Bmatrix} SW(i-1,j-1)+B(p_i,q_j), \\ SW(i-1,j)-d, \\ SW(i,j-1)-d, \\ 0 \end{Bmatrix} \quad (5.54)$$

We have used the notation $SW\ (i, j)$ to emphasize that this score is for the local alignment algorithm due to Smith and Waterman. If you compare this with Equation 5.49, you will see that they are the same except that Equation 5.54 has a 0 in the max function argument list so that the cell entry never becomes negative.

It is expected that values for partial matches provided by the substitution matrix will have a mix of both positive and negative entries. The idea is that we want to derive local alignments and the score for aligning unrelated sequence regions should be negative. We do not want a local alignment to be arbitrarily extended in either direction by partial matches that always contribute a positive increment to the score, which is why negative entries are essential.

To provide a small example, we show the results of a local alignment between the strings: "MNAMRETAW" and "SNAMELDEEN."

This time, we have assumed a gap penalty of $d = -8$. The results are displayed in Figure 5.4.

Working with the trace-back, one gets the following alignment:

```
N A M R
| | | |
N A M E
```

Note that this corresponds to finding the maximum value of 15 and then taking the diagonal path up and to the left until a zero entry is encountered.

		S	N	A	M	E	L	D	E	E	N
	0	0	0	0	0	0	0	0	0	0	0
M	0	0	0	0	5	0	2	0	0	0	0
N	0	1	**6**	0	0	5	0	3	0	0	6
A	0	1	0	**10**	2	0	4	0	2	0	0
M	0	0	0	2	**15**	7	2	1	0	0	0
R	0	0	0	0	7	**15**	7	0	1	0	0
E	0	0	0	0	0	12	12	9	5	6	0
T	0	1	0	0	0	4	11	11	8	4	6
A	0	1	0	4	0	0	3	9	10	7	2
W	0	0	0	0	3	0	0	1	6	7	3

FIGURE 5.4 Local alignment example.

5.11 A SCRIPT TO DO SEQUENCE ALIGNMENT AND THEN STRUCTURAL OVERLAP

In this section, we deal with structural overlaps for which the atom pairings are not immediately obvious but can be determined by means of a local sequence alignment.

We start by describing a class called `SequenceLocalAlignment` and show how it can be used to generate the atom pairings that provide input for the structural overlap class. Here is the script for the class:

```
import chimera, numpy
from numpy import zeros

b62_L = \
   [[ 4],
    [-1, 5],
    [-2, 0, 6],
    [-2,-2, 1, 6],
    [ 0,-3,-3,-3, 9],
    [-1, 1, 0, 0,-3, 5],
    [-1, 0, 0, 2,-4, 2, 5],
    [ 0,-2, 0,-1,-3,-2,-2, 6],
    [-2, 0, 1,-1,-3, 0, 0,-2, 8],
    [-1,-3,-3,-3,-1,-3,-3,-4,-3, 4],
    [-1,-2,-3,-4,-1,-2,-3,-4,-3, 2, 4],
    [-1, 2, 0,-1,-3, 1, 1,-2,-1,-3,-2, 5],
    [-1,-1,-2,-3,-1, 0,-2,-3,-2, 1, 2,-1, 5],
    [-2,-3,-3,-3,-2,-3,-3,-3,-1, 0, 0,-3, 0, 6],
    [-1,-2,-2,-1,-3,-1,-1,-2,-2,-3,-3,-1,-2,-4, 7],
    [ 1,-1, 1, 0,-1, 0, 0, 0,-1,-2,-2, 0,-1,-2,-1, 4],
    [ 0,-1, 0,-1,-1,-1,-1,-2,-2,-1,-1,-1,-1,-2,-1, 1, 5],
    [-3,-3,-4,-4,-2,-2,-3,-2,-2,-3,-2,-3,-1, 1,-4,-3,-2,11],
    [-2,-2,-2,-3,-2,-1,-2,-3, 2,-1,-1,-2,-1, 3,-3,-2,-2, 2, 7],
    [ 0,-3,-3,-3,-1,-2,-2,-3,-3, 3, 1,-2, 1,-1,-2,-2, 0,-3,-1, 4]]
```

```
t2ix_D = \
   {"ALA": 0,"ARG": 1,"ASN": 2,"ASP": 3,"CYS": 4,
    "GLN": 5,"GLU": 6,"GLY": 7,"HIS": 8,"ILE": 9,
    "LEU":10,"LYS":11,"MET":12,"PHE":13,"PRO":14,
    "SER":15,"THR":16,"TRP":17,"TYR":18,"VAL":19}

def b62(resAtype, resBtype):
    ixPair = (t2ix_D[resAtype], t2ix_D[resBtype])
    return b62_L[max(ixPair)][min(ixPair)]


#===================================================================
# SequencesLocalAlignment CLASS

class SequencesLocalAlignment(object):
    def __init__(self, resP_L, resQ_L, penalty = 4):
        # This function computes a local alignment
        # for two residue lists (not character sequences!).
        # Results are stored in the alignment_L list.
        # Gaps are not in the output since we only need
        # the residues that have been aligned.
        self.resP_L = resP_L
        self.resQ_L = resQ_L
        self.penalty = penalty
        self.alignment_L = []

        # Fill in the score matrix and track the largest element.
        sw = zeros((len(self.resP_L) + 1,
                    len(self.resQ_L) + 1), int)
        maxVal = 0;
        for i in range(1, len(self.resP_L) + 1):
            for j in range(1, len(self.resQ_L) + 1):
                sw[i,j] = max(sw[i-1,j-1] +
                              b62(self.resP_L[i-1].type,
                              self.resQ_L[j-1].type),
                              sw[i-1,j] - self.penalty,
                              sw[i,j-1] - self.penalty,
                              0)
                if sw[i,j] >= maxVal:
                    maxVal = sw[i,j]
                    maxLoc = [i, j]

        # Traceback:
        i = maxLoc[0]; j = maxLoc[1]
        while sw[i,j] > 0:
            if sw[i,j] == sw[i-1,j-1] + b62(self.resP_L[i-1].type,
                                            self.resQ_L[j-1].type):
                self.alignment_L.append((resP_L[i-1], resQ_L[j-1]))
                # Note that indexing is shifted for input array
                # relative to indexing of score array.
                i -= 1; j -= 1
            if sw[i,j] == sw[i-1,j] - self.penalty: i -= 1
            if sw[i,j] == sw[i,j-1] - self.penalty: j -= 1

        self.alignment_L.reverse()
```

The script can be imported using the following:

```
from StructBio. StructureOverlap.\
   sequencesLocalAlignment import SequencesLocalAlignment
```

When the class is instantiated, it is given lists of Chimera residue objects for both proteins. After some initialization, the script fills in the score array using Equation 5.54. The `b62` function works with the triangular set of lists (`b62_L`) to provide values from the BLOSUM62 substitution matrix. While filling the array, we compute the maximum entry and subsequently use it for the trace-back. Note that trace-back is done by an evaluation that determines the cell responsible for the entry in the current cell. After trace-back is completed, the residue pairings will have been assembled in the `alignment_L` list. The order of the list is then reversed. This reversal does not affect the residue pairing and is only done for the convenience of the user who might want to list the aligned residues in their sequence order. The aligned residues can be accessed by working with the `alignment_L` attribute of the instantiated `SequencesLocalAlignment` object. With a list of residue pairs, the user can easily extract the alpha carbon atoms that are to be paired in the overlap computation. One can also get the residue types if an alignment is to be displayed in printed form.

By utilizing local sequence alignment to derive atom pairings, we can facilitate protein overlaps that are more challenging than that seen in Exercise 1. See, for example, Exercise 2 at the end of this chapter.

5.12 DEALING WITH WEAKER SEQUENCE SIMILARITY

Section 5.5 covered the standard translate and rotate algorithm that can be used when there is a clear and extensive sequence alignment that defines the pairs of atoms that are to be placed into maximal overlap. The problem of similarity evaluation becomes much more of a challenge when a local sequence alignment becomes sketchy, for example, when the two proteins have a distant evolutionary relationship. Since we know that structure is more conserved than sequence, it is reasonable to strive for algorithms that determine structural alignment with little or no help from a preliminary sequence alignment.

A local sequence alignment with a low score poses difficulties for the structural alignment. The strategy for overcoming these difficulties will depend on the reasons for the low score:

1. Lengthy sequences of mismatches: in some cases, mismatches may reside in loop regions, whereas the hydrophobic core contains residues that show a higher percentage of matches in the sequence alignment. In such cases, it may be reasonable to simply remove the loop region from consideration and try to maximize the overlap of atoms in the hydrophobic core.

2. Presence of gaps in the alignment: gaps in the sequence alignment pose a difficult problem because they indicate a break in continuity of the structural alignment. There are various strategies that can be used:

    a. By breaking up a protein into fragments, we can try to derive separate structural alignments of fragments on either side of the gap.

    b. We can redefine the pairs of atoms that are to be put into maximal overlap.

The second approach is at the heart of a clever algorithm devised by Russell and Barton (see [RB92]). Their algorithm, called STAMP (structural alignment of multiple proteins), uses a sequence alignment but only as an initial step. It is one of the several algorithms that establish a structural alignment with little or no reliance on a sequence alignment. Several algorithms have been developed to compare geometric relationships within the proteins being compared. The advantage of these intramolecular methods is that they avoid the need for an explicit superposition of the two structures. In all cases, the algorithms essentially compare some of the local structures of a protein P with the local structures of a protein Q. The notion of local structure varies from one algorithm to the next. Some examples of such algorithms are as follows:

- DALI: DALI (distance alignment) does an optimal pairwise structural alignment of protein structures based on the similarity of local patterns extracted from distance maps (see [HS93]).

- SSAP: SSAP (secondary structure alignment program) produces a structural alignment using double dynamic programming to generate an alignment of local "views" that are common to both proteins (see [TO89]).

Both algorithms involve the comparison of local structure that deals with individual atoms (DALI working with contact maps and SSAP working with views that are established relative to alpha carbon atoms). This strategy is in contrast with other algorithms that explore the similarity of proteins by dealing with secondary structure elements (SSEs). That is, there is an attempt to derive a set of matching pairs of SSEs in such a way as to maximize the overlap of atom pairs. Several algorithms have used strategies that work in this manner (see secondary structure matching [KH04], vector alignment search tool (VAST) [GM96], and DEJAVU [SK04]).

In this section, we design and implement a class that uses a similar approach to structural alignment.

5.12.1 Structural Alignment by Overlapping Pairs of Secondary Structure Elements

Given two proteins P and Q, we want to design and implement a structural alignment algorithm that meets the following goals:

- The algorithm will not use a sequence alignment of P and Q to get an initial structural overlap. Although a sequence alignment is usually an aid in the determination of atoms that comprise the overlap pairs, we will deliberately avoid this strategy as a challenge to see how far we can get if we totally ignore the primary sequence.
- The algorithm should work even if the SSEs (such as helices and strands) of P represent a "rewiring" of the SSEs in Q. This is frequently seen as a circular permutation of the SSEs [DB07]. Note that this complication would mitigate the need for a preliminary sequence alignment. For other types of nonsequential alignments, see [AV07].
- The success of the structural alignment will be evaluated using the *quality value* (QV) function:

$$\mathrm{QV} = \frac{N_{\mathrm{align}}^2}{\left[1+\left(\mathrm{RMSD}(P,Q)/R_0\right)^2\right]N_P N_Q} \tag{5.55}$$

In Equation 5.55, N_{align}^2 is the number of residues considered to be in the alignment; N_P and N_Q are the number of residues in P and Q, respectively; RMSD (P, Q) is the RMSD of P and Q calculated by using the alpha carbon atoms of the residues in the alignment (see Equation 5.40); and R_0 is an empirical parameter (chosen to be 3 Å) that measures the relative significance of RMSD and N_{align} (see [KH04]). Note that measurement of overlap is more complicated when the amount of overlap is less than the length of the smaller protein. Since there is no sequence alignment to stipulate the atom pairs that would be used in the calculation of RMSD, we typically choose alpha carbon atom pairs that are close to one another as determined by the structural alignment. However, when a limited set of alpha carbon atoms in P is in close structural alignment with a corresponding limited set of alpha carbon atoms in Q, it is to be expected that the remainder of the alpha carbon atoms will be farther apart and will tend to increase the squared deviations of the RMSD so that it gives an overly pessimistic high value. We can lower the RMSD by restricting the calculation so that it covers only those atoms that are very close to one another, but then the measure of overlap is only valid for this limited extent of the overlap and does not represent an accurate appraisal of the alignment. By using the QV formula, we get the number of atom pairings to contribute to the assessment of alignment quality. If you inspect the formula, you can see that a low RMSD gives a higher quality of alignment but the QV also strongly depends on the number of atom pairs in the alignment.

Our strategy will be to try to find the best structural alignment by evaluating several candidate alignments, each involving the overlapping of a pair of SSEs in P with a pair of SSEs in Q (similar to the strategy discussed in [SB97]). If we think of the inertial axes of a pair of SSEs in a protein as defining the opposite sides of a quadrilateral, then we are attempting to find the best overlap involving a quadrilateral in P and a quadrilateral in Q. Figure 5.5 shows a simple example. We have chosen small proteins so that the figures have more clarity.

Figure 5.6a and b shows two proteins with inertial axes placed in the secondary structures. Both proteins have been illustrated using a backbone chain trace so that the axes are easily seen.

If protein P has m_P inertial axes and protein Q has m_Q inertial axes, then we will get $\binom{m_P}{2}$ and $\binom{m_Q}{2}$ quadrilaterals for proteins P and Q, respectively. Since each quadrilateral has four points, we have enough data to define the translation and rotation operations that will give us the best (in a least-squares sense) overlap of a quadrilateral in P with a quadrilateral in Q. Each overlap can be evaluated to check how well other alpha carbon atoms in P align with alpha carbon atoms in Q. While it would be a daunting task to carry out $\binom{m_P}{2}\binom{m_Q}{2}$ overlap operations, in practice

FIGURE 5.5 Quadrilaterals for 1PYI and 1PYC formed by taking secondary structure elements two at a time.

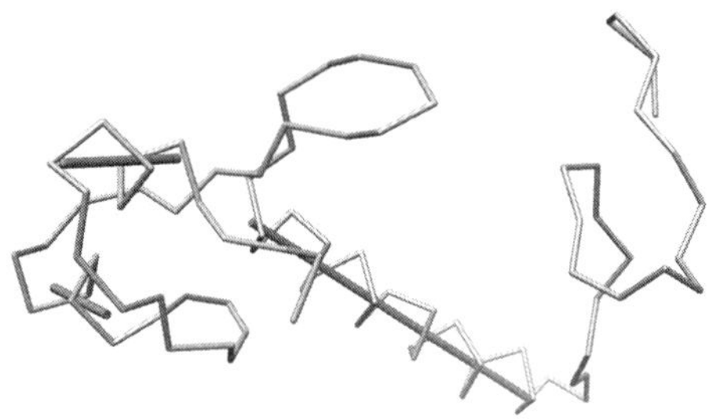

(a)

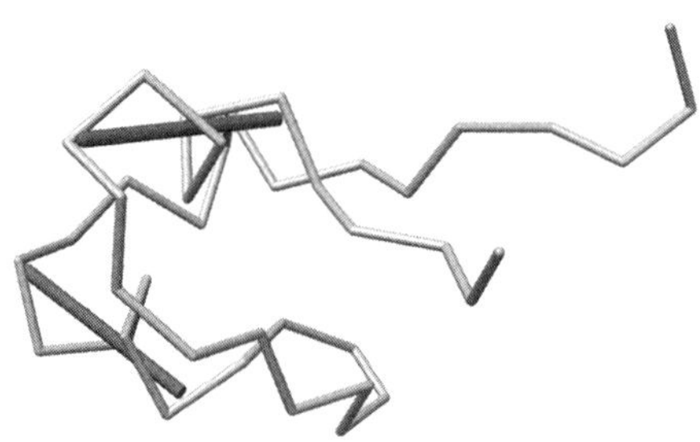

(b)

FIGURE 5.6 (a) Protein 1PYI; (b) protein 1PYC.

we can apply various strategies to filter the number of overlaps that are actually attempted. For example, each inertial axis is labeled with its SSE type, and this allows us to avoid the overlap of a beta strand with an alpha helix. In addition, we can perform tests on the lengths of the sides of the quadrilaterals to avoid an attempt to overlap quadrilaterals with sides that have little hope of initiating an alignment.

Figure 5.7 shows the overlap of the two quadrilaterals that have been chosen to define the structural alignment. In Figure 5.8, we see the same alignment in ribbon form. This alignment has an RMSD of 1.06 across 34 alpha carbon pairs. The best QV, as defined by Equation 5.55, is 0.301.

Figure 5.9 shows the results when this algorithm is applied to the proteins 1VIH (vigilin from *Homo sapiens*, length 71) and 1PYT (procarboxypeptidase A from *Bos taurus*, length 94). The highest QV

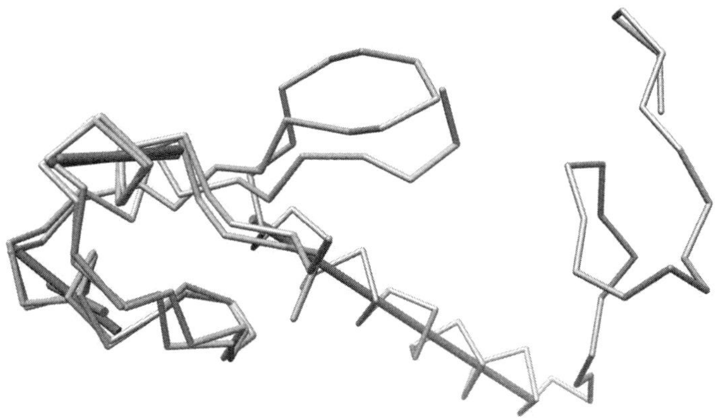

FIGURE 5.7 Overlap of the best matching quadrilaterals gives the structural alignment.

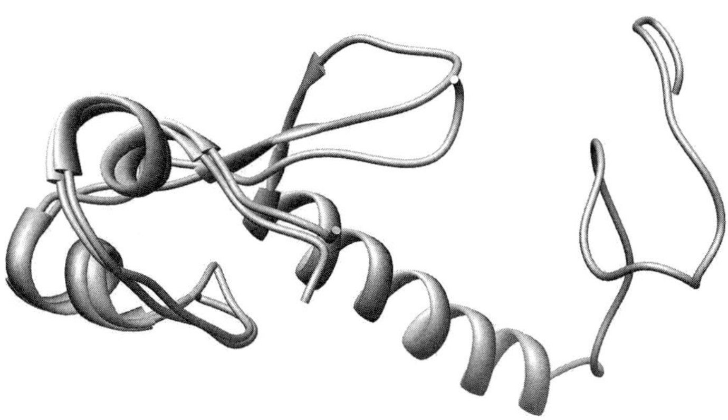

FIGURE 5.8 Structural alignment of 1PYI and 1PYC shown in ribbon form.

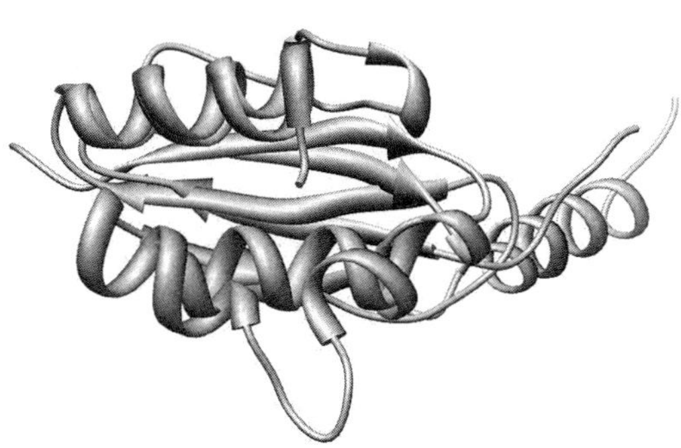

FIGURE 5.9 Structural alignment of 1VIH and 1PYT shown in ribbon form.

is 0.130 over 43 pairs of atoms (RMSD 2.11). This pair of proteins was chosen because SSEs of 1VIH are aligned with SSEs of 1PYT that are not in the same order within the backbone. So, the SSEs of the hydrophobic core in 1VIH are a rewired version of those seen in 1PYT. As reported in [SB98], this similarity of the two structures was not found by DALI [HS93] and VAST [MG95]. The combinatorial extension algorithm discussed in [SB98] finds a structural alignment with an RMSD of 3.2 over 56 atom pairs. By inspecting the alignment, we see that residues Q11 to H34, P51 to K65, and V66 to G72 of 1PYT are in alignment with residues N53 to A75, Y11 to I25, and N31 to K36 of 1VIH, respectively. Figure 5.10 shows the correspondence using dashed lines.

This type of structural alignment shows that the correspondence involves a circular permutation of the secondary structures. Although fairly infrequent, this type of structural alignment has been seen several times (see [JL01] and the references therein). As one might expect in these cases, preliminary processing that calculates a monotonic sequence of residues of protein *P* in local alignment with a monotonic sequence of residues of protein *Q* will only capture a subset of the alpha carbon pairings, and this can lead to an inferior structure alignment.

5.12.2 A Script for Structural Alignment by Overlapping Secondary Structure Elements

In this section, we discuss the `InertialAxisAlignment` class and some of the important issues related to its implementation. The class is accessible using the following:

```
from StructBio. StructureOverlap.inertialAxisAlignment\
   import InertialAxisAlignment
```

The script is fairly short, a bit over two pages, because it makes use of classes that have been developed in earlier sections of this book: `Shell`, `AxesForSSEs`, and `Overlapper`. Instantiation of an `InertialAxisAlignment` object involves the following parameters:

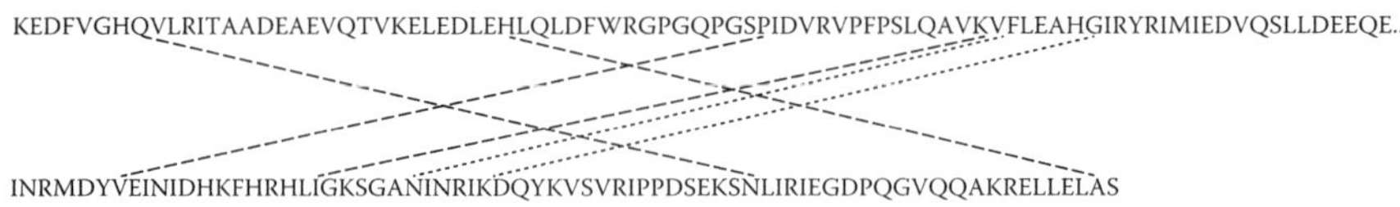

FIGURE 5.10 Correspondence of residues in the structural alignment of 1PYT and 1VIH.

`protP`: the Chimera molecule object (usually a protein chain) defining protein *P*.

`protQ`: the Chimera molecule object (usually a protein chain) defining protein *Q*.

`matchThreshold`: when comparing two quadrilaterals, we ignore the match if potentially overlapped sides differ in length by an amount greater that this threshold (default value is 5.0).

`atomDistThreshold`: two alpha carbon atoms are considered to be close enough to contribute to the QV calculation (Equation 5.55) if they are closer than this threshold (default is 3.0).

`inertialAxisTolerance`: this value is used by the `AxesForSSEs` class and is used to decide whether an SSE should be subdivided to get a better fit between axes and SSE alpha carbon atoms (default value is 3.5).

`axisMinLength`: this threshold determines the minimum axis length for an SSE (default value is 3).

`printingClosestPairs`: if set to True, you will get a list of closest pairs printed in the **Python shell** window (default value is False).

`printProgressiveQ`: if set to True, you will get a three-column table printed in the **Python shell** window. The first column is the number of alpha carbon pairs in alignment, whereas the second and third columns contain the corresponding RMSD and QV values (default value is False).

This initialization function for the `InertialAxisAlignment` class starts by generating a set of inertial axes for each of the protein molecule objects. By using four nested loops, it considers all possible matches of quadrilaterals in *P* with quadrilaterals in *Q*. As described earlier, filtering strategies are used to eliminate incompatible quadrilateral pairs from further consideration. When compatible quadrilaterals are found, the `Overlapper` class is used to perform a structural alignment using the coordinates of the alpha atoms in the quadrilaterals to define the necessary translation and rotation transformations. For each such overlap, the script assesses the success of the overlap by evaluating the extent to which alpha carbon atoms in *P* are brought into close proximity with alpha carbon atoms of *Q*. The overlap with the highest QV evaluation is used to define the best structural alignment. One additional overlap operation is performed, this time based on all the alpha carbon atoms that form close pairs.

Despite the avoidance of any prior sequence alignment, the algorithm provides a very reasonable structural alignment and works fairly quickly when the number of SSEs is low. The algorithm can take considerably longer when, for example, *P* and *Q* are both beta barrels comprising many structurally similar beta strands. In this case, there is a significant combinatorial increase in the number of quadrilaterals to be compared and the execution time may increase to several minutes. On the other hand, one might argue that a beta barrel is obviously similar to another beta barrel with the same number of strands and a structural alignment need not be applied to verify this. The value of the algorithm is mainly for the demonstration of similar structure that is not immediately obvious, and a sequence analysis does not provide a clear correspondence when trying to derive alpha carbon pairings. There is also the possibility that the script can be used as a starting point for other algorithms dealing with structural alignment, for example, loosening our restriction on the avoidance of sequence information and prioritizing the overlap of SSEs that have some reasonable level of sequence similarity. Testing of this class can be found in Script 5_09 InertialAxisAlignmentWithTubes.py on the book's website.

5.13 EXERCISES

The following exercises will make use of the new classes presented in this chapter:

1. Using the `Overlapper` class to superimpose two ligands

   The PDB contains structure files for human oxyhemoglobin (1HHO) and carbonmonoxy hemoglobin (1LJW). For both proteins, use chain A only (delete chain B). In this exercise, you are to use the `Overlapper` class to do translation and rotation operations that will overlap the heme rings of the two proteins. Use the four nitrogen atoms of the heme ring to get four atom pairings that will define the overlap. Now we can compare the structural overlap for other atoms in the heme binding site. In particular, how much overlap do we see for the NE2 atoms of the proximal and distal histidine residues? Refer to Figure 1.25 for a review of these histidine residues. Print out the following values:

a. The RMSD value for the overlapped nitrogen atoms

b. Consider the proximal histidine at position 87. What is the distance between the NE2 atom in this residue in 1HHO and the NE2 atom in this residue in 1LIW?

c. Repeat the calculation in part (b) but use the distal histidine at position 58.

2. Using the `Overlapper` class to superimpose secondary structures in two proteins

   This exercise will use the following PDB files: 1HHP is one chain of HIV protease without a ligand, and 1EBY is HIV protease in complex with the BEA369 inhibitor. The inhibitor has the residue type "BEB." When the protease binds to the inhibitor, there is a shift of the "flaps," which are brought down toward the binding cavity of the protease. The goal of this exercise is to get a visualization and measurement of the conformational changes seen in the flaps when the inhibitor is taken in by the protease. To get an accurate appraisal of flap movement, you are to overlap the two proteins by using atom pairs that come from secondary structures that are reasonably distant from the binding site, for example, the beta strands spanning positions 62 to 73 inclusive and the helices spanning positions 87 to 93 inclusive. The residue at position 49 is a glycine that is near the furthest extent of the flap structure. After the overlap is accomplished, your script should measure the distance traversed by the alpha carbon of GLY 49 when inhibitor binding takes place. Figure 5.11 shows the expected display.

3. Overlap to show the conformational change of helices in a mechanosensitive channel

   Review Exercise 8 of Chapter 4. For this exercise, we want to perform an overlap that exhibits the conformational change of helices within the mechanosensitive channel from *Escherichia coli*. As before, we will use 2VV5 (open conformation) and 2OAU (closed conformation). You should go through the following steps:

   a. Fetch the two PDB files.

   b. Following the same procedure as in Chapter 4, Exercise 8, put in a main axis and a disk as visual components. This is done with

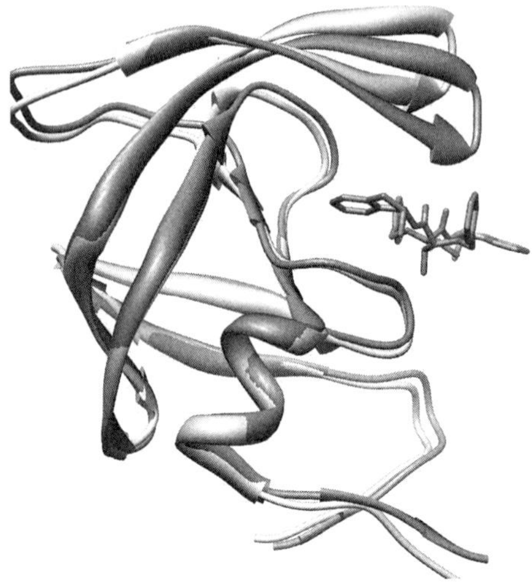

FIGURE 5.11 Overlap showing flap change in HIV protease.

respect to the protein 2VV5. The disk will represent the bottom of the membrane.

c. For clarity, eliminate all chains except for the A chains.

d. Generate an `Overlapper` object that is instantiated for the protein 2OAU.

e. Build a list of alpha carbon pairs from residues at positions 183 to 278 inclusive. These positions avoid the helices that are subject to the conformational change.

f. Perform the overlap by moving 2OAU to superimpose over 2VV5.

g. To provide a visual aid to assess the conformational change put helix tubes into all the helices.

The final display should look like the scene in Figure 5.12. In this figure, the closed conformation is shown in light gray and the open conformation is shown as dark gray.

4. C-loop movement in the nicotinic acetylcholine receptor (nAChR)

Motivation: nAChRs are ligand-gated ion channels [GC01] that can be found embedded in the postsynaptic membrane of a neuromuscular junction. They are drug targets for several neurological

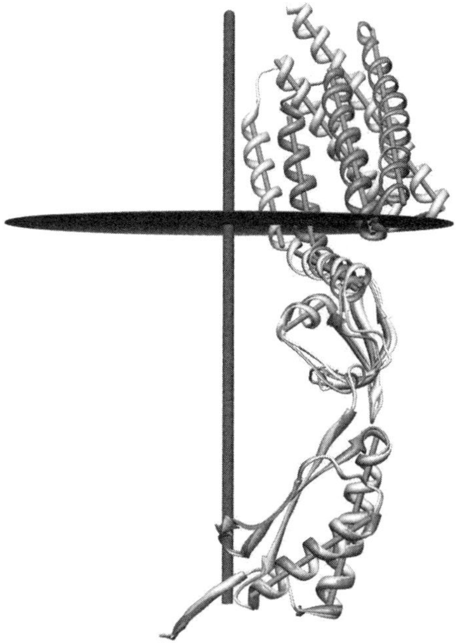

FIGURE 5.12 Overlap to show changes in helices of a mechanosensitive channel.

disorders including myasthenia gravis.* nAChRs can also be found in the postsynaptic membranes of neural synapses.

Architecture of the nicotinic receptor: To get an approximation of the structure of a nAChR, you can use Chimera to fetch 2BG9 (illustrated in Figure 5.13). This is a refined model that is based on electron microscopy. It gives us a reasonable representation of the membrane protein: you can see 5 helices that protrude into the cytosolic environment, and above that there is a group of 20 helices that are situated within the membrane. The part of the structure above this (mostly, beta strands and coils) rises from the membrane to protrude into the synapse. Since the receptor is a membrane protein, we do not yet have a highly accurate representation of its structure as would be available from an x-ray analysis. However, it has been discovered that certain snails have acetylcholine receptors that float freely within their synaptic junctions. These nicotinic acetylcholine binding proteins (nAChBPs) are structurally similar to the synaptic portion of nAChRs. Such a structure is shown in Figure 5.14. These proteins can be crystallized and subjected to x-ray analysis. As evidenced by files in the PDB, they

* http://en.wikipedia.org/wiki/Myasthenia_gravis.

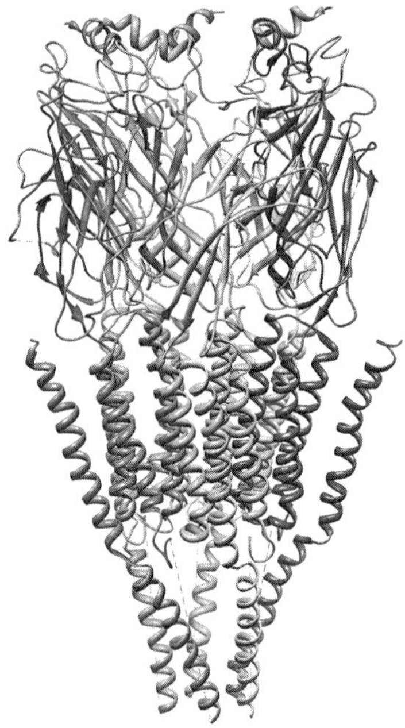

FIGURE 5.13 Side view of the nicotinic acetylcholine receptor.

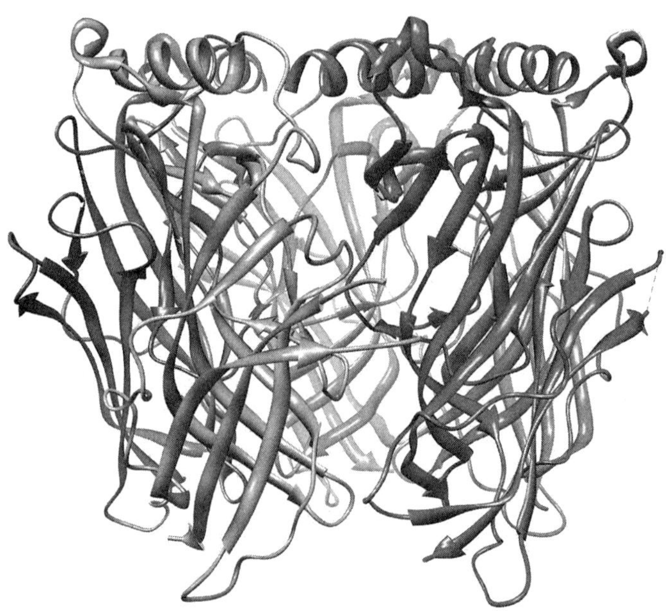

FIGURE 5.14 Side view of the nicotinic acetylcholine binding protein.

have been extensively cocrystallized with a variety of ligands. Although it might seem strange that research on human nicotinic receptors can start with the investigation of proteins from a snail, the structure has been highly conserved over eons of time and it has been estimated that the receptor has been in life forms for at least 1,000,000,000 years.

From Figures 5.15 and 5.16, it is clear that the structure has a pentameric organization. In these figures, each of the five chains has been given a slightly different shade of color to help distinguish the various chains. Figure 5.16 is also a top view, and it shows five nicotine molecules (in sphere representation) held within the five binding sites. Note the slight change in conformation due to the presence of the ligands. Each binding site is actually a crevice that is formed when one chain partially overlaps the adjacent chain, as seen in the side view of Figure 5.17.

In our computational analysis, it will be convenient to deal with only one binding site. In each of the data files used in this exercise, all chains have been removed except for two adjacent chains that surround a single binding site. In Figure 5.17, we are looking at such a binding site from a perspective that is the same as that of Figure 5.14.

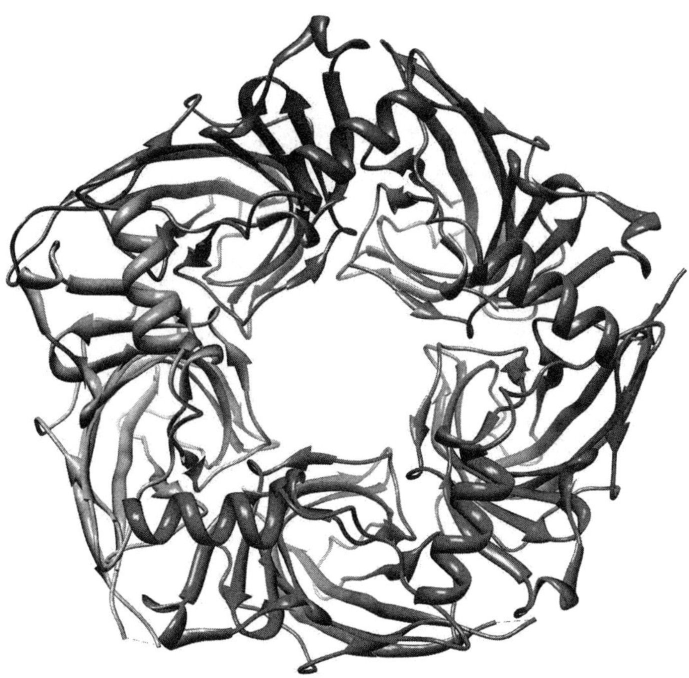

FIGURE 5.15 Nicotinic acetylcholine binding protein in the apo state.

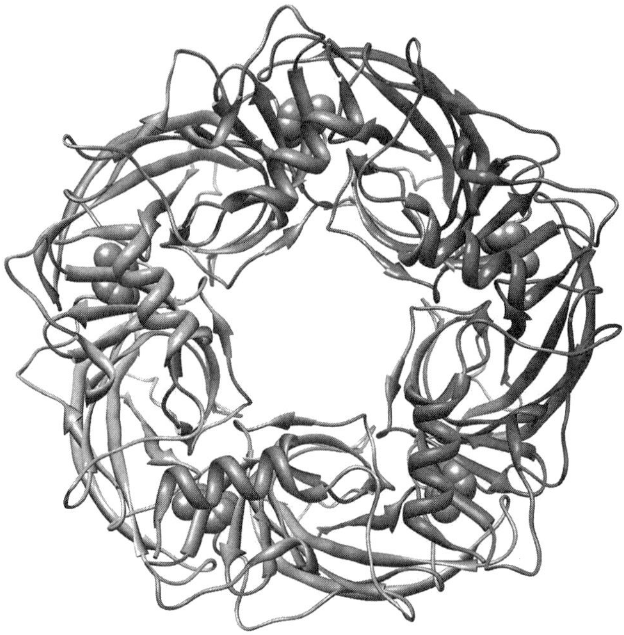

FIGURE 5.16 Nicotinic acetylcholine binding protein with nicotine ligands.

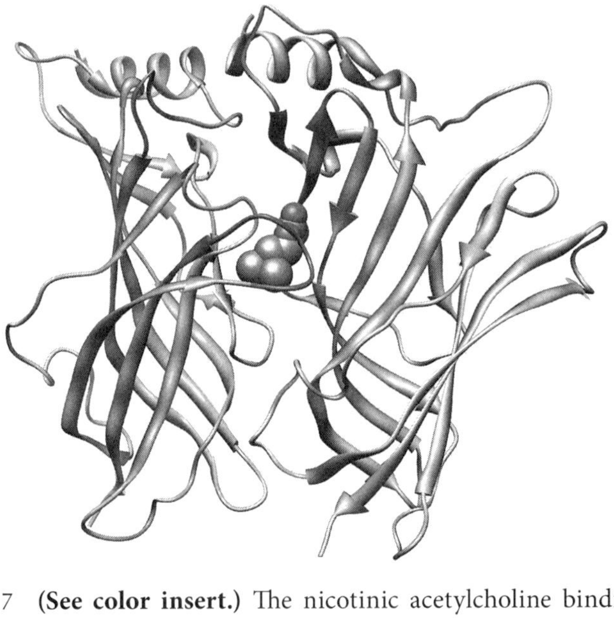

FIGURE 5.17 **(See color insert.)** The nicotinic acetylcholine binding protein binding site.

To proceed with our discussion, some terminology has to be introduced: as mentioned earlier, the binding site is located between two subunits, each one being a chain of the protein. The chain that is somewhat in front of the binding site is called the *principal component* of the binding site. In Figure 5.17, the principal component is colored cornflower blue and a short segment of it is colored green. The green-colored substructure is called the C-loop of the principal component. The movement of the C-loop during ligand binding has been studied with considerable interest since it has been hypothesized that this shift in the C-loop leads to an allosteric change in the conformation of the full receptor (as seen in Figure 5.13) in support of its functionality as a gated ion channel. If you carefully observe the C-loop, you will see that it is a beta strand that ends near the bottom of Figure 5.17. For the nAChBP, this is the usual termination of the chain. In the nAChR protein, this chain continues down into the membrane as a helix. It is believed that when the C-loop is pulled down by a ligand it produces a twisting motion in this helix and this causes the pore to open [CW06].

The other side of the binding site is called the *complementary component*, and it is colored tan in Figure 5.17.

We now discuss important residues in the binding site.

The following files in PDB format have been provided for this exercise:

File Name	Name of Ligand	Primary Citation for the Original PDB File:
2Y7Y_EA.pdb	Apo state (no ligand)	[UA09]
2BYQ_AE.pdb	Epibatidine	[HS05]
2BYR_AE.pdb	Methyllycaconitine	[HS05]
2BYS_AE.pdb	Lobeline	[HS05]
2PGZ_AB.pdb	Cocaine	[HT07]
2WNC_AE.pdb	Tropisetron	[HS09]
3C79_AE.pdb	Imidacloprid	[TH08]
3PMZ_ED.pdb	D-tubocurarine (curare)	[TH10]

The ligands fall into various categories: drugs (lobeline, cocaine, and tropisetron), insecticides (imidacloprid), and toxins (methyllycaconitine, epibatidine, and D-tubocurarine). All ligand names can be found in Wikipedia if you wish to get more information, for example, epibatidine comes from the endangered Ecuadorian frog *Epipedobates tricolor* and D-tubocurarine (used by South American indigenous people to make poisoned arrows) comes from the bark of *Chondrodendron tomentosum*. All data files are for the nAChBPs found in the synapses of *Aplysia californica*, a marine sea slug. Because it has only a few thousand large, easily identified neurons, it is often used for research in neurobiological experiments.

If you access the PDBsum (https://www.ebi.ac.uk/pdbsum/) entries for each protein, you can inspect the ligPlot diagram for each of the ligands. Using this information, we can construct a list of the residues that interact with the ligand in the binding site. Doing this produces the following list of *all* residues that interact with these ligands:

```
holoPrincipalResidues_L = ["THR 91", "TYR 93", "SER 94",
                           "VAL 108", "LYS 143", "PHE
                           144", "GLY 145", "SER 146",
                           "TRP 147", "VAL 148", "TYR
                           149", "GLN 186", "TYR 188",
                           "SER 189", "CYS 190", "CYS
                           191", "TYR 195", "ASP 197"]
holoComplementResidues_L = ["THR 36", "GLN 38",
                           "TYR 55", "GLN 57",
                           "ARG 79", "VAL 108",
                           "MET 116", "ILE 118",
                           "ASP 164", "SER 166",
                           "SER 167"]
```

Notice that the residues appear in two categories. The first batch appears in the principal component, and the second batch appears in the complementary component. These position numbers are valid for all the PDB files corresponding to proteins with ligands (holo-state proteins). Unfortunately, the file for the apo-state protein (2Y7Y_EA.pdb) has the positions shifted somewhat, and so the principal and complementary residues are as follows:

```
apo_PrincipalResidues_L = ["THR 89", "TYR 91",
                          "SER 92", "VAL 106",
                          "LYS 141", "PHE 142",
                          "GLY 143", "SER 144",
                          "TRP 145", "VAL 146",
                          "TYR 147", "GLN 184",
                          "TYR 186", "SER 187",
```

```
                                "CYS 188", "CYS 189",
                                "TYR 193", "ASP 195"]
apo_ComplementResidues_L = ["THR 34", "GLN 36",
                                "TYR 53", "GLN 55",
                                "ARG 77", "VAL 106",
                                "MET 114", "ILE 116",
                                "ASP 162", "SER 164",
                                "SER 165"]
```

(Feel free to copy these lists into your script …)

Using the suffix on the PDB file names: the data files have been extracted from PDB files with the full pentameric structure of each nAChBP, as in Figure 5.15. Unfortunately, (again) researchers have not used a consistent approach for the labeling of the chains in the pentameric structure. Looking at the structure from a top view, some researchers have labeled the chains A, B, C, D, and E in a clockwise fashion, whereas others have done this in a counterclockwise fashion. This can cause confusion when considering the principal and complementary chains for a binding site. Furthermore, some files do not have ligands in each of the five possible binding sites.

To avoid the work of inspecting each and every data file, you can instead use the following rules: the first letter of the suffix specifies the chain identification of the chain acting as the principal component, whereas the second letter specifies the chain identification of the chain acting as the complementary component. So, for example, when your program works with the 2PGZ_AB.pdb file, chain A is the principal component and chain B is the complementary component. These chains will define the single binding site that, in this case, holds a molecule of cocaine (the residue type is "COC"). Note that any residue (consider, e.g., THR 91) will appear in both chains since all chains have the same primary sequence. However, for this particular binding site we only pay attention to a THR 91 residue if it is in the principal component (as specified in our `holoPrincipalResidues_L` list seen earlier). So, for this file we would use the THR 91 residue in chain A and ignore the THR 91 residue in chain B. Note that each chain plays two roles. For example, the chain acting as the complementary subunit for a binding site will also act as the principal subunit for the next binding site.

We can finally state the requirements of the exercise: write a script that reads in the file 2Y7Y_EA.pdb (apo state) and then works through the following list:

```
pdbFiles_L = [("2BYR_AE", "methyllycaconitine"),
              ("3PMZ_ED", "d-tubocurarine"),
              ("2WNC_AE", "tropisetron"),
              ("3C79_AE", "imidacloprid"),
              ("2PGZ_AB", "cocaine"),
              ("2BYQ_AE", "epibatidine"),
              ("2BYS_AE", "lobeline")]
```

For each of these proteins, your script should print the name of the ligand in the Python shell window and then show the apo-state protein overlapped with the holo-state protein. The alpha carbon atoms in the residues of the residue lists should define the overlap transformations, for example, the alpha carbon of THR 91 in the principal component of the holoprotein and the alpha carbon of THR 89 in the principal component of 2Y7Y_EA acting as one of the pairs that specifies the overlap. In other words, the residue lists given in the previous pages are really defining an alignment of residues, one set of residues in 2Y7Y_EA and the other in the holo-state protein. Figure 5.18 shows the results for the overlap between 2PGZ_AB and the apo-state protein 2Y7Y_EA.

After the overlap is done, we can make the holoprotein disappear by executing a statement such as

```
runCommand("sel #" + str(modelNumber) +
           "; ~display sel; ~ribbon sel; ~sel")
```

In this statement, `modelNumber` identifies the model number associated with the Chimera molecular object when you open the file. Since you are fetching 2Y7Y first, it will have model number 0 and then all subsequent holoproteins will have model numbers 1, 2, …, 7. You can allow the user to initiate the overlap with the next holoprotein by asking for user response, for example,

```
raw_input ("Hit Enter to continue.")
```

Note that after all the proteins in the list are brought into overlap, they can be selectively brought back into the scene by executing a statement such as

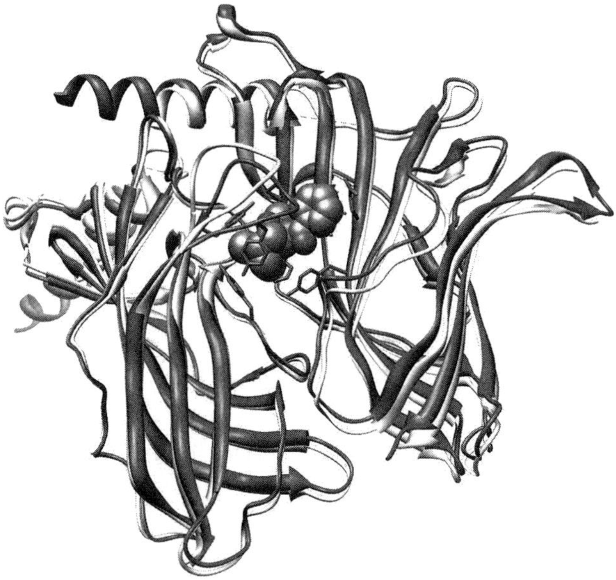

FIGURE 5.18 **(See color insert.)** Overlapping the binding sites of 2Y7Y (white) and 2PGZ (magenta).

```
runCommand("sel #" + str(modelNumber) +
          "; ~display sel; ribbon sel; ~sel")
```

After this is done, the script can be modified to do some calculations on the C-loop. In Figure 5.18, note how the C-loop has been bent toward the cocaine molecule due to the binding of the ligand. For this part of the exercise, we want to measure this shift by calculating the average displacement of all the alpha carbon atoms in the residues of the C-loop. For a holo-state protein, we will define the C-loop to be all the residues between TYR 188 and ILE 196 (both of these residues also included). For the apo-state protein, we will define the C-loop to be all the residues between TYR 186 and ILE 194 (both of these residues also included). Be sure you understand what is meant by displacement. It is the shift in position of the atom due to the binding of the ligand. Here is an example: after we have generated the overlapped versions of the two proteins, we can calculate the displacement of the first alpha carbon in the C-loop by measuring the distance between the alpha carbon of TYR 186 in the principal component of the apoprotein and the alpha carbon of TYR 188 in the principal component of the holoprotein.

Your script should print a table with two columns: the first column will be a list containing the names of the ligands, and the second column will be a list of the corresponding average displacements of the alpha carbon atoms in the C-loop.

5. Testing the `SequencesLocalAlignment` class

   Use the `SequencesLocalAlignment` class to generate a local sequence alignment of two proteins (linear gap model). The script should generate a list of pairs of residues that are in the alignment. You can avoid the mention of those residues that are not paired with any residue from the other protein. In other words, you can ignore the specification of gap locations in the alignment. Test the script using 1MBN and 1JEB. You should also test on 3CS9 and 2HYY, noting that residues in the sequence but missing structural coordinates (3CS9: GLY 224.A to TYR 232.A and GLU 275.A to MET 278.A; 2HYY: VAL 228.A to LYS 234.A, GLU 275.A, and GLU 499.A to SER 599.A) are skipped because they are of no value to the structural alignment.

6. Sequence and structural similarity:

   Write a script that imports the `userSpecificationOfChains` function from the `usefulFunctions` module. This function should be used to bring in n user-specified chains. The first chain will be a target structure, and the remaining $n - 1$ "query" chains are to be overlapped with the target chain. The `SequencesLocalAlignment` class should be used to obtain a local sequence alignment between the target and each of the $n - 1$ query chains. Print out the maximum score value from each dynamic programming score matrix to provide a comparative measure of similarity between the target protein and a query protein. Each sequence alignment should be used to generate alpha carbon pairings that determine the overlap operation. When all the overlaps have been accomplished, the script should repeatedly use a `runCommand` statement such as

```
runCommand("sel #" + str(prot].id) +
           "; ~display sel; ~ribbon sel; ~sel")
```

   to hide all the query chains. After this, use

```
runCommand("sel #" + str(prot.id) +
           "; ~display sel; ribbon sel; ~sel")
```

to selectively bring single query chains back into the display or simply provide a user interaction that will cycle through all the query chains for each depression of the "Enter" key.

To test the script, consider the following questions:

a. Is the myoglobin of pig more similar to that of horse or human? Work with the following protein sequences:

    i. 1PMB: porcine myoglobin (use chain A)

    ii. 1WLA: equine myoglobin

    iii. 2MM1: human myoglobin

   In this case, chain A of 1PMB will act as the target, whereas 1WLA and 2MM1 act as the query proteins. The comparison should address both sequence similarity and structural similarity.

b. Another interesting comparison of structural similarity is to compare the human prion protein with the prion protein from various other mammals (mouse, cattle, and sheep):

    i. 1QM2: *H. sapiens*

    ii. 1AG2: *Mus musculus*

    iii. 1DX0: *B. taurus*

    iv. 1UW3: *Ovis aries*

c. Consider the following two proteins:

    i. Chain B of 1VRU: HIV-1 reverse transcriptase

    ii. Chain B of 1QAI: Moloney murine leukemia virus reverse transcriptase

   Each protein functions as a reverse transcriptase, but the proteins are from different retroviruses. Perform a pairwise sequence local alignment for these proteins. Is there any evidence of significant sequence similarity? What is the percent identity for the longest overlap? Is it in the twilight zone as defined by Rost (see [Ro99])? Use a local sequence alignment to derive alpha carbon pairings that determine a structural overlap.

7. Visualizing domain movements

Read the papers by Brylinski and Skolnick [BS08] and Browner and others [BF92]. These papers deal with proteins that undergo a significant change in conformation when going from the apo to the holo form. In many cases, there is significant hinge bending, shear motion, or even partial unfolding on ligand binding. This is often apparent as a change in the relative position of one domain with respect to another in the protein. The goal of this exercise will be to add scenographic elements to the Chimera display to show the extent of this domain movement. The script should go through the following steps:

a. Execute the following import statements:

```
from StructBio. LeastSquares.inertialAxes\
   import InertialAxes
from StructBio. Scenographics.ellipsoid\
   import Ellipsoid
from StructBio. Scenographics.solids\
   import Solids
from StructBio. StructureOverlap.overlapper\
   import Overlapper
```

b. Fetch both the holo and apo forms of the protein.

c. For each protein, use Table 1 of [BS08] (pages 368 and 369) to determine which chain to use and the positions of residues that define each domain.

d. Use the `Overlapper` class to move the apoprotein so that there is an overlap of domain 1 in the apoprotein with domain 1 in the holoprotein.

e. Work with the `InertialAxes` class to get inertial axes for the following sets of atoms:

   i. Domain 1 in the apo form combined with domain 1 in the holo form.

   ii. Domain 2 in the apo form.

   iii. Domain 2 in the holo form.

f. Use spindles from the `Solids` class to display a visualization of the inertial axes.

g. To accomplish the next step, use the `min_maxProjection` function in the `InertialAxes` class to get the lengths of all the axes.

h. Use the `Ellipsoid` class to place three ellipsoids into the display. There should be an ellipsoid for each set of inertial axes, as described in step e. Each ellipsoid should have its own color. Work with the `displayMesh` function to display the ellipsoidal mesh in the Chimera window.

You should end up with a display that is similar to that shown in Figure 5.19.

8. Visualizing domain movements

Build a graphical user interface (GUI) for the application described in Exercise 7. As before, the user will specify the apo- and holo-proteins along with designations for the chains and domains. The interface should allow the user to envelop the ellipsoids with a transparent surface or mesh or both (color to be specified as well). See Figure 5.20 for an example of such an interface. Note the last frame and the reporting of domain 2 movement by calculating

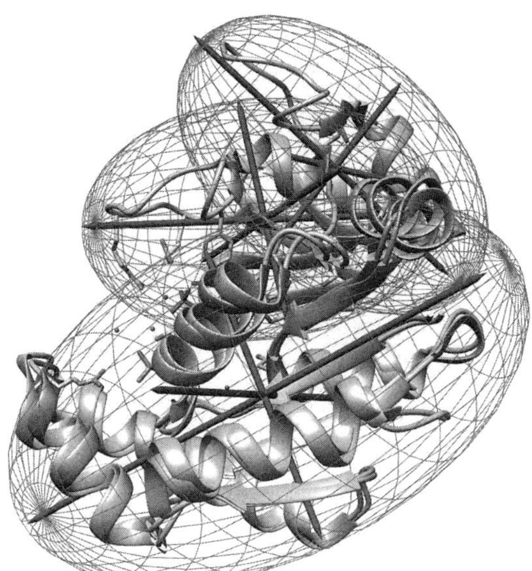

FIGURE 5.19 **(See color insert.)** Domain movement for guanylate kinase (1EX6B and 1EX7A).

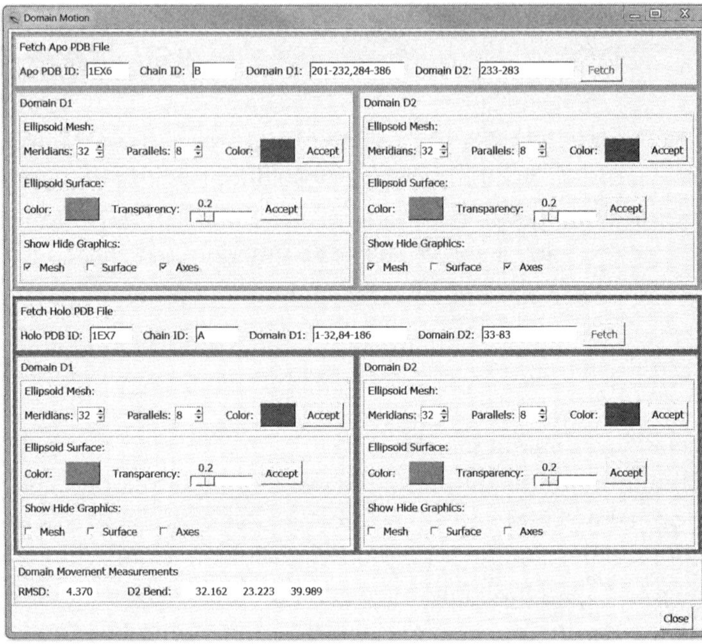

FIGURE 5.20 Suggested modeless dialog for domain movement visualization.

RMSD and domain bending. This interface makes all the controls simultaneously visible and so it is easy to build such a GUI, but it is rather demanding for screen space. A more sophisticated version can be constructed to use tabs or some strategy that incorporates modal dialogs that would be instantiated from this modeless dialog.

REFERENCES

[AV07] A. Abyzov and V. A. Ilyin. A comprehensive analysis of non-sequential alignments between all protein structures. *BMC Structural Biology*, **7** (2007), 78.

[Be54] R. Bellman. The theory of dynamic programming. *Bulletin of the American Mathematical Society*, **60** (1954), 503–516.

[BO96] N. P. Brown, C. A. Orengo, and W. R. Taylor. A protein structure comparison methodology. *Computers and Chemistry*, **20** (1996), 359–380.

[BF92] M. F. Browner, E. B. Fauman, and R. J. Fletterick. Tracking conformational states in allosteric transitions of phosphorylase. *Biochemistry*, **31** (1992), 11297–11304.

[BS08] M. Brylinski and J. Skolnick. What is the relationship between the global structures of apo and holo proteins? *Proteins*, **70** (2008), 363–377.

[CW06] X. Cheng, H. Wang, B. Grant, S. M. Sine, and J. A. McCammon. Targeted molecular dynamics study of C-loop closure and channel gating in nicotinic receptors. *PLOS Computational Biology*, **2** (2006), 1173–1184.

[Do81] R. F. Doolittle. Similar amino acid sequences: chance or common ancestry? *Science*, **214** (1981), 149–159.

[DB07] J. Dundas, T. Binkowski, B. DasGupta and J. Liang. Topology independent protein structural alignment. *BMC Bioinformatics*, **8** (2007), 388. See also: *Lecture Notes in Computer Science*, **4645** (2007), 171–182.

[DE98] R. Durbin, S. Eddy, A. Krogh, and G. Mitchison. *Biological sequence analysis*, Cambridge University Press, Cambridge, UK, 1998.

[GM96] J.-F. Gibrat, T. Madej, and S. H. Bryant. Surprising similarities in structure comparison. *Current Opinion in Structural Biology*, **6** (1996), 377–385.

[GC01] T. Grutter and J.-P. Changeux. Nicotinic receptors in wonderland. *Trends in biochemical sciences*, **26** (2001), 459–463.

[HS05] S. B. Hansen, G. Sulzenbacher, T. Huxford, P. Marchot, P. Taylor, and Y. Bourne. Structures of *Aplysia* AChBP complexes with nicotinic agonists and antagonists reveal distinctive binding interfaces and conformations. *EMBO Journal*, **24** (2005), 3635–3646.

[HT07] S. B. Hansen and P. Taylor. Galanthamine and non-competitive inhibitor binding to ACh-binding protein: evidence for a binding site on non-alpha-subunit interfaces of heteromeric neuronal nicotinic receptors. *Journal of Molecular Biology*, **369** (2007), 895–901.

[HH92] S. Henikoff and J. Henikoff. Amino acid substitution matrices from protein blocks. *Proceedings of National Academy of Science USA*, **89** (1992), 10915–10919.

[HS09] R. E. Hibbs, G. Sulzenbacher, J. Shi, T. T. Talley, S. Conrod, W. R. Kem, P. Taylor, P. Marchot, and Y. Bourne. Structural determinants for interaction of partial agonists with acetylcholine binding protein and neuronal alpha7 nicotinic acetylcholine receptor. *EMBO Journal*, **28** (2009), 3040–3051.

[HS93] L. Holm and C. Sander. Protein structure comparison by alignment of distance matrices. *Journal of Molecular Biology*, **233** (1993), 123–138.

[JL01] J. Jung and B. Lee. Circularly permuted proteins in the protein structure database. *Protein Science*, **10** (2001), 1881–1886.

[KH04] E. Krissinel and K. Henrick. Secondary-structure matching (SSM), a new tool for fast protein structure alignment in three dimensions. *Acta Crystallographica*, D**60** (2004), 2256–2268.

[MG95] T. Madej, J. F. Gibrat, and S. H. Bryant. Threading a database of protein cores. *Proteins*, **23**, (1995), 356–369.

[MO78] S. Meier and S. Özbek. A biological cosmos of parallel universes: does protein structural plasticity facilitate evolution? *BioEssays*, **29** (2007), 1095–1104.

[NW70] S. B. Needleman and C. D. Wunsch. A general method applicable to the search for similarities in the amino acid sequence of two proteins. *Journal of Molecular Biology*, **48** (1970), 443–453.

[OJ03] C. A. Orengo, D. T. Jones, and J. M. Thornton. *Bioinformatics: genes, proteins and computers*, BIOS Scientific Publishers, Oxford, UK, 2003.

[Ro99] B. Rost. Twilight zone of protein sequence alignments. *Protein Engineering*, **12** (1999), 85–94.

[RB92] R. B. Russell and G. J. Barton. Multiple protein sequence alignment from tertiary structure comparison: assignment of global and residue confidence levels. *Proteins*, **14** (1992), 309–323.

[SB06] M. Sagermann, W. A. Baase, and B. W. Matthews. Sequential reorganization of β-sheet topology by insertion of a single strand. *Protein Science*, **15** (2006), 1085–1092.

[SB97] A. P. Singh and D. L. Brutlag. Hierarchical protein structure superposition using both secondary structure and atomic representations. *Proceedings of the International Conference on Intelligent Systems for Molecular Biology ISMB-97*, Halkidiki, Greece, (1997), 284–293.

[SK04] M. L. Sierk and G. J. Kleywegt. Déjà vu all over again: finding and analyzing protein structure similarities. *Structure*, **12** (2004), 2103–2111.

[SB98] I. N. Shindyalov, and P. E. Bourne. Protein structure alignment by incremental combinatorial extension (CE) of the optimal path. *Protein Engineering*, **11** (1998), 739–747.

[SW81] T. F. Smith and M. S. Waterman. Identification of common molecular subsequences. *Journal of Molecular Biology*, **147** (1981), 195–197.

[TH08] T. T. Talley, M. Harel, R. E. Hibbs, Z. Radic, M. Tomizawa, J. E. Casida, and P. Taylor. Atomic interactions of neonicotinoid agonists with AChBP: molecular recognition of the distinctive electronegative pharmacophore. *Proceedings of National Academy of Science USA*, **105** (2008), 7606–7611.

[TH10] T. T. Talley, M. Harel, J. G. Yamauchi, Z. Radic, S. Hansen, T. Huxford, and P. W. Taylor. The curare alkaloids: analyzing the poses of complexes with the acetylcholine binding protein in relation to structure and binding energies. To appear.

[TO89] W. R. Taylor and C. A. Orengo. Protein structure alignment. *Journal of Molecular Biology*, **208** (1989), 1–22.

[UA09] C. Ulens, A. Akdemir, A. Jongejan, R. Van Elk, S. Bertrand, A. Perrakis, R. Leurs, A. B. Smit, T. K. Sixma, D. Bertrand, and I. J. De Esch. Use of acetylcholine binding protein in the search for novel alpha7 nicotinic receptor ligands. In silico docking, pharmacological screening, and x-ray analysis. *Journal of Medical Chemistry*, **52** (2009), 2372–2383.

[VB96] I. R. Vetter, W. A. Baase, D. W. Heinz, J. P. Xiong, S. Snow, and B. W. Matthews. Protein structural plasticity exemplified by insertion and deletion mutants in T4 lysozyme. *Protein Science*, **5** (1996), 2399–2415.

[ZB07] M. Zvelebil and J. Baum. *Understanding Bioinformatics*. Garland Science, Oxford, UK, 2007.

CHAPTER 6

Potential Energy Functions

6.1 MOTIVATION

The central problem of molecular dynamics is to determine the relative motion of all atoms in a set of molecules. It is essentially the n-body problem that Newton pondered when dealing with a collection of planetary bodies but it is much more formidable in its complexity. This extra complexity is caused by the following issues:

1. *Complexity in modeling forces*: Although the classical mechanics model can be used to describe both planetary motion and molecular motion, the latter involves a much more complicated description of the forces between two bodies. Moreover, forces between two atoms can be both attractive and repulsive (described by Equation 6.10).
2. *Number of interacting bodies*: The system contains not just a few interacting bodies but perhaps hundreds or even thousands of bodies (atoms).
3. *Proximity of interacting bodies*: For a planetary system, one can derive a very reasonable approximation by modeling the motion of a planet using a two-body system, namely the planet and the sun with its very large gravitational effect. The gravitational influence of other planets can be neglected. For a biological system, several atoms may be in close proximity and consequently it is not as easy to eliminate their influence in the calculations.

The problem may be stated as follows: How can we best represent the forces that act on an atom within a molecular ensemble?

6.1.1 Empirical Observations

There are many observables that can be extracted from a molecular system. This includes the empirical assessments of the various force constants that are used in the force-field formulas described in Section 6.1.2. As noted in [MA98], data may be collected from the gas phase using electron diffraction, microwave, infrared, and Raman techniques to name just a few. We can also use the results of x-ray analysis. In the case of a ligand bound to a protein, an x-ray analysis of the cocrystallized molecules can be used to derive the three-dimensional coordinates of all atoms in the molecules. In particular, we can determine the distance between the various atoms and these distances can be used in the mathematical model that calculates the potential energy of the atomic ensemble.

6.1.2 Mathematical Modeling

Our most accurate description of an atomic ensemble is via quantum mechanics but this does not facilitate a tractable computational strategy. Instead, our molecular mechanical analysis will consider the atoms to be a collection of masses interacting with each other through various forces. In fact, we will assume that the total energy of the system is minimized when the atoms assume a stable position that is characterized by the minimization of

$$E_{\text{total}} = E_{\text{bond}} + E_{\text{angle}} + E_{\text{dihedral}} + E_{\text{vdw}} + E_{\text{coulomb}} \tag{6.1}$$

Other formulas may be developed using more or fewer terms, depending on the needs of the application. As we will see later, there are several useful models that can be used in the modeling of the energy function.

In general, the terms on the right-hand side of the equation fall into two categories:

1. *Bonded atoms*: These are energy terms related to atoms that share a covalent bond.
2. *Nonbonded atoms*: These are energy terms related to atoms that are not bonded but are close enough to interact.

6.1.2.1 Energy Terms for Bonded Atoms

E_{bond} is composed of all the various bond-stretch terms. Each is described by a simple quadratic function:

$$K^b(r-\bar{r})^2 \tag{6.2}$$

where r is the distance between two bonded atoms. This formula is derived from Hooke's law and it models the potential energy as if the two bonded atoms were connected by a spring. If the bond is not *strained* it will have an equilibrium length of $\bar{r}$. If the bond is stretched or compressed, the deviation from $\bar{r}$ measured as $r-\bar{r}$ will increase the potential energy in a quadratic manner. The bond stretch force constant K^b has been determined by empirical observations and it will take on various values depending on the atom types specified for the two atoms that are involved. For example, as reported by [CC95],* $\bar{r}=1.507$ Å and K^b = 317 kcal/(mol Å^2) when the bond is C–C, whereas $\bar{r}=1.273$ Å and K^b = 570 kcal/(mol Å^2) when the bond is C=N.

Because E_{bond} will be the sum of all such quadratic contributions, we can write

$$E_{\text{bond}} = \sum_{(i,j)\in B} K_{ij}^b\left(r_{ij}-\bar{r}_{ij}\right)^2 \tag{6.3}$$

where:

K_{ij}^b is the bond stretch force constant that would be used when *atom* (*i*) is bonded to *atom* (*j*)

r_{ij} is the distance between *atom* (*i*) and *atom* (*j*)

$\bar{r}_{ij}$ is the distance between *atom* (*i*) and *atom* (*j*) at equilibrium

B is the set of pairs: $\{(i,j)\,|\,atom(i)\,\text{is bonded to}\,atom(j)\}$

E_{angle} is the sum of all bond angle bending terms and each is modeled as a quadratic spring function with an angle-bending force constant K^a:

$$K^a\left(\theta-\bar{\theta}\right)^2 \tag{6.4}$$

where θ is the bond angle formed by three atoms and $\bar{\theta}$ the value of the unstrained or equilibrium angle. Like K^b, the value of the force constant will depend on the type of atoms involved. For example, again reported by [CC95], $\bar{\theta}=112.4^\circ=1.96\,\text{radian}$ with K^a = 63 kcal/(mol radian^2) when the

* Table 14 of this article presents well over 200 of the many parameters that can be used in this calculation.

angle is formed by C–C–C, while $\bar{\theta} = 121.9^\circ = 2.13$ radian and $K^a = 50$ kcal/(mol radian^2) when the angle is formed by C–N–C.

Since E_{angle} is the sum of all such quadratic contributions, we can write

$$E_{\text{angle}} = \sum_{(i,j,k)\in A} K_{ijk}^{a} \left(\theta_{ijk} - \bar{\theta}_{ijk}\right)^2 \tag{6.5}$$

where:

K_{ijk}^{a} is the force constant used for the bond angle formed when *atom* (*i*) and *atom* (*k*) are bonded to *atom* (*j*)

θ_{ijk} is the bond angle made when *atom* (*i*) and *atom* (*k*) are bonded to *atom* (*j*)

$\bar{\theta}_{ijk}$ is the bond angle made when θ_{ijk} is at equilibrium

A is the set of triplets: $\{(i,j,k) \mid atom(i) \text{ and } atom(k) \text{ are bonded to } atom(j)\}$

E_{dihedral} involves the contributions made by the dihedral angle terms. This is a bit more complicated and is modeled as

$$\frac{K^d}{2}\left[1+\cos(n\phi-\gamma)\right] \tag{6.6}$$

where ϕ is the dihedral angle formed by a chain of four consecutive bonded atoms, n is the periodicity of the dihedral angle (see [CC95]), and γ is the phase offset. For example, $K^d/2 = 14.50$ kcal/mol, $n = 2$, and $\gamma = 180^\circ$ when the dihedral angle is formed by the four consecutive atoms X–C–CA–X, where X is any atom, C is a carbonyl sp^2 carbon, and CA is an aromatic sp^2 carbon. As before, we sum over all contributions:

$$E_{\text{dihedral}} = \sum_{(i,j,k,l)\in D} \frac{K_{ijkl}^{d}}{2}\left[1+\cos\left(n_{ijkl}\phi_{ijkl} - \gamma_{ijkl}\right)\right] \tag{6.7}$$

where

K_{ijkl}^{d} is the force constant used for the dihedral angle formed by the four consecutive atoms indexed by *i*, *j*, *k*, and *l*

n_{ijkl} is the periodicity of this dihedral angle

ϕ_{ijkl} is the dihedral angle formed by the consecutive atoms indexed by *i*, *j*, *k*, and *l*

γ_{ijkl} is the phase offset for this dihedral angle

D is the set of quadruplets that specify all the dihedral angles, namely, $\{(i,j,k,l) \mid i,j,k, \text{ and } l \text{ are the indices of four consecutive bonded atoms}\}$

6.1.2.2 *Energy Terms for Nonbonded Atoms*

E_{vdw} is due to the attractive/repulsive forces that are characterized as van der Waals interactions. The van der Waals interaction between two atoms is given by the *12-6 potential*:

$$\frac{A}{r^{12}} - \frac{B}{r^{6}} \tag{6.8}$$

where r is the distance between the two atoms. We assume that the atoms are in different molecules or if they are in the same molecule, then they are separated by at least three bonds. From the formula, it is clear that for very small r, the r^{-12} term contributes a large increase in the potential energy but it has very little impact for larger values of r, where the potential is negative but very close to zero. Values for constants A and B also depend on the atom types of the two interacting atoms. The derivation of their values is somewhat involved and we will omit these details that can be found in [CC95, Table 14 footnote]. Adding up all contributions we get

$$E_{vdw} = \sum_{j \in F} \sum_{[i<j]} \left[\frac{A_{ij}}{r_{ij}^{12}} - \frac{B_{ij}}{r_{ij}^{6}} \right] \tag{6.9}$$

where

A_{ij} and B_{ij} are the Lennard–Jones 12-6 parameters (see [CC95]) for *atom*(i) interacting with *atom*(j)

r_{ij} the distance between *atom*(i) and *atom*(j)

F the full set of all indices for the atoms, and $[i < j]$ means that index i is less than index j and also that *atom*(i) and *atom*(j) are either in separate molecules or if in the same molecule they are separated by at least three bonds.

$E_{coulomb}$ is due to the electrostatic forces that exist between two charged atoms. As before, these nonbonded interactions are only calculated for atoms that are separated by at least three bonds. For two atoms indexed by i and j, this is characterized by the formula:

$$K_c = \frac{q_i q_j}{\varepsilon r_{ij}} \tag{6.10}$$

where ε is the dielectric constant and variables q_i and q_j the atomic charges of the interacting atoms that are separated by a distance of r_{ij}. This formula can be derived from Coulomb's law that expresses the electrostatic force as

$$F \propto \frac{q_i q_j}{r_{ij}^2} \tag{6.11}$$

The force is positive (indicating repulsion) if the two atoms have the same charge, otherwise it is negative and produces an attractive interaction. The K_c constant is a conversion factor needed to obtain energies in units of kcal/mol. Using the electrostatic charge unit (esu) to define a unit of charge as in [Sc02, p. 254], we can derive the value of the conversion factor as $K_c = 332$ (kcal/mol)(Å/esu^2). Partial charge values for various atoms in both amino acids and nucleotides are reported in [CC95] based on [CCB95]. For now, we will assume that $\varepsilon = 1$. Other values of ε will be discussed near the end of Section 6.1.3. Adding up all contributions, we get

$$E_{\text{coulomb}} = K_c \sum_{j \in F} \sum_{[i<j]} \frac{q_i q_j}{\varepsilon_{ij} r_{ij}} \tag{6.12}$$

where

K_c is 332 (kcal/mol)(Å/esu^2)

q_i, q_j are the partial charges for *atom* (i) and *atom* (j), respectively

r_{ij} is the distance between *atom* (i) and *atom* (j)

ε_{ij} is the dielectric constant for the *atom* (i) and *atom* (j) *environment* (assumed to be 1 for now)

F is the full set of all indices for the atoms

$[i<j]$ means that index i is less than index j and also that *atom* (i) and *atom* (j) are either in separate molecules or if in the same molecule they are separated by at least three bonds

6.1.2.3 *Total Potential Energy*

Combining all terms just defined into one grand total we get

$$\begin{aligned} E_{\text{total}} &= E_{\text{bond}} + E_{\text{angle}} + E_{\text{dihedral}} + E_{\text{vdw}} + E_{\text{coulomb}} \\ &= \sum_{(i,j)\in B} K_{ij}^b \left(r_{ij} - \bar{r}_{ij}\right)^2 + \sum_{(i.j.k)\in A} K_{ijk}^a \left(\theta_{ijk} - \bar{\theta}_{ijk}\right)^2 \\ &+ \sum_{(i,j,k,l)\in D} \frac{K_{ijkl}^d}{2} \left[1 + \cos\left(n_{ijkl}\phi_{ijkl} - \gamma_{ijkl}\right)\right] \\ &+ \sum_{j\in F} \sum_{[i<j]} \left[\frac{A_{ij}}{r_{ij}^{12}} - \frac{B_{ij}}{r_{ij}^{6}}\right] + K_c \sum_{j\in F} \sum_{[i<j]} \frac{q_i q_j}{\varepsilon_{ij} r_{ij}} \end{aligned} \tag{6.13}$$

As mentioned earlier, this empirical energy calculation depends on many parameters and it is typical that the program performing the calculation will have rather extensive tables to provide values for $K_{ij}^{b}, \bar{r}_{ijk}, K_{ijk}^{a}, \bar{\theta}_{ijk}, K_{ijkl}^{d}, n_{ijkl}, \phi_{ijkl}, \gamma_{ijkl}, A_{ij}$, and B_{ij}, all of which depend on the atom types assigned to the atoms involved. There are several versions of these tables. The tables included in [CC95] can give the reader a general idea about the information that is required. There are several parameter sets that are now available for various applications. They have names such as AMBER, CHARMM, GROMOS, MMFF, and OPLS.

6.1.3 Computational Issues

Although somewhat daunting, the empirical force field just described would seem to be a rather straightforward calculation. The program would do all the appropriate table lookups, make the necessary substitutions, and do the indicated calculations. Unfortunately, there are some issues that complicate the situation:

1. *Repetitive energy calculations*: There are various applications, such as molecular dynamics, for which it is necessary to repeatedly do this computation across several, perhaps billions of time steps. Each small change in the conformation or position of a molecule requires a new evaluation of the potential energy.

2. *Many atoms in the molecule ensemble*: A large number of atoms in a molecular ensemble will greatly increase computation time especially for the double sums associated with E_{vdw} and E_{coulomb}. Note that if we do the computation as indicated, then the number of computations for these terms increases as the square of the number of atoms in the system.

Because of these issues, computational chemists have proposed several strategies designed to reduce the computational load. Before proceeding with a description of typical strategies, it should be noted that the given formula for E_{total} is itself a simplified version of other more complicated formulas that strive for higher accuracy. For example, some "Class II force fields," such as MMFF, use

$$K^{b_2}(r-\bar{r})^2 + K^{b_3}(r-\bar{r})^3 + K^{b_4}(r-\bar{r})^4 \tag{6.14}$$

instead of $K^{b_2}\left(r-\bar{r}\right)^2$ with a similar enhancement for $K^{a}\left(\theta-\bar{\theta}\right)^2$. It is also possible to replace $\frac{K^d}{2}\left[1+\cos\left(n\phi-\gamma\right)\right]$ with $\left[\mathrm{K}^{d_1}(1-\cos\phi)+K^{d_2}(1-\cos 2\phi)+K^{d_3}(1-\cos 3\phi)\right]/2$.

In an effort to simplify E_{total} even further, some research studies will assume that bond lengths and bond angles are constant. In these applications, the *change* in potential energy will be due to changes in molecular positions or conformational changes in molecules that produce variations in dihedral angles. Consequently, the modified energy calculations will be due to $E_{dihedral}$, E_{vdw}, and $E_{coulomb}$ only.

However, neglecting E_{bond} and E_{angle} will not significantly reduce computation time. Using the notation of asymptotic analysis, the number of bonds and the number of bond angles are both $\Theta(n)$, where n is the number of atoms. The most important time savings are related to computational strategies that try to reduce the calculation times of E_{vdw} and $E_{coulomb}$ since both are $\Theta(n^2)$. This is usually done by using a *spherical cut-off* strategy for E_{vdw} and $E_{coulomb}$. When the distance r_{ij} between two atoms exceeds a particular threshold, say 10 Å, then their interaction contribution to E_{vdw} or $E_{coulomb}$ is considered to be zero. The approach works quite well for E_{vdw} because the 12-6 Lennard–Jones potential approaches zero very rapidly and can be safely neglected when r_{ij} is over 9 or 10 Å. The handling of $E_{coulomb}$ requires more caution. To understand the computational trouble stemming from the $E_{coulomb}$ calculation, consider a particular atom A near the center of a large globular protein. For the sake of argument, we can consider the other atoms that interact with A to reside in concentric spherical shells of thickness 1 Å. Although the Coulomb potential falls off as the inverse of the distance, the average number of charged atoms per shell increases with the volume of the shell, a quantity that is proportional to r^2. Consequently, even though the potential of each interacting atom decreases as you get farther from A, the number of these charges goes up. As noted in [SS92], the cumulative sum over the shells does not converge to a final value in a monotonic manner. Instead, as the radius of the shells grows, the electrostatic contributions oscillate in sign. It should be stressed that even though the number of atoms in a shell increases, their net resultant is usually reduced, on average, due to cancellation effects. Consider, for example, atom A situated between two atoms of same charge but on opposite sides within the same shell. With these issues in mind, the electrostatic potential is typically regarded as a slowly decaying potential and the strategies for reducing computation time are at times somewhat controversial. So, it is not surprising to see the 1992 paper entitled "Cutoff size does strongly influence molecular dynamics results on solvated polypeptides" [SS92] followed by the 2005 paper entitled "Cutoff size need not strongly influence molecular dynamics results for solvated polypeptides" [BA05].

From a computational perspective, the value of the dielectric constant ε also requires more discussion. It is a dimensionless factor that has the value 1 when the interacting atoms are in a vacuum. When the interacting atoms are in a polarizable medium such as water, the Coulombic force is reduced because the intervening water molecules have a screening effect that weakens the electrostatic force between the two atoms. This reduction is usually modeled by increasing the value of ε to 80 when the solvent is water.* The value of 80 for the dielectric constant should be used when the space between the two charges is so large that it admits many water molecules. As this space decreases, for example, in the simulation of a ligand binding to a cleft in a protein, it is reasonable to expect that two charges would eventually have vacuum between them and so, at that time, $\varepsilon = 1$. But when and how should this change take place? Researchers typically adopt one of two methodologies for the modeling of electrostatic charge in a solvent:

1. *Explicit solvent*: The molecular ensemble explicitly includes *all* the water molecules. In this case we simply use $\varepsilon = 1$ but the calculation must include all the atoms and the extra water molecules will lead to much longer computation times.
2. *Implicit solvent*: This approach is also referred to as *continuum modeling*. The idea is to remove the water molecules, thus reducing the atom count, and this leads to a decrease in computation time. However, the dielectric effect of the water molecules is retained by replacing ε with a dielectric function (also called a screening function).

As mentioned earlier, a coarse version of the dielectric function would be $\varepsilon = 80$, but this is inappropriate for very small r_{ij} values. A more realistic screening function has a sigmoid shape. This approach has been used by Hingerty et al. [HR85] and the sigmoid function is given as

$$\varepsilon(r) = 78 - 77\left(\frac{r}{2.5}\right)^2 \frac{e^{r/2.5}}{\left(e^{r/2.5} - 1\right)^2} \tag{6.15}$$

A plot of the sigmoid dielectric function is given in Figure 6.1.

* Some research articles use 78 instead of 80.

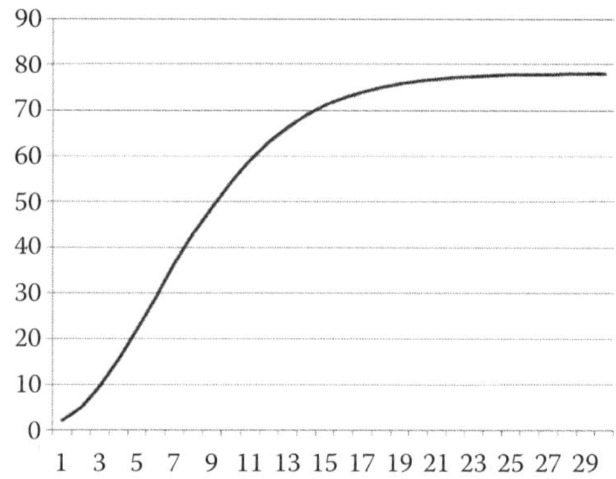

FIGURE 6.1 The dielectric function used by Hingerty et al.

6.2 SOME SIMPLE SCRIPTS TO DO ENERGY CALCULATIONS

Section 6.1 makes it clear that energy calculations are complicated and time consuming. In particular, there are several issues that relate to the trade-offs between execution time and accuracy of the computations. When a high level of accuracy is needed, one should consider the use of a force-field program such as Amber [PC03] (http://ambermd.org/). Metaphorically, this is like "flying business class" and one should be prepared to pay license fees along with the time investment necessary to develop the expertise to properly use the Amber codes. If you can manage with a less accurate energy calculation, then there are other approaches that are both fast and simple. In the following subsections, we cover some strategies that have appeared in the research literature.

6.2.1 Linear Repulsive Energy Terms

In some situations, it is suitable to use a steric *pseudo-energy function* that is just a simple linear repulsive energy term. This approach is used by Canutescu et al. [CS03], in a research study that deals with prediction of protein side-chain conformations. The function can be used for atom pairs with one atom in a side chain and the other atom in the protein backbone or in a different side chain. The definition is as follows:

$$E(r)=\begin{cases} 10 & r<0.8254R_{ij} \\ 57.273\left(1-r/R_{ij}\right) & 0.8254R_{ij}\le r\le R_{ij} \\ 0 & R_{ij}<r \end{cases} \tag{6.16}$$

where r is the interatomic distance and R_{ij} is the sum of the hard-sphere radii for atoms i and j. Typically, Cβ is treated as a backbone atom. The energy is capped at a value of 10.0 kcal/mole. They used the following radii: carbon, 1.6 Å; oxygen, 1.3 Å; nitrogen, 1.3 Å; and sulfur, 1.7 Å. A more detailed set of radii are used in the related paper by Bower et al. [BC97] (see Table 5, p. 1278). As explained in Figure 8 of [BC97], this piecewise linear function is intended to be a rough approximation of the standard Lennard–Jones potential. In particular, it is intended for atom pairs where the atoms are separated by three or more bonds. Piecewise linear energy functions can be accessed via

```
from StructBio.EnergyFunctions.LinearEnergyFunctions\
   import atomPairEnergy
from StructBio.EnergyFunctions.LinearEnergyFunctions\
   import residuePairEnergy
```

6.2.2 An Energy Function Based on the Lennard–Jones Formula

We now discuss a force-field calculation that incorporates a Lennard–Jones formula with parameters that are based on the research done by Cornell et al. [CC95]. It is a more accurate energy evaluation than that described in Section 6.2.1, but it is still somewhat incomplete because it ignores E_{bond}, E_{angle}, $E_{dihedral}$, and $E_{coulomb}$ in Equation 6.1. Consequently, it would be inappropriate for applications such as molecular dynamics simulations, but could be used for side-chain interaction energies (Chapter 7) when one wishes to have an energy function that is somewhat more sophisticated than mere collision detection (see [KI05]). The `LJ _ evaluation` class can be accessed using:

```
from StructBio.EnergyFunctions.LJ import LJ_evaluation
```

A method within this class called `atomPairEnergy(a,b)` can be used to evaluate the Lennard–Jones energy for two atoms `a` and `b`.

We now describe the computations done within the script for the `atomPairEnergy` function. Let us assume that we are evaluating E_{vdw} defined in Equation 6.9 and we need to calculate the value of

$$\frac{A_{ij}}{r_{ij}^{12}} - \frac{B_{ij}}{r_{ij}^{6}} \tag{6.17}$$

for two atoms indexed by i and j. According to the footnotes for Table 14 in [CC95], the numerators are defined using

$$A_{ij} = \varepsilon_{ij}\left(R_{ij}\right)^{12} \tag{6.18}$$

and

$$B_{ij} = 2\varepsilon_{ij}\left(R_{ij}\right)^{6} \tag{6.19}$$

where

$$\varepsilon_{ij} = \sqrt{\varepsilon_i \varepsilon_j} \tag{6.20}$$

and

$$R_{ij} = R_i + R_j \tag{6.21}$$

The R_i value represents the van der Waals radius of atom i in Å and have been recorded in the data structure `radiusMap _ D`. The ε_i value represents the van der Waals well depth for atom i in kilocalorie per mole. These values are stored in the `epsMap _ D` data structure. Both sets of values were taken from the final entries in Table 14 of [CC95]. Because the atom names used in [CC95] are somewhat different from the atom labels used by Chimera, it was necessary to define a map that would convert the latter to the former. This is done by means of the dictionary `elementNameMap _ D`.

6.2.3 Computing Energy Using Amber Parameters in Chimera

Chimera allows the user to do an energy minimization for molecular structures. This can be invoked by using the menu item: **Tools… Structure Editing… Minimize Structure**. This will initiate a steepest descent minimization to eliminate steric clashes followed by a conjugate gradient minimization to reach a local energy minimum (not necessarily the global energy minimum).

To evaluate the energy of a molecule, one can do a "zero-step minimization." This will calculate the initial energy of the system and will not alter the positions of any atoms in the structure. Minimization routines are provided by the Molecular Modeling Toolkit (MMTK) [Hi00], which is included with the Chimera download. The code uses Amber parameters (http://ambermd.org/) for standard residues. For nonstandard residues, the code uses Amber's Antechamber module [WW06], which is also included with Chimera.

The function evaluateAmberEnergy can be found in the file AmberForceField.py (Chapter 6 scripts), which provides a simple example. It fetches a protein in Protein Data Bank (PDB) format and performs an energy calculation. The script is as follows:

```
import chimera
from MMMD import base

def evaluateAmberEnergy(molecules_L, excludedResidues = set()):

    def run(minimizer):
       minimizer.run()

    minimizer = base.Minimizer(molecules_L, nsteps = 0, cgsteps = 0,
                     nogui = True, addhyd = True,  callback = run,
                     exclres = excludedResidues, cache = False,
                     prep = True)

    return minimizer._mi.universe.energy()

#=====================================================
# Mainline:

pdbID = raw_input("Type 4 character PDB ID: \n")
prot = chimera.openModels.open(pdbID, type="PDB")[0]
energyAmber = evaluateAmberEnergy([prot])
print energyAmber
```

Here are some notes to clarify the parameter settings for the base.Minimizer call:

- molecules _ L is a list of molecule objects. Note that the call to evaluateAmberEnergy uses [prot] and not simply prot.
- The settings for both nsteps and cgsteps should be 0. Note that we set nsteps = 0 because we only want the initial energy evaluation.
- nogui set to True will suppress the appearance of tool dialogs that might otherwise appear.
- addhyd set to True will add hydrogen atoms to the molecules. This is necessary for an accurate assessment of the energy.
- callback = run will run the minimizer as a final step in the instantiation of the Minimizer object.
- exclres = excludedResidues specifies a set of excluded residues. In the given script, the default value for this is an empty

set, which would specify that no residues will be excluded from the energy calculation. It is established as an argument in the call to `evaluateAmberEnergy` in case the user wishes to exclude residues from the energy evaluation.

- `cache = False` will speed up the calculation because it prevents the caching of parameters such as partial charges. Since we are not allowing the calculations to progress over several steps of minimization, caching is not required.
- `prep = True` will initiate a call to Dock Prep. This will delete water molecules, repair truncated side chains, add in the hydrogen atoms, and assign partial charges. For more details you may consult the Dock Prep documentation on the Chimera website: http://www.cgl.ucsf.edu/chimera/docs/ContributedSoftware/dockprep/dockprep.html.

The energy calculation provided by `evaluateAmberEnergy` will be much more realistic than those provided by the two functions described in Sections 6.2.1 and 6.2.2. Evaluation is fairly fast but occasionally you may experience lengthy execution times, especially when the scene contains nonstandard residues (e.g., 1HIV). It should be noted that the order of hydrogen atom addition is dependent on the use of a Python dictionary (essentially a hash table) within Chimera and so the order of hydrogen atom placement is not completely deterministic. Consequently, you might run this script repeatedly and get slightly different results. For example, three successive runs, each applied to the protein 1CRN, gave energy evaluations of –3471.23, –3470.99, and –3467.21.

6.3 EXERCISES

6.3.1 Comparing Lennard–Jones and Piecewise Linear Energy Functions

Doing this exercise will give you some practice in the implementation of a modeless dialog that contains a Tkinter canvas with a plot. Using `matplotlib` will give you more experience in the generation of plots. Recall that Section A.7 of Appendix A gives some guidelines for the embedding of a plot into a modeless dialog.

Consider the following table that gives estimates of radii for various atom types:

Atom Type	Clash Radius	LJ Radius
Backbone N	1.1	1.82
Backbone CA	1.3	1.91
Backbone C	1.3	1.91
Backbone O	1.1	1.66
ASP, ASN OD1	0.7	1.66
LYS NZ	0.4	1.88
ARG CZ	0.8	1.91
ARG NH2	0.4	1.82

Entries for the middle column are taken from Table 5 in [BC97]. The rightmost column has van der Waals radii taken from Table 14 in [CC95].

Given two atom types, we can calculate R_{ij} in Equation 6.16 as the sum of two radii taken from the middle column. We can then plot the piecewise linear energy function $E(r)$ defined by Equation 6.16. In a similar manner, we can use two radii taken from the last column to compute the R_{ij} value defined by Equation 6.21 and subsequently plot the Lennard–Jones energy function

$$E(r) = \frac{A_{ij}}{r^{12}} - \frac{B_{ij}}{r^{6}} \tag{6.22}$$

which is simply Equation 6.17 with an atom separation distance of r. As noted in Section 6.2.2, the `LJ _ evaluation` class contains the dictionaries that can be used to calculate both A_{ij} and B_{ij}.

The goal of this exercise is to write a script that displays a graphical user interface with two columns of radio buttons allowing a user to select two atom types. After both atom types are chosen, the dialog should display a plot of the two energy functions as shown in Figure 6.2.

The plot should display the functions using the format shown in Figure 8 of [BC97]. More precisely, we want to use the ratio $r/r0$ as the x axis for the plot as shown in Figure 6.2. The value of $r0$ is the sum of the radii used in the Lennard–Jones calculation (see earlier note about Equation 6.21). Here is a suggested approach:

1. Define a list of ratios using: `ratio_L = arange(0.1, 2.01, 0.01)`
2. The r values for the atomic separation distances can be taken from a list that is computed as: `r_L = ratio_L * r0LJ` where `r0LJ`

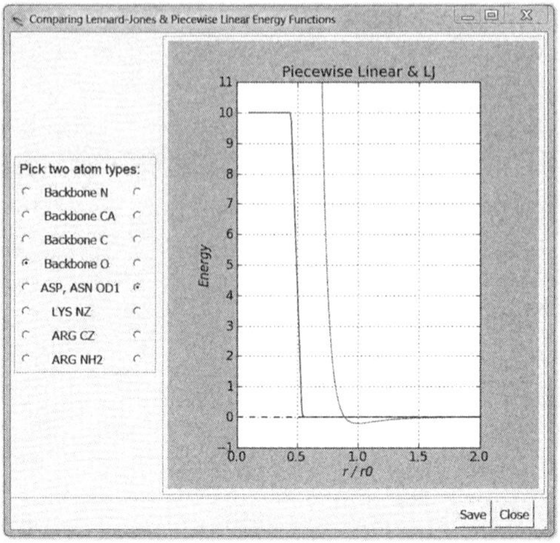

FIGURE 6.2 The user interface for Exercise 6.1.

is the *r*0 value corresponding to the sum of the radii used in the Lennard–Jones calculation (same as *r*0 in the previous paragraph).

3. To generate the plot, these *r* values are used as the values of the independent variable for *both* energy functions.

REFERENCES

[BA05] D. A. C. Beck, R. S. Armen, and V. Daggett. Cutoff size need not strongly influence molecular dynamics results for solvated polypeptides. *Biochemistry*, **44** (2005), 609–616.

[BC97] M. J. Bower, F. E. Cohen, and R. L. Dunbrack Jr. Prediction of protein side-chain rotamers from a backbone-dependent rotamer library: a new homology modeling tool. *Journal of Molecular Biology*, **267** (1997), 1268–1282.

[CC95] W. D. Cornell, P. Cieplak, C. I. Bayly, I. R. Gould, K. M. Merz Jr., D. M. Ferguson, D. C. Spellmeyer, T. F. Fox, J. W. Caldwell, and P. A. Kollman. A second generation force field for the simulation of proteins, nucleic acids, and organic molecules. *Journal of the American Chemical Society*, **117** (1995), 5179–5197.

[CCB95] P. Cieplak, W. D. Cornell, C. I. Balyl, and P. A. Kollman. Application of the multimolecule and multiconformation RESP methodology to biopolymers: derivation for DNA, RNA and proteins. *Journal of Computational Chemistry*, **16** (1995), 1357–1377.

[CS03] A. C. Canutescu, A. A. Shelenkov, and R. L. Dunbrack Jr. A graph-theory algorithm for rapid protein side-chain prediction. *Protein Science*, **12** (2003), 2001–2014.

[Hɪ00] K. Hinsen. The molecular modeling toolkit: a new approach to molecular simulations. *Journal of Computational Chemistry*, **21** (2000), 79–85.

[HR85] B. E. Hingerty, R. H. Ritchie, T. L. Ferrell, and J. E. Turner. Dielectric effects in biopolymers: the theory of ionic saturation revisited. *Biopolymers*, **24** (1985), 427–439.

[Kɪ05] C. L. Kingsford. *Computational approaches to problems in protein structure and function*. PhD thesis, Princeton University, Princeton, NJ, 2005, AAI3188669.

[Mᴀ98] A. D. MacKerell Jr. Protein force fields. In: *Encyclopedia of Computational Chemistry*. P. v. R. Schleyer, N. L. Allinger, T. Clark, J. Gasteiger, P. A. Kollman, H. F. Schaefer III, and P. R. Schreiner (editors). John Wiley, Chichester, UK, 1998.

[Pᴄ03] J. W. Ponder and D. A. Case. Force fields for protein simulations. *Advances in Protein Chemistry*, **66** (2003), 27–85.

[Sᴄ02] T. Schlick. *Molecular Modeling and Simulation: An Interdisciplinary Guide*. Springer-Verlag, New York, 2002.

[SS92] H. Schreiber and O. Steinhauser. Cutoff size does strongly influence molecular dynamics results on solvated polypeptides. *Biochemistry*, **31** (1992), 5856–5860.

[WW06] J. Wang, W. Wang, P. A. Kollman, and D. A. Case. Automatic atom type and bond type perception in molecular mechanical calculations. *Journal of Molecular Graphics and Modelling*, **25** (2006), 247–260.

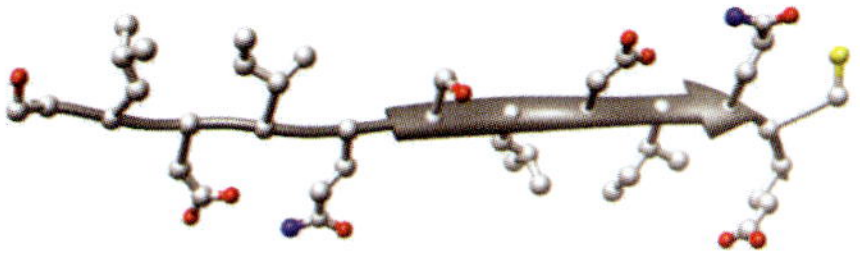

FIGURE 1.36 What is the Protein Data Bank ID?

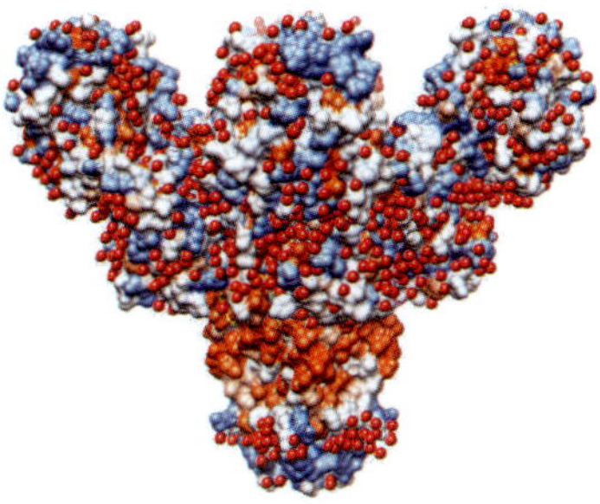

FIGURE 2.8 Potassium channel 1K4C with all chains in place.

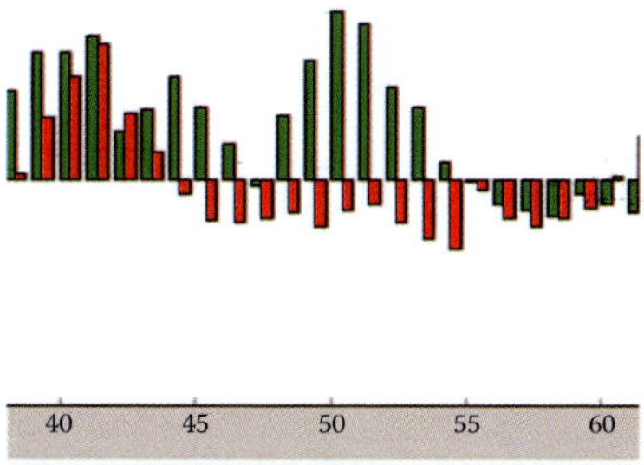

FIGURE 2.9 A section of the plot comparing the B-factors.

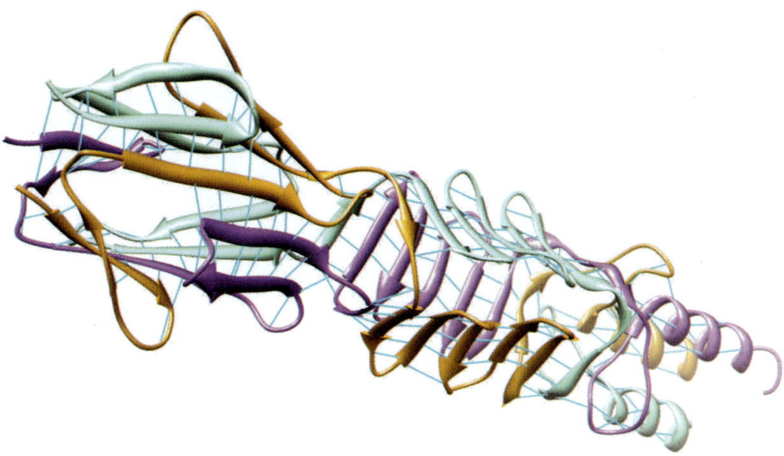

FIGURE 2.13 The autotransporter adhesin protein from Bartonella (PDB ID = 3D9X).

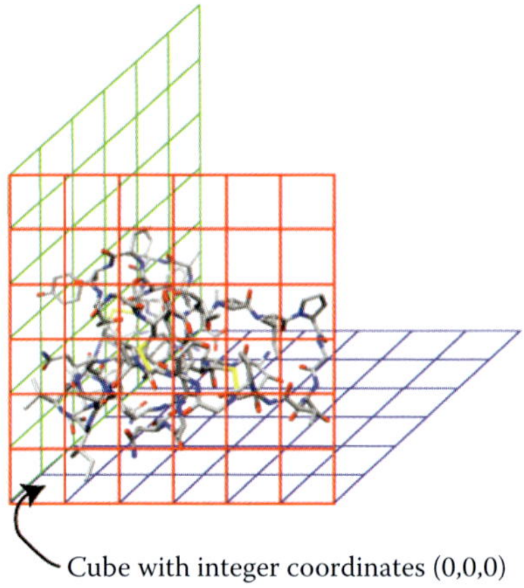

FIGURE 3.2 Three sides of the bounding box.

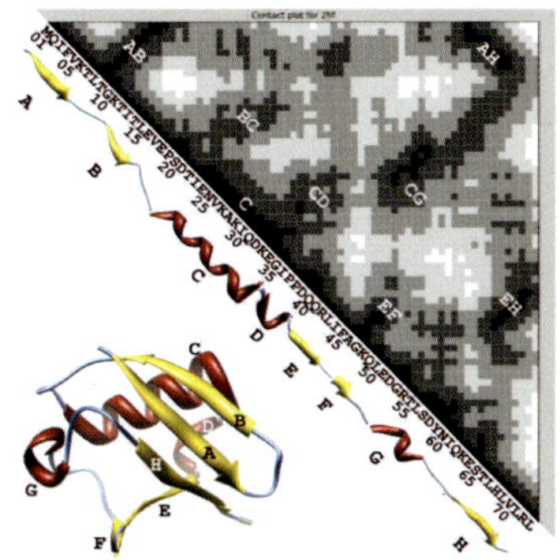

FIGURE 3.4 Distance map for chain A of 2FIF.

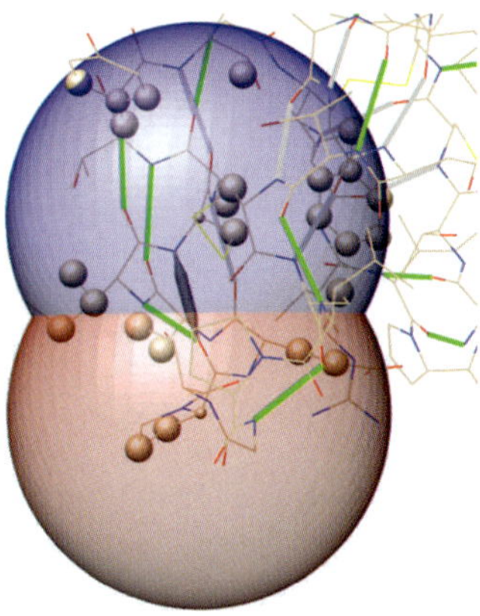

FIGURE 3.8 Well-wrapped hydrogen bond.

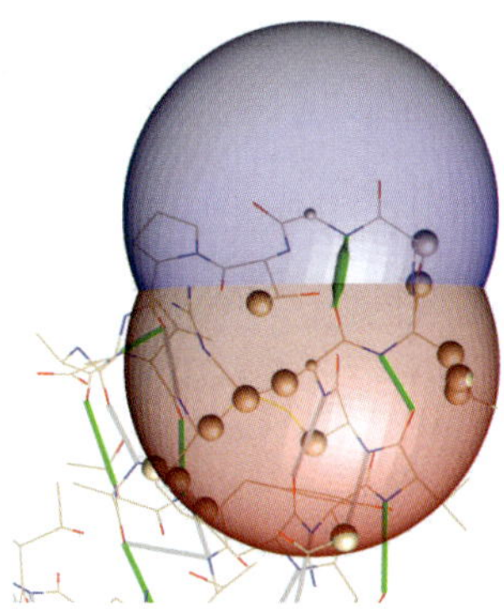

FIGURE 3.9 Dehydron: a deficiently wrapped hydrogen bond.

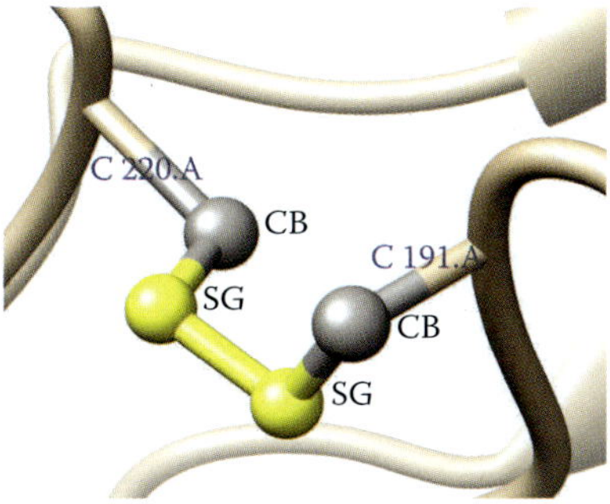

FIGURE 3.10 Disulfide bridge in chain A of 1BQY.

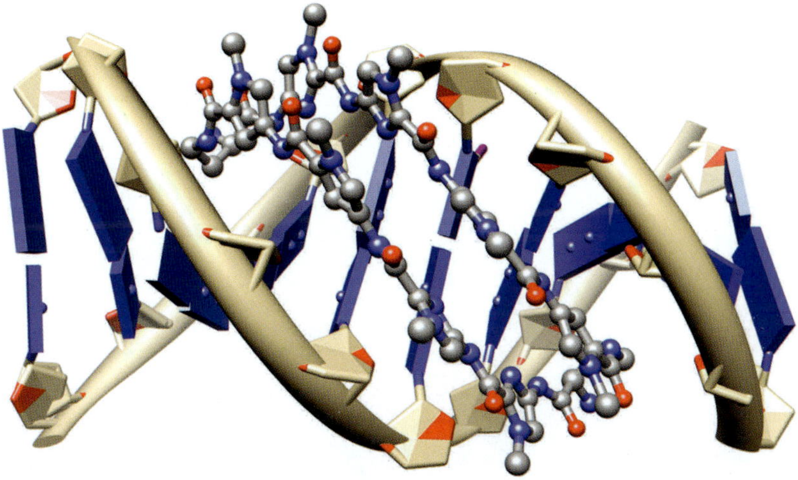

FIGURE 3.12 DNA segment and polyamide ligand (Protein Data Bank identifier: 3I5L).

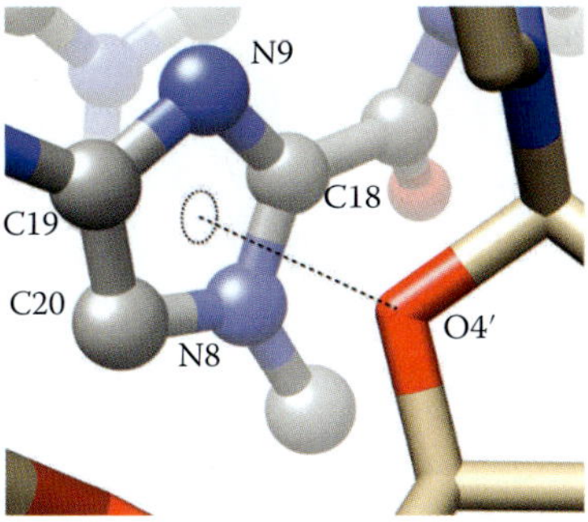

FIGURE 3.13 Distance from centroid of imidazole ring to O4′.

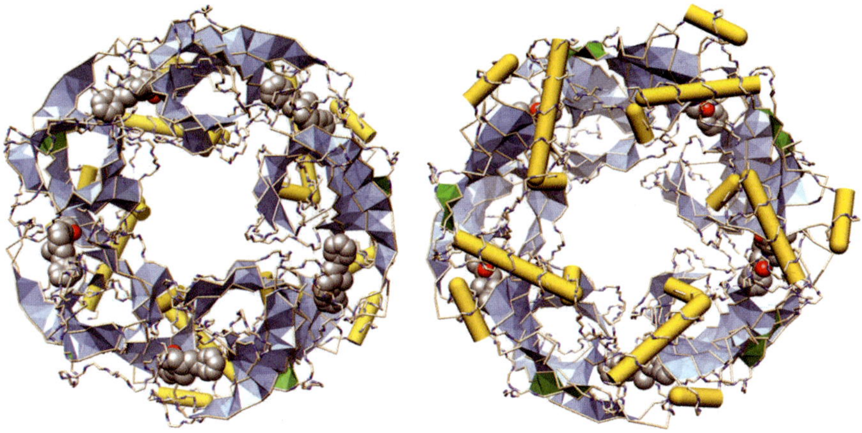

FIGURE 3.17 Plates and tubes for 2BYS.

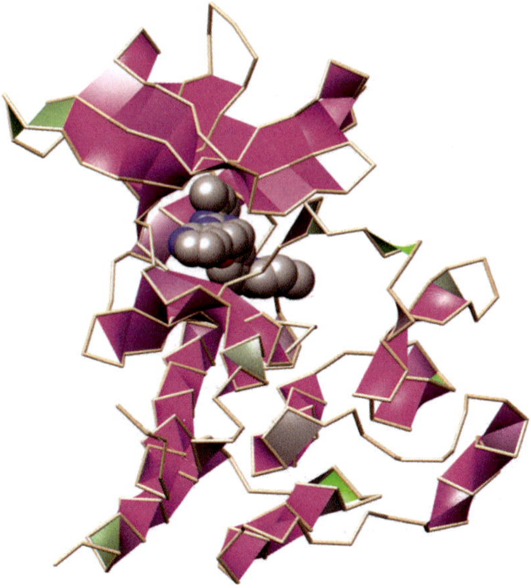

FIGURE 3.21 Hydrogen bond plates colored with respect to wrapper atom count (1T46).

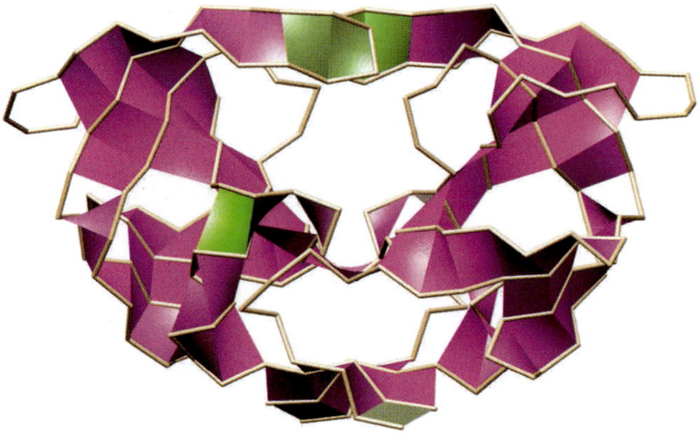

FIGURE 3.22 HIV protease without substrate (1HIV with ligand removed).

FIGURE 3.23 HIV protease with substrate (1F7A).

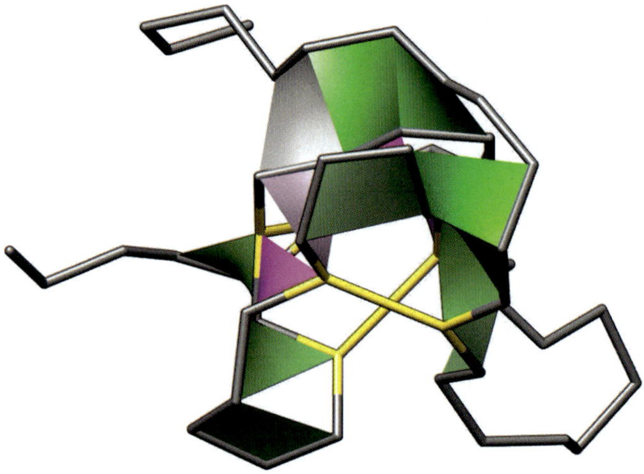

FIGURE 3.24 Hirudin (1HIC).

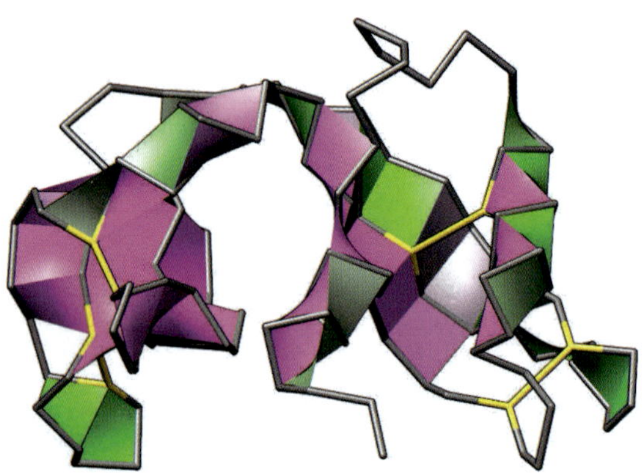

FIGURE 3.25 Ribonuclease A (1RHB).

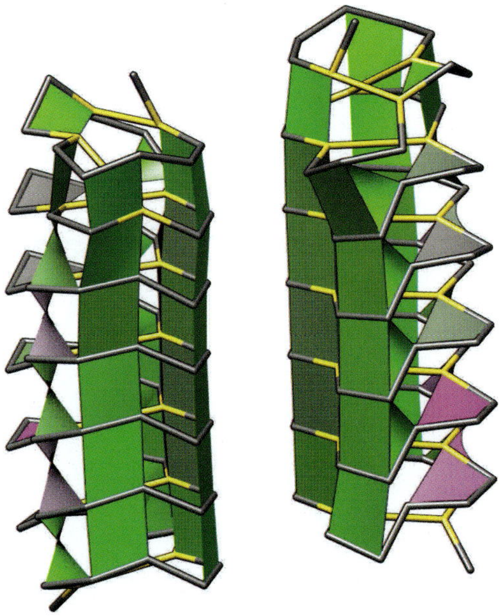

FIGURE 3.26 *Tenebrio molitor* antifreeze protein (1EZG).

FIGURE 3.27 Snow flea antifreeze protein (2PNE).

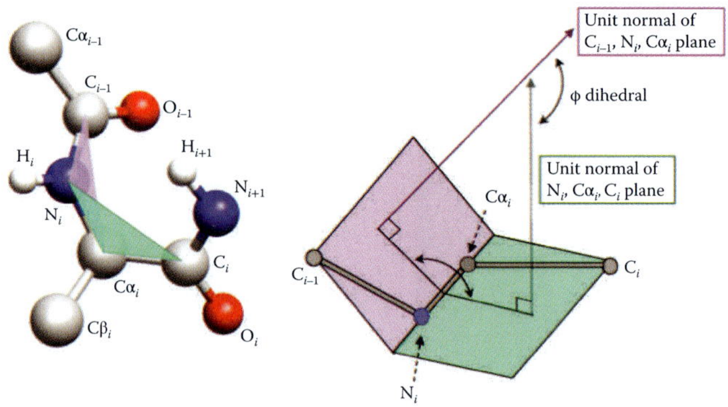

FIGURE 4.2 Phi dihedral angle.

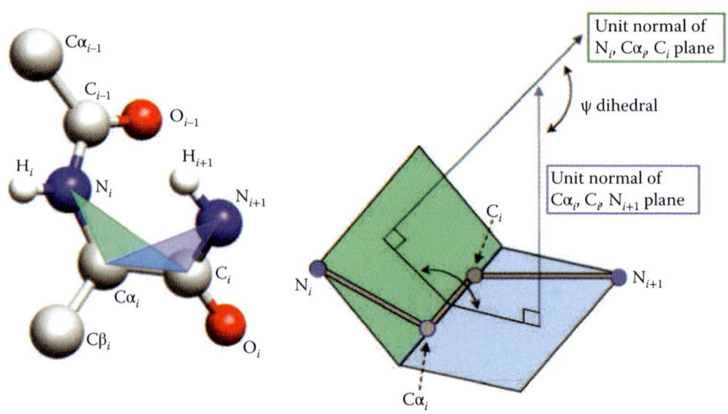

FIGURE 4.3 Psi dihedral angle.

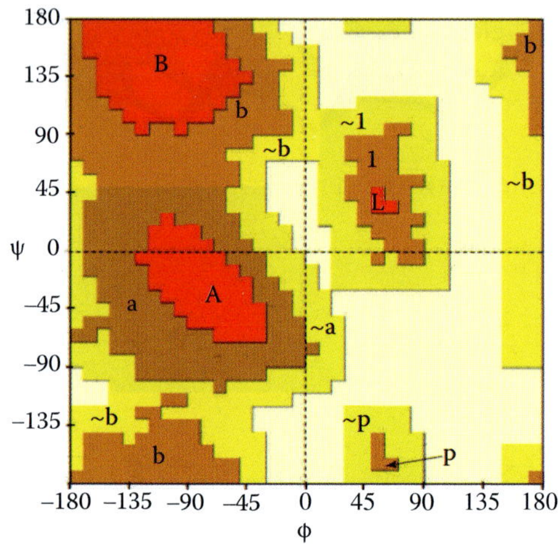

FIGURE 4.7 Ramachandran plot for 121,870 residues (463 protein structures).

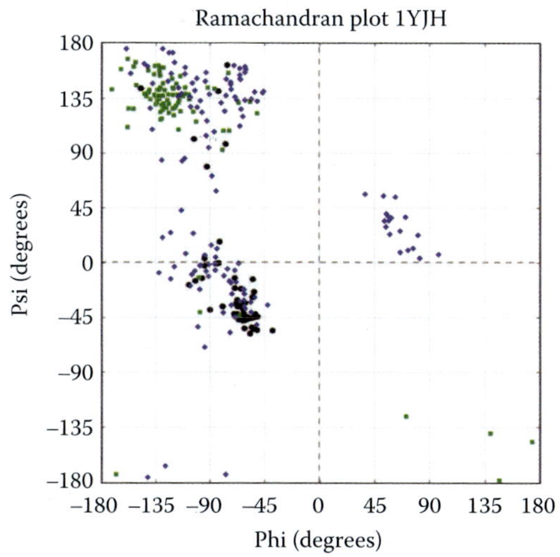

FIGURE 4.13 Ramachandran plot for 1YJH.

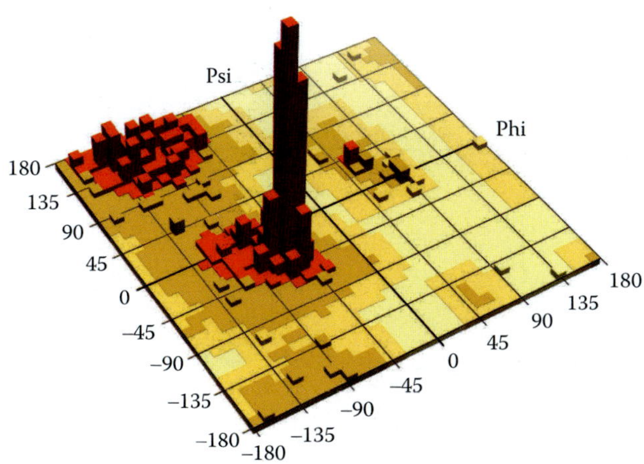

FIGURE 4.15 Three-dimensional Ramachandran plot for 1PTM_A.

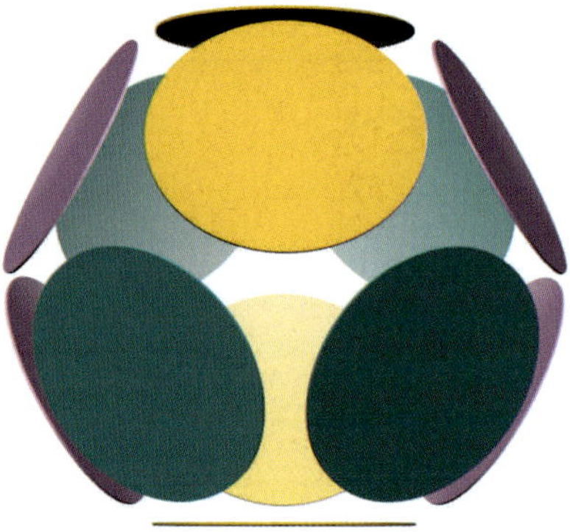

FIGURE 4.22 Dodecahedral disks (testing the addDisk function in the Solids class).

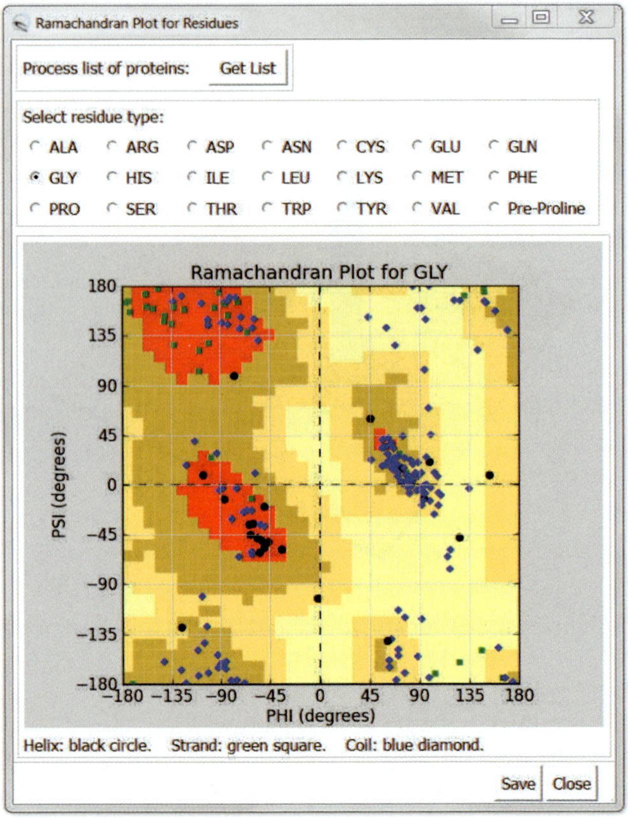

FIGURE 4.27 A GUI for Ramachandran plots applied to residues.

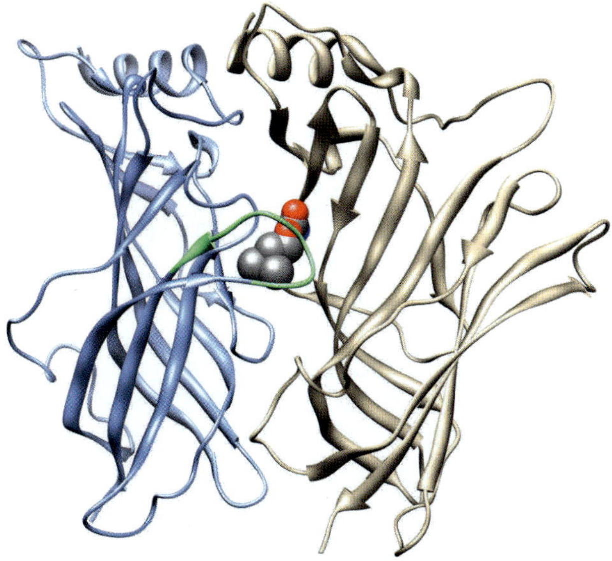

FIGURE 5.17 The nicotinic acetylcholine binding protein binding site.

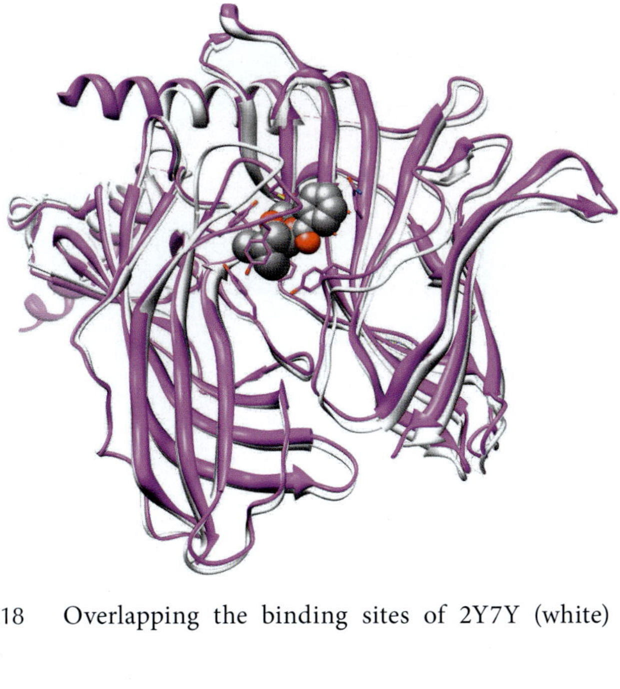

FIGURE 5.18 Overlapping the binding sites of 2Y7Y (white) and 2PGZ (magenta).

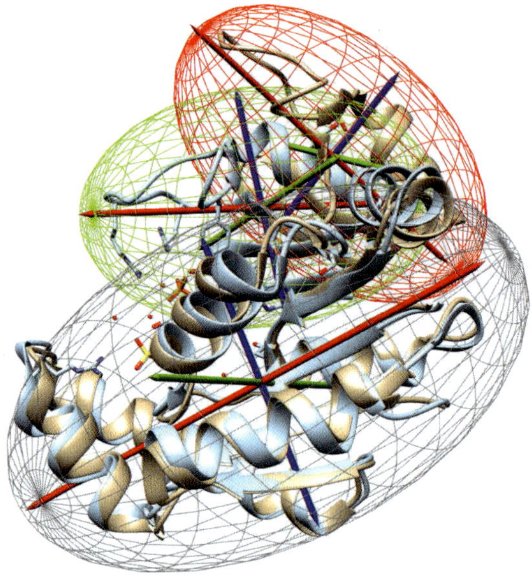

FIGURE 5.19 Domain movement for guanylate kinase (1EX6B and 1EX7A).

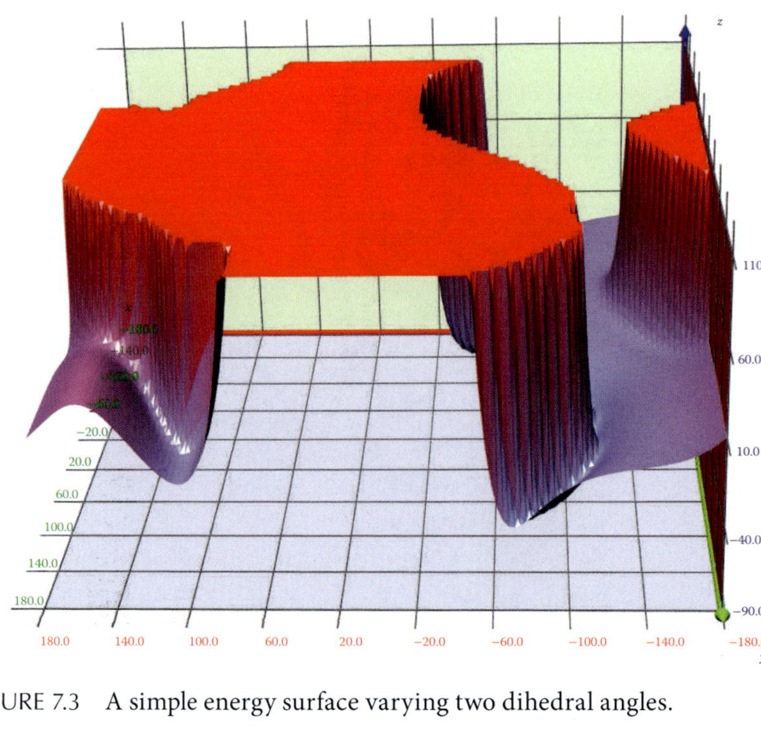

FIGURE 7.3 A simple energy surface varying two dihedral angles.

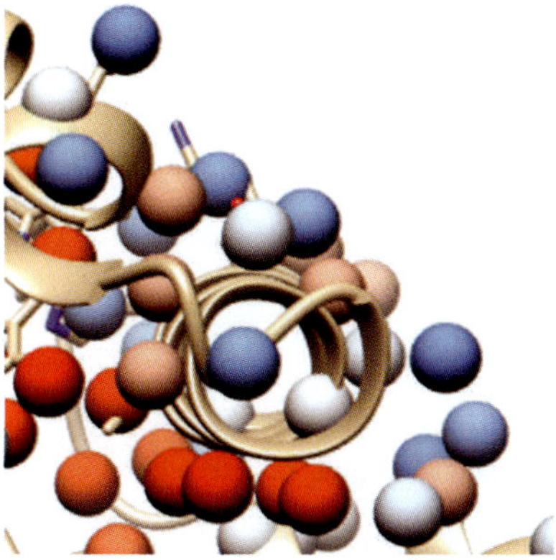

FIGURE 8.2 Amphiphilicity.

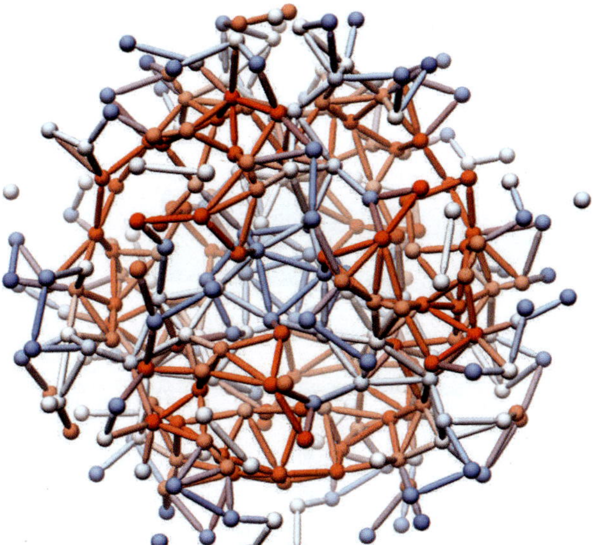

FIGURE 8.3 Network of neighboring residues (1ZNJ).

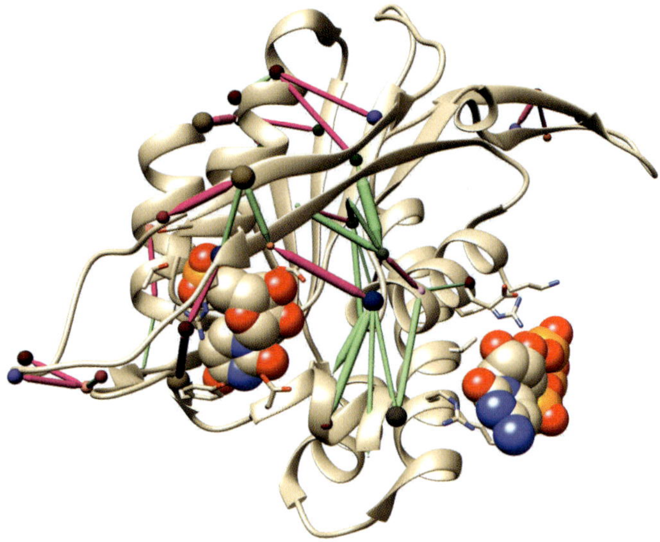

FIGURE 8.8 Graph visualization of a contact rearrangement network (1XTU and 1XTT).

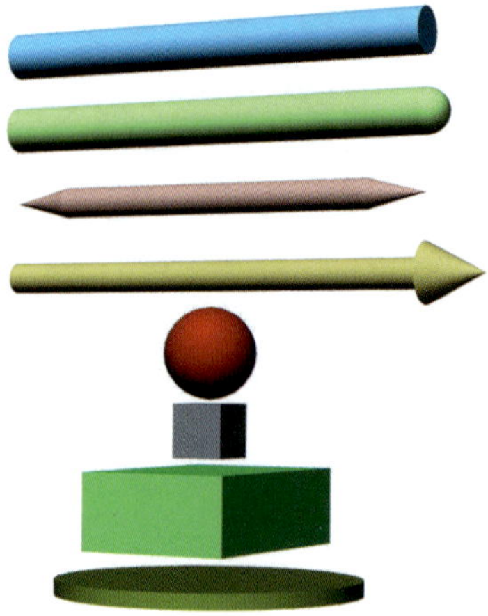

FIGURE B.1 Some examples of the available three-dimensional (3D) objects from the Solids class.

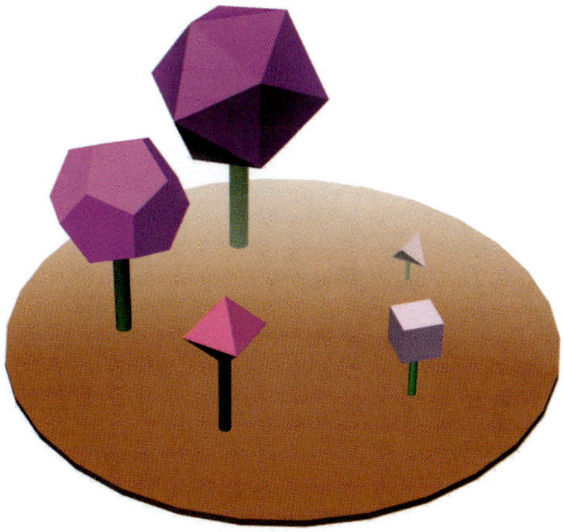

FIGURE B.2 A Platonic garden.

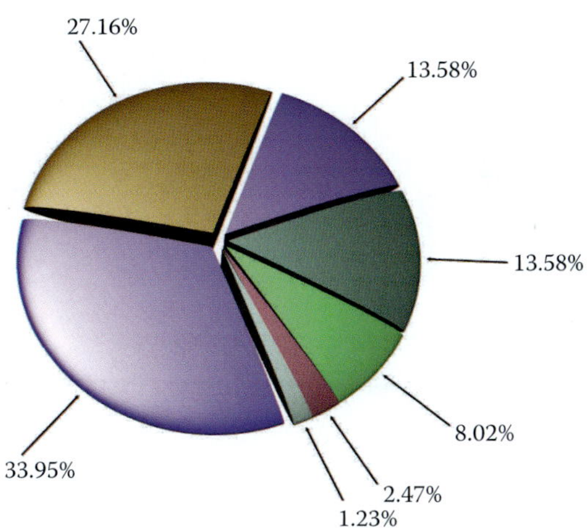

FIGURE B.11 Pie base curves.

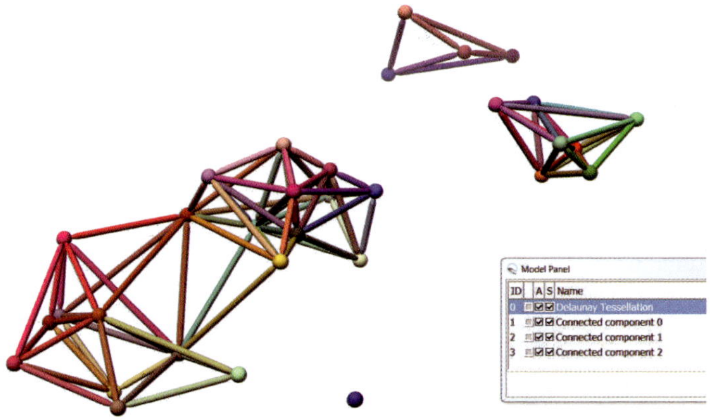

FIGURE C.6 Connected components produced by Script C.5.

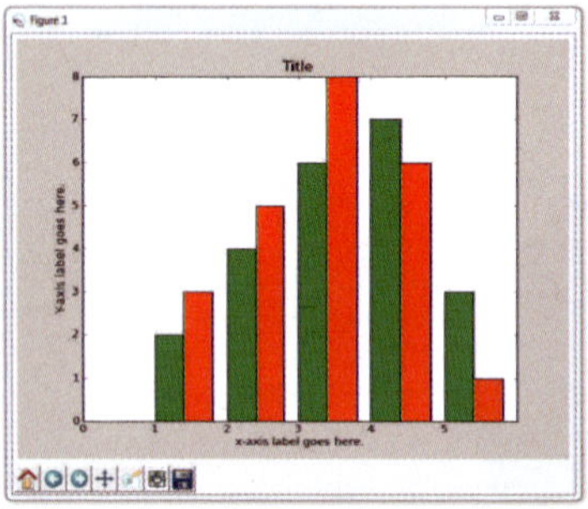

FIGURE D.1 Bar chart from Script D.01.

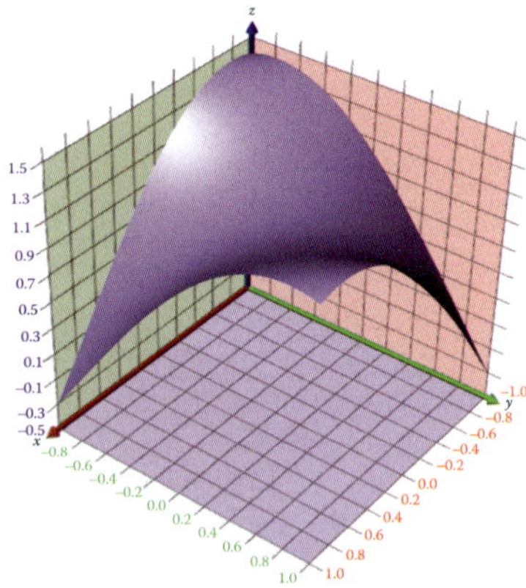

FIGURE D.9 Surface plot in the Chimera display.

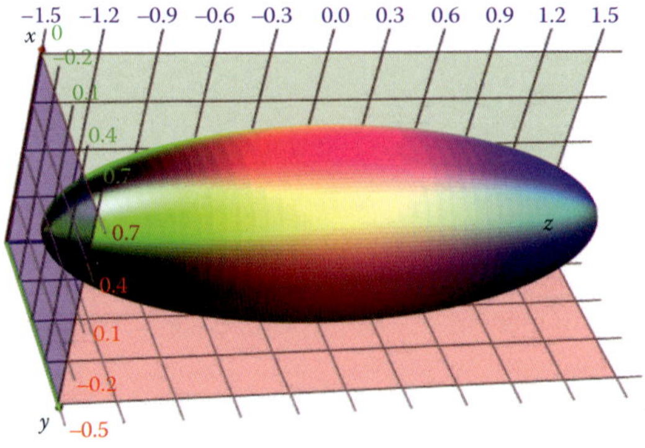

FIGURE D.10 Surface plot of an ellipsoid in the Chimera display.

CHAPTER 7

Rotamers and Side-Chain Conformation

7.1 MOTIVATION

The difficult and recalcitrant problem of protein structure prediction is often considered to be the "Holy Grail" of structural bioinformatics. The problem can be stated in simple terms: Given the primary sequence of a protein, predict the tertiary conformation that achieves a global energy minimum. There are several approaches to the problem. For an overview of the various techniques, the reader may consult Tramontano [TR05]. Some of the more successful strategies rely on a *fold recognition* step. This starts by comparing the given primary sequence with protein sequences in the Protein Data Bank (PDB). The idea is to find one or more protein folds that can act as possible structural templates for the given protein with unknown structure. Various tactics can be used to select the most promising fold followed by techniques for structural modifications. If all goes well, the researcher will get an accurate evaluation of the three-dimensional (3D) coordinates of the backbone atoms. The final step is to determine the positions of atoms in the side chains. This is often referred to as the *side-chain packing problem* (see [DT05] and the references therein). Side-chain packing is important for other applications such as ligand docking, protein design, and studies dealing with protein allostery.

7.2 SIDE-CHAIN PACKING: COMPUTATIONAL ISSUES

The problem is known to be NP-hard [AK97]. Consequently, the algorithms for side-chain packing are typically heuristic in nature and do not guarantee a conformation that will achieve a global energy minimum for the entire protein.

Let us restate the problem in terms of the inputs and expected output of an algorithm that attempts to solve the problem. Since the 3D conformation of the backbone is known, we would have the coordinates of all the alpha carbon atoms, each acting as a point of attachment for a particular side chain. A reasonable approximation is that both the bond lengths and geminal distances within a side chain can be assessed by simply copying these values from side-chain structures previously determined by x-ray analysis. With these assumptions in play, deriving the conformation of a side chain reduces to the determination of the dihedral angles* within the side chain. In summary, the algorithm should determine the settings of the dihedral angles so that the side-chain atoms have 3D positions that will (ideally) yield a global energy minimum.

The number of dihedral angles ranges from 0 to 5 inclusive. Arginine is the only side chain with more than four dihedral angles. This is illustrated in Figure 7.1, which shows the positions of all five dihedral angles and their labels: χ_1, χ_2, χ_3, χ_4, and χ_5. Chimera allows one to access dihedral angles of a residue, but only the first four are available. In a Python script, these would be referenced as `res.chi1`, `res.chi2`, `res.chi3`, and `res.chi4` for a given residue object `res`. Since the `chi1` dihedral angle is defined by four atoms (usually N, CA, CB, and CG), it is to be expected that glycine and alanine do not have dihedral angles. Table 7.1 specifies the atoms used for dihedral angle definitions.

If the minimum energy of a protein is dependent on a particular setting of all the dihedral angles, then it should be clear that the domain of investigation is a space with a very large dimension. Consider a small protein such as 1CRN with 46 residues in its primary sequence: TTCCPSIVARSNFNVCRLPGTPEAICATYTGCIIIPGATCPGDYAN. There are 65 dihedral angles that would have to be determined. An exhaustive search, for example, trying out all possible angle settings in 6° increments would require $60^{65} = 3.8*10^{115}$ evaluations that would not

* We will use the terms *dihedral* and *torsional* interchangeably.

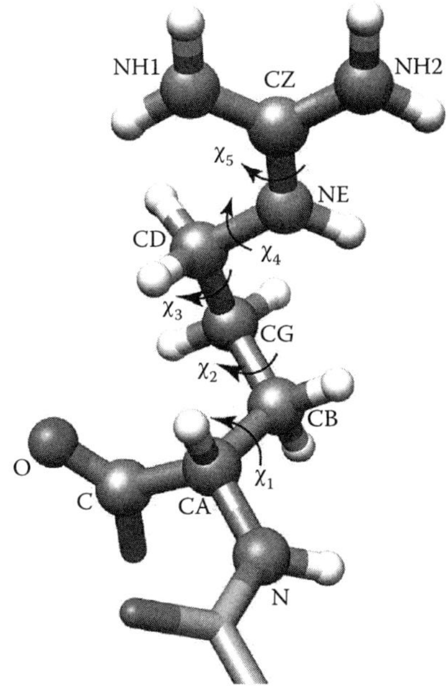

FIGURE 7.1 Side-chain dihedral angles for arginine.

be possible. Some type of gradient descent algorithm working with the energy function is another reasonable attack, but the energy surface is a very rough landscape and there is no guarantee that the global energy minimum will be found.

7.3 ROTAMERICITY

Even though our assumptions about predetermined bond lengths and bond angles have somewhat simplified the problem, we will use a further approximation in an effort to reduce the computational load. This approximation is a discretization of side-chain conformations that capitalize on the phenomenon of rotamericity: it has been observed that a side chain will typically have a tendency to adopt a conformation that is close to a member of small set of possible conformations. Each such conformation is referred to as a *rotamer.*

Various studies [BC97], [DK93], and [SD11a] have produced libraries that store the rotamer sets for each of the amino acids (see also [Du02]). The number of rotamers in a set will depend on the number of dihedral angles in the side chain. The more complicated side chains have rotamer sets with as many as 81 members. A library entry will specify both the

TABLE 7.1 Atoms Defining the Dihedral Angles in Side Chains

	χ_1	χ_2	χ_3	χ_4	χ_5
ARG	N-CA-CB-CG	CA-CB-CG-CD	CB-CG-CD-NE	CG-CD-NE-CZ	CD-NE-CZ-NH1
ASN	N-CA-CB-CG	CA-CB-CG-OD1			
ASP	N-CA-CB-CG	CA-CB-CG-OD1			
CYS	N-CA-CB-SG				
GLN	N-CA-CB-CG	CA-CB-CG-CD	CB-CG-CD-OE1		
GLU	N-CA-CB-CG	CA-CB-CG-CD	CB-CG-CD-OE1		
HIS	N-CA-CB-CG	CA-CB-CG-ND1			
ILE	N-CA-CB-CG1	CA-CB-CG1-CD1			
LEU	N-CA-CB-CG	CA-CB-CG-CD1			
LYS	N-CA-CB-CG	CA-CB-CG-CD	CB-CG-CD-CE	CG-CD-CE-NZ	
MET	N-CA-CB-CG	CA-CB-CG-SD	CB-CG-SD-CE		
PHE	N-CA-CB-CG	CA-CB-CG-CD1			
PRO	N-CA-CB-CG	CA-CB-CG-CD			
SER	N-CA-CB-OG				
THR	N-CA-CB-OG1				
TRP	N-CA-CB-CG	CA-CB-CG-CD1			
TYR	N-CA-CB-CG	CA-CB-CG-CD1			
VAL	N-CA-CB-CG1				

conformation and the frequency of its occurrence, usually expressed as a probability of occurrence.

The notion of rotamericity rests on the evidence that the side-chain dihedral angles occur in tight clusters around certain values. Typically, the dihedral angle space is subdivided into bins and a statistical analysis is applied to determine the average conformation in each bin. There is a computational trade-off involved in this discretization. If the statistical binning uses a larger number of rotamers, we will get a higher level of accuracy for the final calculated conformation but there will be a much higher computational cost that grows in a combinatorial fashion with the sizes of the rotamer sets.

Rotamer libraries may be backbone independent or backbone dependent. In the latter case, choice of a rotamer set will be guided by taking the local backbone conformation into account. Otherwise, no backbone information is used and the rotamer set is chosen only by the specification of the side-chain type.

7.4 ACCESSING ROTAMERS USING A PYTHON SCRIPT

Chimera allows one to access a rotamer library provided by the Dunbrack Lab* [SD11b]. Let us start with a simple script that will fetch a protein file and subsequently print out, for each residue, its position in its chain, the residue type, the number of rotamers available for this side chain, and whether the rotamers are backbone dependent (Script 7.01 in the file: `Script 7_01 Rotamer counts for each side-chain position.py`). For example, the first 10 lines for 1CRN would be

```
Residue     Type Rotamer Backbone
position            count    dependent
      1     THR       3         False
      2     THR       3         True
      3     CYS       3         True
      4     CYS       3         True
      5     PRO       2         True
      6     SER       3         True
      7     ILE       9         True
      8     VAL       3         True
      9     ALA       0
     10     ARG      81         True
```

* Chimera only offers the older Dunbrack library due to licensing restrictions on the new one.

Note that the rotamer count for alanine is zero (as expected), and since there are no rotamers for this residue there is no indication as to whether they are backbone dependent. The script is as follows:

```
import chimera, Rotamers
from Rotamers import getRotamers

targetPath = raw_input("Type 4 character PDB ID: \n")
prot = chimera.openModels.open(targetPath, type="PDB")[0]
print "Residue  Type Rotamer Backbone"
print "position       count  dependent"
for res in prot.residues:
    if res.type in ("ARG","ASN","ASP","CYS","GLN","GLU",
                    "HIS","ILE","LEU","LYS","MET","PHE",
                    "PRO","SER","THR","TRP","TYR","VAL"):
        (bbd_flag, rot_L) = getRotamers(res)
        print "  %3d    " % res.id.position, res.type, \
              "  %2d     " % len(rot_L), bbd_flag
    if res.type in ("GLY", "ALA"):
        print "  %3d    " % res.id.position, res.type, \
              "  0     "
```

Let us examine the most relevant lines. The first line imports the `Rotamers` module and the next line imports, from that module, the function that is needed to bring in the rotamers. Note that the script will only attempt to get rotamers if the side chain is a standard residue and not glycine or alanine. A list of rotamers is accessed by means of the statement:

```
(bbd_flag, rot_L) = getRotamers(res)
```

Note that the return value of the `getRotamers` function is a tuple. The first entry is a Boolean value indicating whether the rotamers are backbone dependent, and the second entry is a list of molecule objects representing the rotamers.

Now suppose we want to see the various conformations of the rotamers for a particular residue site. The script in the file named `Script 7_02 Iterating through rotamers.py` will ask the user to specify a PDB ID and then to select a residue from the display:

```
import chimera, Rotamers
from chimera import runCommand, selection
from Rotamers import getRotamers, useRotamer, NoResidueRotamersError

# ------------------------------------------------------------------
def userSelectsResidue(focus = False):
```

```
    print  "Select an atom or a standard residue in the display."
    raw_input("Then bring the cursor back to this line and hit Enter.")
    atm_L = selection.currentAtoms()
    if len(atm_L) == 0: return None
    res = atm_L[0].residue
    if focus:
        runCommand("sel :" + str(res.id.position) + "." +
                   res.id.chainId + " za<8.0; focus sel; ~sel")
    return res

#==================================================================
# Mainline:

targetPath = raw_input("Type 4 character PDB ID: \n")
prot = chimera.openModels.open(targetPath, type="PDB")[0]

runCommand("set bg_color white; ~ribbon; repr stick; show")

selR = userSelectsResidue(focus = True)
# Use first residue if user does not do a select:
if not selR: selR = prot.residues[0]
resStr = ":" + str(selR.id.position) + "." + selR.id.chainId

runCommand("repr bs " + resStr)

print "Use the Enter key to step through the rotamers."
print "Type any character then Enter to terminate the script.\n"
print selR.type, "residue " + resStr + " selected."
print "Rotamer  Rotamer "
print "index:   probability:"

try:
    rot_L = getRotamers(selR)[1]
    numRots = len(rot_L)
    nullChar = ""
    rotIx = 0
    while nullChar == "":
        useRotamer(selR, [rot_L[rotIx]])
        printLine = "%2d    %7.5f" % (rotIx, rot_L[rotIx]. rotamerProb)
        nullChar = raw_input(printLine)
        rotIx += 1
        rotIx = rotIx%numRots
        if rotIx == 0: print ""

except NoResidueRotamersError:
    print selR.id.position, selR.type, "No rotamers"
```

The script introduces more rotamer code and a small function that may be useful for other applications as well. The mainline part of the script fetches the PDB file specified by the user. The display is set up to show all the atoms in a stick representation. Relevant notes:

- By way of the `userSelectsResidue` function, the script issues a request for the user to select a residue in the display. The user can do this by selecting a residue from the sequence window, by selecting a residue via the **Select** menu, or by simply selecting an atom. To avoid any complicated interaction with a user who fails to select a single residue, the function simply extracts the first atom from the set of selected atoms and works with the residue that contains this atom. Note that the user will be interacting with both the shell window and the display window. To reduce any confusion that this might entail, the script provides some advice about cursor placement so that the appropriate window is active for further user interaction. If the function is called with `focus` set to `True`, the display changes so that the selected residue is centered in the display. Additional rotations of the molecule can be performed to get a better view of the residue site with the understanding that we must subsequently make the shell window active to resume user interaction.

- To make the rotamer clearly discernible from other nearby residues, the script changes the display representation of the selected residue to be `bs` (ball and stick).

- As in the previous script, we use `getRotamers` to fetch a list of rotamers for the site of the selected residue. Note that `getRotamers` is imported from the `Rotamers` module.

- The `getRotamers` call and subsequent processing of the fetched rotamers appear in `try: except:` clauses. This allows a graceful handling of exceptions when rotamers do not exist for some particular side chain. This might happen for residues other than GLY and ALA, for example, the CSO residue at position 67 in 1HIV. Note that `NoResidueRotamersError` is also imported from the `Rotamers` module.

- The `while` loop runs until the user types an arbitrary character before depressing the Enter key. The body of the loop uses `rotIx` (incremented modulo the number of rotamers) as an index for the list of rotamers fetched via `getRotamers`. Each iteration of the while loop will set up a different rotamer by means of the `useRotamer` function (also imported from the `Rotamers` module). The first argument of this function specifies the residue that will be replaced by a rotamer and the second argument specifies which rotamer should be used. The second argument has to be a list of rotamers, so even though `rotIx` is designating a single rotamer, that rotamer is passed to the function within list brackets. Notice that using rotamers for a selected cysteine side-chain will not maintain a disulfide bridge if one exists.

- As the loop progresses, note how the conformation of the rotamer changes in the display. The **Shell** window will give you the rotamer index and the probability associated with this rotamer. The list returned by `getRotamers` is such that rotamers appear in order of decreasing probability. A cautionary note: Running the script with large proteins requires longer processing times for each iteration of the `while` loop.

7.5 ROTAMERS AND DIHEDRAL ANGLES

As one might expect, the `useRotamer` function will modify the dihedral angles of a residue. The changes are usually small. The following lines of output are typical results if you run `Script 7_03 Rotamers and torsion angles.py`:

```
Using rotamer having index 0 probability 0.195628002286
chi angles for rotamer before useRotamer call:
chi1: -70.600
chi2: 174.900
chi3: -3.900
chi angles for rotamer after useRotamer call:
chi1: -69.749
chi2: 175.018
chi3: -3.900
```

This output was seen when residue 23 was selected in 1CRN.

7.6 A SINGLE SIDE CHAIN AND ENERGY CONSIDERATIONS

Before any further discussion of the side-chain packing problem, it would be worthwhile to get a feel for the complexity of the problem. If you run the script in the file `Script 7_04 Side chain control panel.py`, you will get a blank version of the dialog shown in Figure 7.2. Either before or just after launching the script, you should fetch a protein for the Chimera display. Put the Chimera arrow cursor over an atom, hold down your **Ctrl** key, and do a left mouse click. Going back to the **Side Chain Control Panel** dialog, click on the **Accept** button and you should see a string identifying the chosen residue in the associated label field. In Figure 7.2, residue 41 in chain A of 1EX7 was

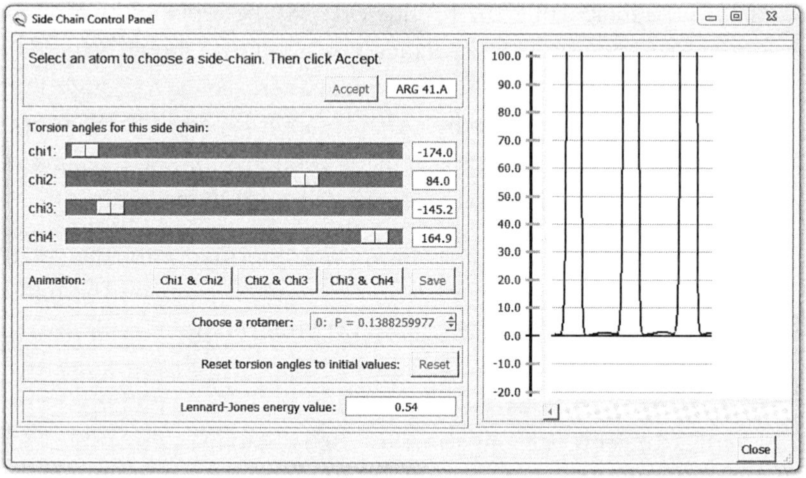

FIGURE 7.2 Side Chain Control Panel.

selected in this manner. You can also use the menu invocation **Favorites... Sequence** to get a sequence dialog that will allow you to select a residue. Avoid the Animation frame until after trying out the other widgets.

Once the residue is selected, 0, 1, 2, 3, or 4 of the slider widgets will be enabled. There will be one slider activated for each dihedral angle in the chosen residue. If you move the sliders, you should notice that the dihedral angles of the chosen residue in the Chimera display will change in accordance with the settings of the sliders. The chosen angle can be read from the label widget to the right of the scale. If at any time you wish to go back to the PDB original values, click on the Reset button. Notice that the last frame of the dialog shows the Lennard–Jones energy value. If you wish to replace the side chain with a rotamer from the Dunbrack library, then use the spin box in the frame just above the Reset button. Again, you can go back to the original angle settings by clicking on Reset.

Clicking on one of the Animation buttons will produce a series of changes in the display. The dihedral angles specified on the button will assign all possible settings between −180° and +180° in increments of 3° and 6°, respectively. Since the energy evaluation for each setting is fairly costly, the time for the total animation may be quite extended depending on the power of your computer. If you wish to use a coarser set of increments, you can change the global constants `CHI_A_DEG_INCR`, `CHI_B_DEG_INCR` that can be found near the beginning of the script. As you can see from the labels on the animation buttons, you have three choices, each choice working with two consecutive dihedral angles of the chosen residue. If a button is not enabled, then it is because that residue does not have the dihedral angle shown on the button label.

While the animation progresses, the energy values associated with that residue will be computed as the sum of all pairwise atom interactions between this residue and its neighbors. This energy will be displayed in the Lennard–Jones energy value label in the bottom frame and simultaneously it will be displayed in the strip chart in the right side of the dialog. Information about scripts that can display strip charts can be found in Section D.9 of Appendix D.

As the graph displaying the energy values is played out in the strip chart, you may notice that the curve has several local minima. It should be reasonably clear that the energy landscape is quite rough even though the curve represents the energy corresponding to the setting of only two consecutive dihedral angles. A truer and more biologically sound appraisal of the situation would lead you to understand that the energy "surface" is really rough and moreover is actually a space that is not just two but several hundred dimensions for most proteins. Searching for a global energy minimum in this high-dimensional space is computationally intractable and the combinatorial explosion of local minima speaks to the NP-hardness of the side-chain packing problem.

Since the energy values were computed by going through all dihedral angle pairs each member of the pair taken from a discrete set of angles, we can consider each pair to define a point in the Euclidean plane with the associated energy values defining a surface hovering over this plane. The data for this surface can be obtained by clicking on the Save button of the Side Chain Control Panel dialog after the strip chart recording is finished. Surface data will be saved to the directory containing the script for the dialog. The data are saved as a "pickled" version of a surface plot bundle. These terms are explained in Section D.11 of Appendix D. If you run the script in the file `Script D_14 Chimera surface plot with GUI.py`, the pickled surface data can be displayed in the Chimera window. To do this, go through the following steps:

- Clear the Chimera display by using the menu invocation **File... Close Session**.
- Launch `Script D_14 Chimera surface plot with GUI.py`.
- Click on the Fetch button and use the file open dialog to find and fetch the pickled surface plot bundle saved by the Side Chain Control Panel.

You should get a display that looks something like that in Figure 7.3.

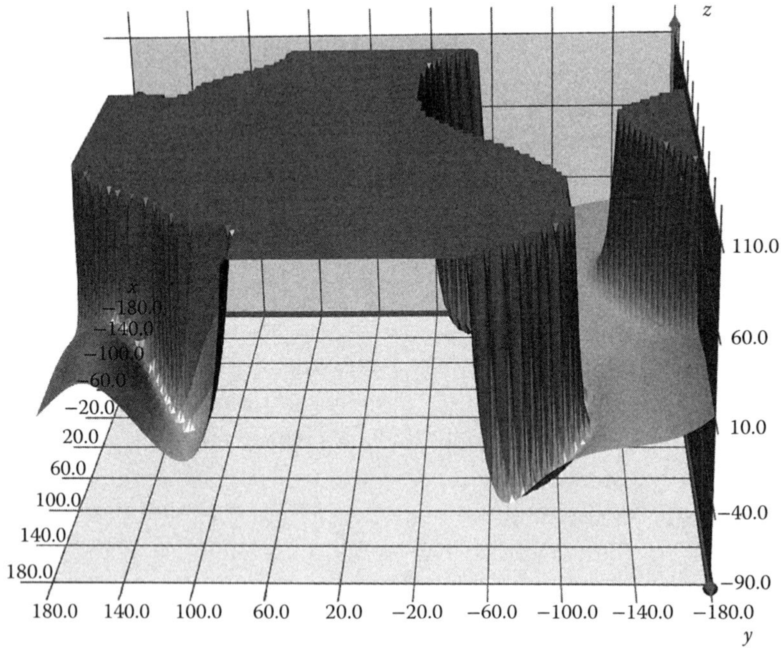

FIGURE 7.3 (**See color insert.**) A simple energy surface varying two dihedral angles.

NOTES:

- The energy values have been multiplied by 10 so that there is more surface detail in the regions holding local minima.
- The maximum energy value that is displayed tops out at 10 (seen in the display as 100).
- When the surface is first displayed, the maximum energy values have a transparent rendering. This was done to show the utility of using multiple transparency settings for one surface.
- To get a colored version of the surface (as shown in Figure 7.3), you can use the buttons in the Surface Color Selection frame.
- Typically, the high plateau in the figure will represent angle settings that have introduced a steric clash.

7.7 SIDE-CHAIN PACKING AND DEAD-END ELIMINATION

Ideally, the side-chain packing problem is solved by finding the conformations of all the side chains as they would appear in the minimum energy conformation of the molecule. Since the problem is intractable one expects that attempts to derive an approximate solution will depend on the usage of various heuristics that give reasonable low-energy conformations while not guaranteeing the achievement of a global energy minimum.

A popular heuristic is to consider the "solution" to be an optimal selection of rotamers. We select one rotamer from each rotamer set corresponding to a residue position and we try to ensure that the selected rotamers give a final structural conformation that minimize the total potential energy of the protein. Here is a more detailed description of the problem:

- We assume the protein backbone is fixed in the 3D space.
- Each protein residue (apart from glycine and alanine) will have a set of rotamers that can be accessed from a rotamer library. The residue at position k will have a rotamer set, call it S_k containing n_k members. Each rotamer is represented by the symbol r_k^i, where k denotes the residue position and i denotes the index of this rotamer within the rotamer set S_k. Hence: $S_k = \{r_k^1, r_k^2, \ldots, r_k^{n_k}\}$
- The solution depends on making an optimal choice from each S_k. More precisely, find an optimal set of integers $\{c_1, c_2, \ldots, c_k\}$ such that $1 \leq c_k \leq n_k$ for all k and the energy of the protein is minimized when rotamer $r_k^{c_k}$ is used for the residue at position k.

At the very least, selection of rotamers will be done so as to avoid steric clashes that cause the energy level to spike into very high values, so we must avoid choices that involve conflicts between neighboring rotamers. In addition to this we can make the following observation: A rotamer can be eliminated from the set S_k if it can be shown that there is another member of that same set that always gives a lower energy value *independent* of the rotamer choices for the nearby or neighboring residue sites. There are various approaches to find and eliminate such rotamers; all strategies striving to reduce the size of the rotamer sets associated with each residue site. These strategies are generally referred to as DEE (*dead-end elimination*) strategies [DM92].

We now consider the approach described by Goldstein in [Go94].

7.7.1 Goldstein's DEE: Basic Strategy

To avoid any misunderstandings that might occur later, we reaffirm that side-chain atoms of a rotamer are all the atoms in the residue except for the backbone atoms (N, Cα, C, O). In Figure 7.4, these atoms, for rotamers r_k^i and r_m^j, are shown in the two ellipses in the upper part of the figure. The ellipses in the lower part of the figure enclose backbone atoms corresponding to each rotamer.

We now introduce two functions that will be used in our energy calculations. The function $E_{\text{self}}\left(r_k^i\right)$ will be used to denote the rotamer r_k^i energy contribution that is independent of the rotamer choices made at any other residue site. Expressed as an equation:

$$\begin{aligned} E_{\text{self}}\left(r_k^i\right) &= E_{\text{frame}}\left(r_k^i\right) + E_{\text{intrinsic}}\left(r_k^i\right) \\ &= E_{\text{frame}}\left(r_k^i\right) + c_k \log\left(\frac{p\left(r_k^*\right)}{p\left(r_k^i\right)}\right) \end{aligned} \tag{7.1}$$

The term $E_{\text{frame}}\left(r_k^i\right)$ represents the sum of the energies due to the pairwise interactions between all side-chain atoms of rotamer r_k^i and all the backbone atoms of the other nearby residues. In Figure 7.4, the dashed line indicates the contribution that would be provided by the side-chain atoms and the backbone atoms corresponding to a single reside. The term $E_{\text{intrinsic}}\left(r_k^i\right)$ is the r_k^i *intrinsic* energy, which is due to all the pairwise interactions between atoms belonging exclusively to r_k^i. Since some of these atom pairs correspond to bonded atoms, the energy computation is complicated and typically approximated by using the logarithmic approximation specified in Equation 7.1. In Equation 7.1, the term $p\left(r_k^i\right)$ represents the probability of the occurrence of this rotamer, whereas $p\left(r_k^*\right)$ represents the probability

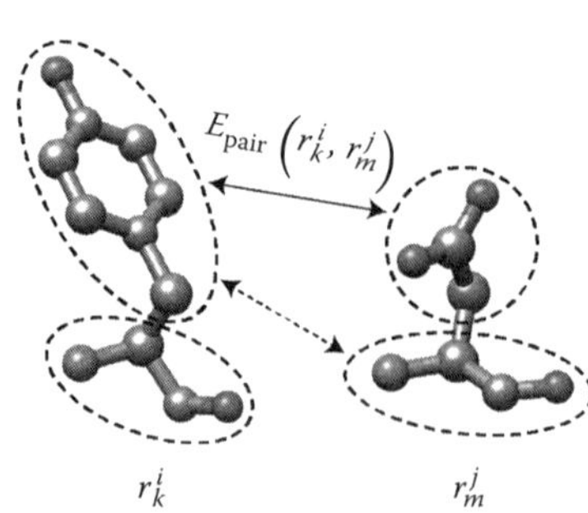

FIGURE 7.4 Energy computations for a rotamer.

of the most frequent rotamer for the residue at position k. In Chimera, the most probable rotamer has an index of 0 and so we would simply let $p(r_k^*) = p(r_k^0)$. The value of c_k provides the appropriate weighting coefficient so that the energy computed via the logarithm is properly commensurate with the other energy terms. We use the value of 10 for all c_k as suggested in Kingsford's thesis [Ki05]. The second function that we need is $E_{\text{pair}}(r_k^i, r_m^j)$, defined as the sum of all energy contributions made by atom pairs such that one atom is in the side chain of r_k^i and the other atom is in the side chain of r_m^j. This is indicated in Figure 7.4 by means of the solid arrow.

We will refer to rotamers at the same residue site as *sibling* rotamers. The energy contribution made by a rotamer will be its intrinsic energy plus the sum of all its interaction energies with nearby rotamers situated at other residue sites that are nearby in 3D space. If the "best" energy contribution of r_m^i is **larger** than the "worst" energy contribution of a sibling r_m^j, then rotamer r_m^i can be eliminated because its sibling will always be better at reducing the total energy independent of the conformations of nearby side chains. Equation 7.2 is a mathematical expression for the previous statement. We can remove r_m^i from further consideration if we can find a sibling r_m^j such that the following inequality is true:

$$E_{\text{self}}(r_m^i) + \sum_{k \in N(m)} \min_p \left\{ E_{\text{pair}}(r_m^i, r_k^p) \right\} > E_{\text{self}}(r_m^j) + \sum_{k \in N(m)} \max_p \left\{ E_{\text{pair}}(r_m^j, r_k^p) \right\} \tag{7.2}$$

where $\min_p \left\{ E_{\text{pair}}(r_m^i, r_k^p) \right\}$ is the minimum (i.e., best) energy possible between rotamer r_m^i of the side-chain site at residue[m] and **any** rotamer r_k^p of the side-chain site at residue[k]. In other words, we extract the minimum interaction energy as p ranges across all the siblings for residue[k]. A similar definition holds for the max function term. We use the notation $k \in N(m)$ to mean that k is an integer in the set of integers designated as $N(m)$. Here $N(m)$ is the set of indices of residues that are *nearby* residue[m]. For our purposes, a nearby residue will have its alpha carbon within 10 Å of the alpha carbon of residue[m].

The test is applied at each residue site and rotamers are eliminated whenever possible. Care should be exercised in the organization of the computations so that you do not repeat various energy calculations. As the algorithm progresses and rotamers are removed, the neighbor sets will change and so a rotamer that is not removed on the first pass through all the residues sites might eventually be eliminated by a future pass. The algorithm is finished when a full pass through all the residue sites is done without any elimination of rotamers.

7.7.2 Goldstein's DEE: A More Efficient Strategy

Inspection of Equation 7.2 leads to the following observation: the value of p producing the minimum on the left-hand side of the inequality is not necessarily the value of p producing the maximum on the right-hand side of the inequality. Consequently, the elimination criterion is overly stringent and somewhat inefficient because of the need for the extraction of both a minimum and maximum value. A better test is the following:

$$E_{\text{self}}\left(r_m^i\right) - E_{\text{self}}\left(r_m^j\right) + \sum_{k \in N(m)} \min_p \left\{ E_{\text{pair}}\left(r_m^i, r_k^p\right) - E_{\text{pair}}\left(r_m^j, r_k^p\right) \right\} > 0 \quad (7.3)$$

where the variables have the same significance as in Equation 7.2. As before, we can remove r_m^i from further consideration if we can find a sibling r_m^j such that Equation 7.3 is true.

7.7.3 Side-Chain Packing

After the completion of dead-end elimination, you will be left with a reduced set of rotamers at each site and the side-chain packing algorithm is ready for the next phase. One can find in the literature a wide range of algorithms to get a final solution. Covering these is beyond the scope of this text, but if you need to follow this research further, then you can access these papers and the references therein: [BC13], [KS09], and [XU05]. The introduction contained in [KS09] is particularly noteworthy since it outlines various difficulties associated with the problem and provides several references related to various heuristics that have been used to solve the problem.

7.8 EXERCISES

7.8.1 Dead-End Elimination

Write a script that performs the following steps:

1. Fetch a protein file from the PDB.
2. For each residue in the protein, determine its neighboring residues. We will consider two residues to be neighbors if their beta carbon atoms have a separation that is less than or equal to 22 Å. This is easily done by using the `Shell` class imported from `StructBio.Utilities.shellNeighbors`.

3. Working with a rotamer library (e.g., the Dunbrack library) determine the rotamer sets for each protein residue site. While doing this you should calculate the self-energy for each rotamer.

4. For each rotamer, use the linear repulsive energy function (described in Section 6.2.1 of Chapter 6) to calculate the energy interactions between this rotamer and all the rotamers for neighboring residues. We will say that two rotamers are *interacting* if they are derived from neighboring residues.

5. Use the Goldstein dead-end elimination algorithm to reduce the number of rotamer members in each set.

6. Generate a list that gives the energy for each pair of interacting rotamers. Hopefully, after dead-end elimination, this list will be a small subset of the interaction pairs generated in Step 4.

7. For each residue in the original protein determine whether it is "close" to some remaining member in its rotamer set. Two conformations of a side chain will be considered as close if their Chi 1 dihedral angles differ by 40° or less and, if so, then their Chi 2 angles must also differ by 40° or less.

8. Calculate the percentage of residues that pass this evaluation.

It is interesting to take the results of the last step and compare them with the corresponding results of algorithms that start with dead-end elimination and then do side-chain packing as a subsequent procedure.

REFERENCES

[Ak97] T. Akutsu. NP-hardness results for protein side-chain packing. *Genome Informatics*, **8** (1997), 180–186.

[DT05] K. C. Dukka Bahadur, E. Tomita, J. Suzuki, and T. Akutsu. Protein side-chain packing problem: a maximum edge-weight clique algorithmic approach. *Journal of Bioinformatics and Computational Biology*, **3** (2005), 103–126.

[BC97] M. J. Bower, F. E. Cohen, and R. L. Dunbrack, Jr. Prediction of protein side-chain rotamers from a backbone-dependent rotamer library: a new homology modeling tool. *Journal of Molecular Biology*, **267** (1997), 1268–1282.

[BC13] F. J. Burkowski, Y.-L. Cheung, and H. Wolkowicz. Efficient use of semidefinite programming for selection of rotamers in protein conformations. *INFORMS Journal on Computing* (2014).

[DK93] R. L. Dunbrack, Jr. and M. Karplus. Backbone-dependent rotamer library for proteins: application to side-chain prediction. *Journal of Molecular Biology*, **230** (1993), 543–574.

[DM92] J. Desmet, M. de Maeyer, B. Hazes, and I. Lasters. The dead-end elimination theorem and its use in protein side-chain positioning. *Nature*, **356** (1992), 539–542.

[Du02] R. L. Dunbrack, Jr. Rotamer libraries in the 21st century. *Current Opinion in Structural Biology*, **12** (2002), 431–440.

[Go94] R. F. Goldstein. Efficient rotamer elimination applied to protein side-chains and related spin 7 glasses. *Biophysical Journal*, **66** (1994), 1335–1340.

[Ki05] C. L. Kingsford. *Computational approaches to problems in protein structure and function*. PhD thesis, Princeton University, Princeton, NJ, USA, 2005.

[KS09] G. G. Krivov, M. V. Shapovalov, and R. L. Dunbrack, Jr. Improved prediction of protein side-chain conformations with SCWRL4. *Proteins*, **77** (2009), 778–795.

[SD11a] A. D. Scouras and V. Daggett. The dynameomics rotamer library: amino acid side chain conformations and dynamics from comprehensive molecular dynamics simulations in water. *Protein Science*, **20** (2011), 341–352.

[SD11b] M. S. Shapovalov and R. L. Dunbrack, Jr. A smoothed backbone-dependent rotamer library for proteins derived from adaptive kernel density estimates and regressions. *Structure*, **19** (2011), 844–858.

[Tr05] A. Tramontano. *The Ten Most Wanted Solutions in Protein Bioinformatics*. Chapman & Hall/CRC, Boca Raton, FL, 2005.

[Xu05] J. Xu. *Rapid Protein Side-Chain Packing via Tree Decomposition*. RECOMB, Boston, MA, 2005.

CHAPTER 8

Residue Networks

8.1 MOTIVATION

Recently, the investigation of interactions between neighboring residues has been a "hot topic" in protein structure research. In these studies, the notion of "neighboring" is usually determined by some type of distance criterion. For example, two residues R_a and R_b may be designated as *contact neighbors* if some atom in R_a is within a short distance threshold of another atom in R_b. A simpler criterion might require that the alpha carbon atoms in R_a and R_b be within some distance threshold.

A residue network is a graph theory representation of these interactions. In these graphs, a node corresponds to a residue in the protein. Typically, the requirements of the study will dictate that the node will correspond to some particular subset of the atoms in the residue. This could be all the residue atoms or perhaps some subset of the residue atoms, for example, the side-chain atoms (excluding the backbone atoms) or just the single alpha carbon. Edges in the graph will exist between two nodes when the corresponding residues are reckoned to be neighbors. Labels on nodes typically designate the residue type and chain position. Labels on edges may specify an edge weight that has some particular relevance to the research study.

Several research papers have dealt with residue networks. As a starting point you may consult the following: Krishnan et al. [KZ08] introduce residue networks as an extension of contact maps. They discuss the significance of paths in the graphs derived from residue networks.

In [DK11], the authors describe software for analyzing and visualizing residue networks. A more recent paper by Csermely et al. [CS12] is a good starting point—the title provides a comprehensive description of the content: "Disordered proteins and network disorder in network descriptions of protein structure, dynamics and function: Hypothesis and a comprehensive review." Another review paper [BK07] presents a topological analysis of protein structures using networks with an emphasis on the *small-world* characterization of these networks. More advanced applications and analysis of residue networks can be found in [BS05] (residue networks used in the study of protein oligomerization) and [MF09] (Residue Interaction graphs).

Modeling tools arising from the study of residue networks can be applied to the analysis of protein allostery. This topic will be discussed at the end of this chapter.

8.2 THREE-DIMENSIONAL (3D) VISUALIZATION OF RESIDUE NETWORKS

Visualization of these networks can be shown by means of a "flat" two-dimensional (2D) graph drawing. While this is often the basis for a figure that is submitted for publication, the properties of the graph are usually difficult to discern because a typical protein has a large number of residues leading to a corresponding large number of nodes. In turn this produces a huge number of graph edges that will be crisscrossing to such an extent that local features of the graph are obliterated.

An objective of this section is to demonstrate that this confusion can be somewhat reduced by drawing the graph as a 3D structure that is superimposed on the 3D structure of the protein. This has an immediate advantage: the density of nodes and edges is reduced because the graph is not flattened onto a 2D plane. More significantly, by placing the graph nodes near their corresponding residues, the viewer can recognize the significance of the node and its relationship to the overall structure of the protein.

The exercises at the end of this chapter act as a starting point for these visualizations. Nodes are represented by spheres, and edges are represented by spindles. The reader should consult Appendix B, which describes a Python class containing functions that can provide these 3D elements. Exercises 1, 2, and 3 pose problems that are to be done using these representations.

8.3 ALLOSTERY AND CONTACT REARRANGEMENT NETWORKS

Recall that allosteric regulation is a control mechanism in which an effector molecule binds at one site of a protein and this event induces a conformational change that alters the binding behavior at another binding site called the substrate site. It is reasonable to assume that there is some sort of communication between the two sites, but the phenomenon is not yet fully understood. There are several models that attempt to characterize this communication. For example, the paper by Daily et al. [DU08] presents a technique for elucidating the communication path between effector and substrate site by quantifying the changes in contacts between neighboring residues. While their approach does not explain the exact physical causes of the communication, it does help to reveal its presence. To review their work we start with the following definitions:

Given two residues R_i and R_j in the same protein, we will say that they are in *contact* if we can find an atom in R_i and another atom in R_j such that the distance between these two atoms is less than some threshold value d (we will work with $d = 5$ Å). Given two residues R_i and R_j that are in contact, we could inspect each atom in R_i to determine if that atom is within a distance d from an atom in R_j. If the distance test is satisfied, then both atoms are placed into a set called the *contact interface* C_{ij}. Naturally, if R_i and R_j are not in contact, then C_{ij} is an empty set.

Each allosteric analysis will involve a protein that has two different conformational states: the inactive (I) state and the active (A) state. Each state corresponds to an entry in the Protein Data Bank (PDB). For example, phosphofructokinase can be in the I state (PDB ID 6PFK) or in the A state (PDB ID 4PFK).

We will use the notation C_{ij}^I to represent the contact interface C_{ij} when the protein is in the I state and C_{ij}^A to represent the contact interface C_{ij} when the protein is in the A state. For example, the top diagram in Figure 8.1 shows all the atoms in C_{ij}^I within the enclosure defined by the dotted line. The dotted line enclosure of the lower diagram shows all the atoms in C_{ij}^A. Let us consider some changes that may or may not occur in C_{ij} when the protein makes a state transition, for example, it goes from the I state to the A state. First, let us consider the possibility that the two residues R_i and R_j are in contact and $C_{ij}^A = C_{ij}^I$. This would mean that the contact interface does not change in the transition; in other words, the two residues are in contact but their contact with each other is not affected

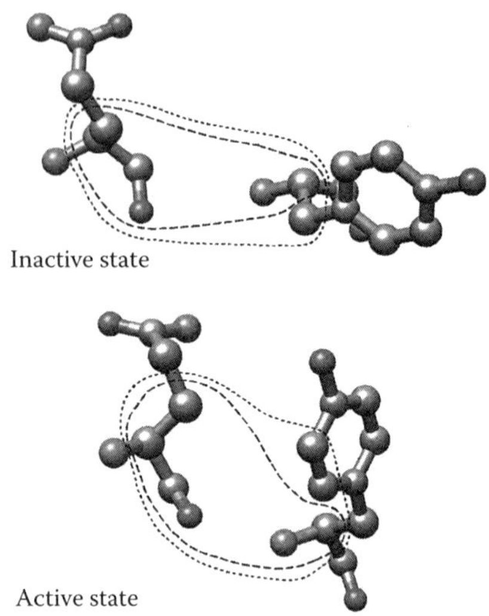

FIGURE 8.1 Atom sets in the contact rearrangement interfaces.

by the allosteric change in conformation. Note that, in this case, we have $C_{ij}^{A} \cap C_{ij}^{I} = C_{ij}^{A} = C_{ij}^{I}$.

Note that the dashed line in each diagram contains the intersection $C_{ij}^{A} \cap C_{ij}^{I}$ and, of course, it is the same for both diagrams.

Now consider the set difference: $C_{ij}^{I} - \left(C_{ij}^{A} \cap C_{ij}^{I}\right)$. This set contains those atoms in the contact interface of the I state that do not stay in the contact interface when there is a transition from I to A. Similarly, $C_{ij}^{A} - \left(C_{ij}^{A} \cap C_{ij}^{I}\right)$ is the set that contains those atoms in the contact interface of the A state that do not stay in the contact interface when there is a transition from A to I. Note that both of these sets are empty when $C_{ij}^{A} \cap C_{ij}^{I} = C_{ij}^{A} = C_{ij}^{I}$. In Figure 8.1, the set $C_{ij}^{I} - \left(C_{ij}^{A} \cap C_{ij}^{I}\right)$ contains the single atom CB of the TYR side chain and the set $C_{ij}^{A} - \left(C_{ij}^{A} \cap C_{ij}^{I}\right)$ contains the two atoms CD1 and CE1 of the TYR side chain.

In the strategy used by Daily et al., the number of atoms in these difference sets indicates the amount of loss or gain in a contact interface during a transition. More significantly, this count can *quantify* the rearrangement of contacts in the contact interface during a transition. They use these considerations to define the *rearrangement factor* $RF(i, j)$ as follows:

$$RF(i,j) = \frac{\max\left\{\left|C_{ij}^{I} - \left(C_{ij}^{A} \cap C_{ij}^{I}\right)\right|, \left|C_{ij}^{A} - \left(C_{ij}^{A} \cap C_{ij}^{I}\right)\right|\right\}}{N_i + N_j}.$$

Notice the normalization computation done via division by $N_i + N_j$, which is the sum of the number of atoms in R_i and R_j, respectively.

Daily et al. use the $RF(i, j)$ values to define a graph called the contact rearrangement network (CRN). In this graph, the nodes correspond to the residues of the protein. Two nodes labeled R_i and R_j are connected by an edge if their rearrangement factor $RF(i, j)$ is above some threshold T (they use $T = 0.3$). The CRN is considered to be a *small-world network*, and further calculations are performed to evaluate various small-world network parameters.

The final exercises in this chapter deal with scripts to make these calculations leading to the generation of a network graph.

8.4 EXERCISES

The following exercises will take you through various examples that involve the construction of residue networks and their display in the Chimera window:

8.4.1 Spheres Representing Residues

Write a script that computes the "center" of a side chain by computing the centroid of all atoms in the side chain excluding backbone atoms CA, C, N, and O. The script should treat glycine as a special case with the center taken to be the coordinates of the CA atom. Eventually, we will want to pass this center to various other scripts that work with atom coordinates, for example, methods in the `Shell` class. Consequently, you should implement a class called `ResidueCenter` that has scripts to implement two methods with definition headers as follows:

```
def __init__(self, resObj):
def coord(self):
```

This means that `ResidueCenter` will be an object with a method to return its coordinates. We can then pass it to any other function that deals with atoms, as long as that function only uses the `coord()` method for the object. In this sense, the center will act as a *virtual* atom.

Use the `addSphere` functions in the `Solids` class to put a small sphere (`radius = 1.0`) at each center. The sphere should be colored according to its Kyte–Doolittle hydrophobicity value. Software for this book has a function to generate this color. Use the import statement:

```
from StructBio.Utilities.miscFunctions import residueKDhColor
```

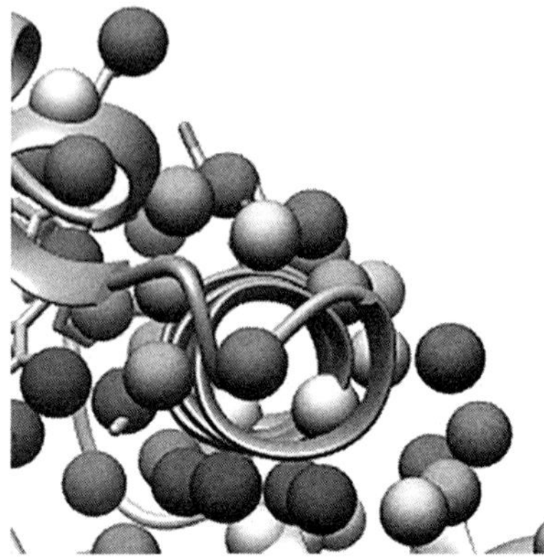

FIGURE 8.2 **(See color insert.)** Amphiphilicity.

to get the function from the `StructBio` package. If you move the structures, so that you are looking along the axis of a helix, you can usually see the amphiphilicity [EW82] of the helix as a tendency for hydrophobic residues to be on the side of the helix that is closer to the center of the protein (see Figure 8.2).

Test your script on the following PDB files: the winged helix protein 1SMT, the beta barrel 2POR, and insulin 1ZNJ.

8.4.2 Spheres and Spindles for Residue Networks

The visualization generated by the previous exercise gives the viewer a realistic assessment of the distribution of hydrophobic and hydrophilic residues within the protein. However, the visual representation still has a somewhat jumbled appearance because it is difficult to assess when two spheres correspond to nearby residues. This can be improved somewhat by drawing a link between the spheres corresponding to neighboring residues. Even though the scene has more visual elements, there is a better portrayal of the neighborhood structure and so more information is conveyed to the viewer.*

Extend the capabilities of the previous script by linking spheres that represent neighboring residues. We will consider two residues to be neighbors if their centers have a separation that is less than or equal to a value set by the parameter `resDistThreshold`.

This is an excellent opportunity to use the `Shell` class with virtual atom objects instead of atom objects.

---

* This type of visualization will be seen again when we deal with residue networks (see Exercise 3).

Use the `Solids` class given in Appendix B to place spheres at the residue centers (as done in the previous exercise) and then link the spheres corresponding to neighboring residues by putting a spindle between these spheres. As in Exercise 1, spheres should be colored according to their Kyte–Doolittle hydrophobicity values. The color of a spindle may be derived by taking the RGB tuples for each sphere and calculating the average. The scene should have an appearance similar to that shown in Figure 8.3.

Note that the spindles can be categorized into three groupings that are characterized by the hydrophobicity attributes of the spheres being linked.

1. HP: hydrophobic-hydrophilic
2. HH: hydrophobic-hydrophobic
3. PP: hydrophilic-hydrophilic

If the various categories of spheres and spindles are associated with separate objects instantiated from the `Solids` class, then you can use the `Model Panel` to selectively display hydrophobic or hydrophilic networks. Spheres are shown only if they are at the ends of spindles that are being displayed. For example, Figure 8.4 shows spindles linking spheres corresponding to hydrophobic residues (`resDistThreshold = 8`).

If you refer to Figure 1.32, you can recall that the hexameric structure of 1ZNJ is really a trimer of dimers. Each of the three dimers has a

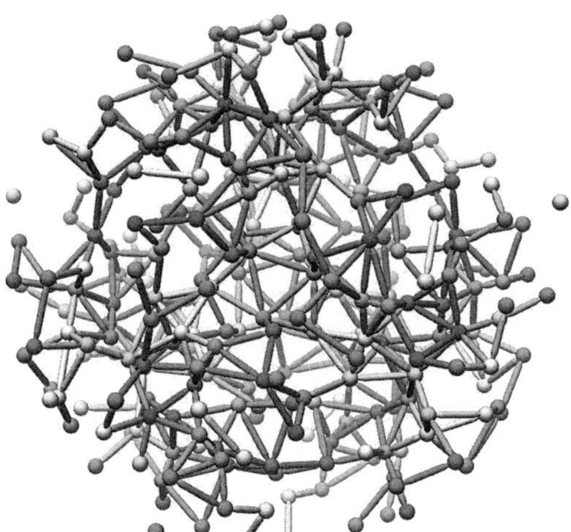

FIGURE 8.3 **(See color insert.)** Network of neighboring residues (1ZNJ).

hydrophobic core as shown in Figure 8.4. Figure 8.4 shows how the three cores combine to surround an innermost core that is hydrophilic but not shown. Figure 8.5 illustrates this hydrophilic network clearly showing a hydrophilic cage-like structure that contains the zinc atoms (one of which can be seen at the exact center of the structure).

Work with the porin protein 2POR to get another compelling visualization. You should see a "hydrophobic girdle" around this beta barrel with various hydrophilic residues on the inside of the pore as might be expected

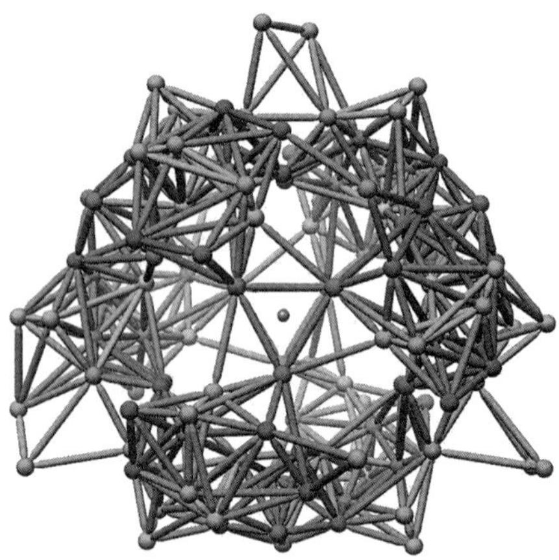

FIGURE 8.4 Network of hydrophobic residues (1ZNJ).

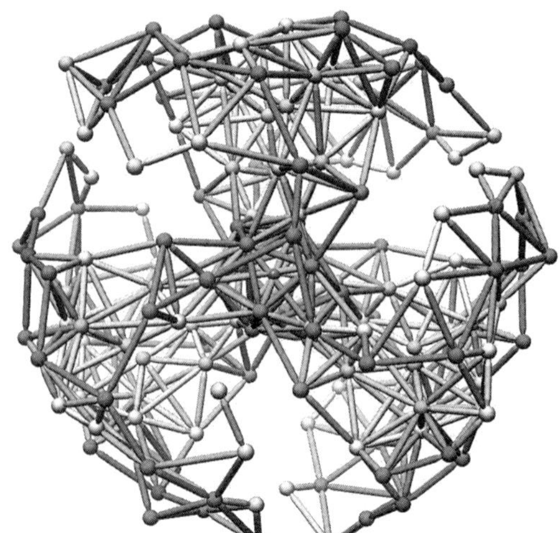

FIGURE 8.5 Network of hydrophilic residues with zinc atoms (1ZNJ).

for a bacterial outer membrane protein. The antifreeze protein 2PNE is another interesting case.

8.4.3 Spheres and Spindles for Residue Networks: Graphical User Interface (GUI) Implementation

The previous script requires various specifications and Model Panel interactions with the user. These would be handled better with a suitable GUI. Study the material in Appendix A so that you can generate two dialogs to be used by the program. Execution begins with a modal dialog that contains two widgets: an entry box that allows the user to specify the PDB ID of the protein that is to be fetched and a scale widget to define the value of `distThreshold`. After the protein is fetched and data structures established, the script should launch a modeless dialog that allows the user to specify the various types of spindles to be displayed. Since the dialog is modeless, the dialog should allow the user to change this specification at will, each new choice causing the display to be updated.

Figures 8.6 and 8.7 give the appearance of the two dialogs.

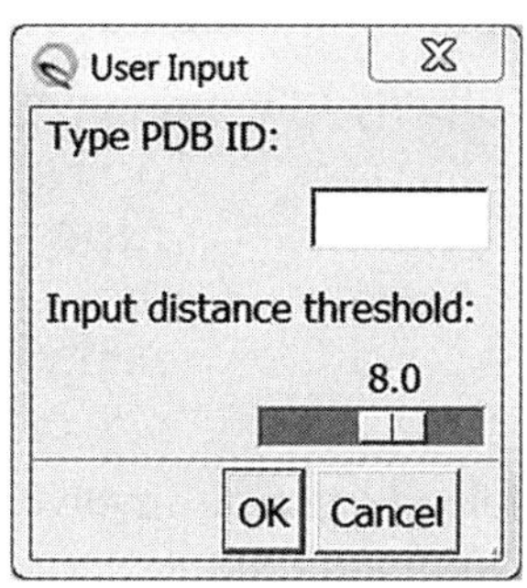

FIGURE 8.6 Dialog window to select the Protein Data Bank (PDB) file and set `distThreshold`.

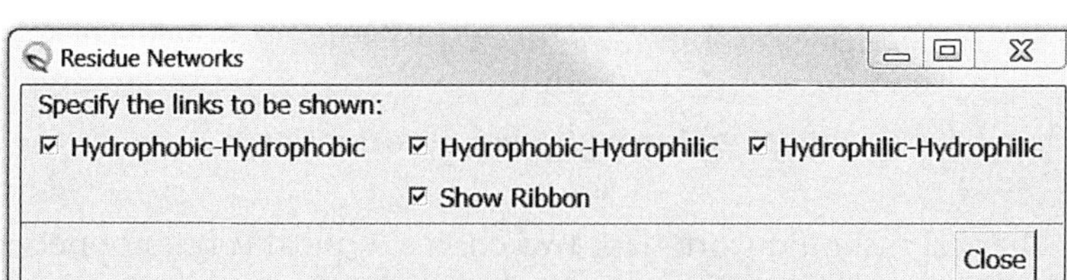

FIGURE 8.7 Dialog window to select visual elements.

8.4.4 Graphs for Residue Networks: GUI Implementation

The previous exercise can be handled with greater ease by working with the graph classes that can be imported by using:

```
from StructBio.Graphs.graph import Graph, Node, Edge
```

Get familiar with the contents of Appendix C, if you have not already done so, and modify the previous script so that the graph classes are used. Showing and hiding the various links can now be done by calling functions that work with subgraphs.

8.4.5 Contact Rearrangement Networks: Evaluating *RF*(*i*, *j*)

These final exercises deal with CRNs. The overall problem has been broken down into a series of smaller steps to make the development easier. We start with the evaluation of the *RF*(*i*, *j*) function. To test the exercises, we will work with the inactive and active states of the following six proteins:

Inactive	Active
1HH4	1MH1
1XTQ	1XTS
1XTU	1XTT
1CD5	1HOT
1WPL	1IS7
1KAO	2RAP

In all cases, when multiple chains are involved, you should work with Chain A only. Any other chains should be deleted.

Start with

```
from StructBio.Utilities.miscFunctions \
   import userSpecificationOfChains
from StructBio.StructureOverlap.sequencesLocalAlignment \
   import SequencesLocalAlignment
```

and then have the script go through the following steps:

1. The script should work with two chains, typically, but not necessarily, from two different proteins. User determination of the chains can be done with a GUI implementation or by working with the

function `userSpecificationOfChains`. These two protein chains will represent the structural information for the inactive and active states.

2. For each protein, extract a list of standard residues with the requirement that corresponding entries from the lists must be the same residue types. One might think that this is easily accomplished because each protein file represents a different state of the same protein. Ideally, both files have the same number of residues and the same set of position numbers to specify the locations of residues in a chain. Unfortunately, this ideal situation might not hold because the inactive and active versions of the protein may have been deposited in the PDB by two different research laboratories. In fact, it is possible that some parts of a protein chain may simply be missing in one file while being completely intact in the other file! To get around this situation, you should read both files and then do a sequence alignment to get corresponding pairs of residues that have all their structural information present. Use a simple alignment that relies on *identical* matching. In other words, do *not* use matching scores such as BLOSUM 62. Consider working with the function `SequencesLocalAlignment`.

3. Calculate $RF(i, j)$ *for all pairs of residues that are in contact.* You should *not* calculate or store values of $RF(i, j)$ when R_i and R_j are not in contact. Since the $RF(i, j)$ values are not to be stored in a 2D array, you will likely use a dictionary or list of tuples each containing values such as: i, j, $RF(i, j)$ for $RF(i, j) \geq T$. Note that i and j are the indices for the lists created in the previous step. They may or may not correspond to the position numbers of a residue in its chain and they may or may not correspond to the indices of a residue in the Chimera residue lists for the proteins (because of the sequence alignment and extraction of corresponding residues done in the previous step).

Implementation notes:

- Use the parameter settings stipulated in Section 8.3 ($d = 5.0$, $T = 0.3$).
- The script should execute in time $O(n)$, where n is the number of residues in the protein. So when evaluating residue contact, you should not use a loop within a loop with each loop ranging over all possible residues. Use the `Shell` class.

- In the calculation of $C_{ij}^{I} - \left(C_{ij}^{A} \cap C_{ij}^{I}\right)$, it is natural to use the `Set` class in Python but you have to be careful. If C_{ij}^{I} and C_{ij}^{A} contain atom objects, then the intersection will always be empty because Chimera will have different atom objects for the two proteins. So the sets should contain integers, each integer representing a tracking identifier that uniquely specifies an atom within a particular residue of the list created in Step 2. For example, atom "CA" in the residue at some particular index i, for example, $i = 22$ would be given a unique tracking identifier but it would be valid for both the inactive and active residue lists. Of course, similar considerations hold for $C_{ij}^{A} - \left(C_{ij}^{A} \cap C_{ij}^{I}\right)$.

To complete this exercise, you should have your program compute the $RF(i, j)$ values that are equal to or greater than T and then print them out. Since the $RF(i, j)$ function is symmetric with respect to i and j, you should only print out the $RF(i, j)$ values for which $i < j$. For example, here are five lines from the 1HH4 and 1MH1 case:

```
37 PHE     58 THR     0.833
38 ASP     57 ASP     0.750
37 PHE     59 ALA     0.750
36 VAL     57 ASP     0.846
21 ILE     37 PHE     0.684
```

Residues are specified by position in the chain of the protein followed by residue type. This output is of limited use, but you can use it to test progress thus far. The main purpose of this exercise is to serve as a preamble for the next exercise.

8.4.6 Displaying the Residue Network for Rearranged Contacts

Note that the list of $RF(i, j)$ values computed in the previous exercise contains enough information for you to build a graph, where each node n_i corresponds to a residue R_i and there is an edge between node n_i and node n_j if $RF(i, j) \geq T$. In the paper by Daily et al., this graph is considered to be a *small-world network* and various graph-related operations can be performed to assess the properties of the network.*

* For example, you could extract the subgraph that is induced by all the nodes having degree larger than some given threshold or you could use a depth-first search to derive all the connected components of the graph.

Use the Graph, Node, and Edge classes to place a residue network into the display. Start with the import statements:

```
from StructBio.StructureOverlap.overlapper \
   import Overlapper
from StructBio.Utilities.shellNeighbors \
   import Shell
from StructBio.Utilities.miscFunctions \
   import userSpecificationOfChains, colorByName
from StructBio.Utilities.miscFunctions \
   import residueKDhColor, residueFletterickColor
from StructBio.StructureOverlap.sequencesLocalAlignment \
   import SequencesLocalAlignment
from StructBio.Graphs.graph \
   import Graph, Node, Edge
```

and do the same three steps as in the previous exercise followed by

4. Work with the Overlapper class to derive an overlap of the inactive and active versions of the protein. The sequence alignment done in the previous exercise can be used to get lists of the corresponding alpha carbon atoms that will define the overlap. Since a transition between the inactive and active states is often accompanied by an extensive conformational change, the overlap may show a poor structural alignment across its entire length. It is better to have a tight structural alignment over a smaller range of residues with the rest of the chains then showing a more extensive movement relative to the tighter overlapped regions. Consequently, it is recommended that the initial overlap operation be followed by a scan that selects a set of alpha carbon pairs that have the shortest interatomic distance (e.g., the first third of the pairs after sorting the pairs to get ascending order by interatomic distance). Use these pairs to redo the overlap.

5. Use the list of $RF(i, j)$ values computed in the previous exercise to build a graph, where each node n_i corresponds to a residue R_i and there is an edge between node n_i and node n_j if $RF(i, j) \geq T$. Work with the Graph, Node, and Edge classes to construct nodes and edges for the display. Each chain should have its own graph. The nodes should be spheres that are centered over the alpha carbon atoms of the residues in a chain. To provide more visual information, the radius of a sphere can be made proportional to the number of atoms in the corresponding residue. The color of the sphere can

be determined by using the `residueFletterickColor` function, which will give it a unique color for each residue type. The spindles that act as links between nodes can be given a radius that is proportional to the *RF*(*i*, *j*) value corresponding to that edge. Color the spindle light green if the corresponding contact interface is going to gain atoms in the transition to the other state and color it deep pink if the corresponding contact interface is going to lose atoms in the transition to the other state.

6. There will be several nodes in the graphs that have either no edges or are members of connected components that are too small to be of interest. This final step should decompose each graph into a set of connected components each with a size that is at least 3.

The Model Panel can be used to selectively determine the visual elements to be shown in the scene. You may wish to streamline this user interaction by designing a GUI that is more appropriate for this application. Depending on the amount of structural movement in a transition, one graph may look better than the other. Figure 8.8 shows a graph visualization for proteins 1XTU and 1XTT (inactive and active forms of the protein uracil phosphoribosyltransferase). With close inspection, you

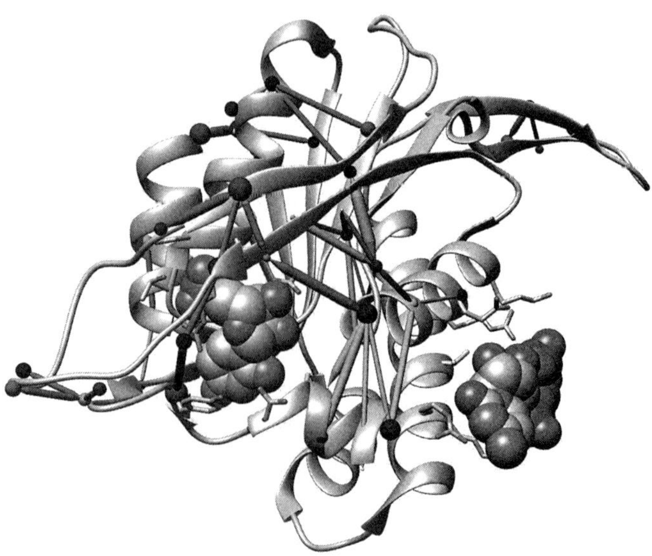

FIGURE 8.8 **(See color insert.)** Graph visualization of a contact rearrangement network (1XTU and 1XTT).

should be able to see the various spindles that have been drawn between nodes that correspond to residues that undergo contact rearrangement in a transition. Note that these residues tend to be close to the ligands and, in fact, represent a reasonable path of rearrangement between the two binding sites helping to elucidate the allosteric phenomenon.

REFERENCES

[BK07] C. Böde, I. A. Kovács, M. S. Szalay, R. Palotai, T. Korcsmáros, and P. Csermely. Network analysis of protein dynamics. *FEBS Letters*, **581** (2007), 2776–2782.

[BS05] K. V. Brinda, A. Surolia, and S. Vishveshwara. Insights into the quaternary association of proteins through structure graphs: a case study of lectins. *Biochemical Journal*, **391** (2005), 1–15.

[CS12] P. Csermely, K. S. Sandhu, E. Hazai, Z. Hoksza, H. J. M. Kiss, F. Miozzo, D. V. Veres, F. Piazza, and R. Nussinov. Disordered proteins and network disorder in network descriptions of protein structure, dynamics and function: hypothesis and a comprehensive review. *Current Protein and Peptide Science*, **13** (2012), 19–33.

[DU08] M. D. Daily, T. J. Upadhyaya, and J. J. Gray. Contact rearrangements form coupled networks from local motions in allosteric proteins. *Proteins*, **71** (2008), 455–466.

[DK11] N. T. Doncheva, K. Klein, F. S. Domingues, and M. Albrecht. Analyzing and visualizing residue networks of protein structures. *Trends in Biochemical Science*, **36** (2011), 179–182.

[EW82] D. Eisenberg, R. M. Weiss, and T. C. Terwilliger. The helical hydrophobic moment: a measure of the amphiphilicity of a helix. *Nature*, **299** (1982), 371–374.

[KZ08] A. Krishnan, J. P. Zbilut, M. Tomita, and A. Giuliani. Proteins as networks: usefulness of graph theory in protein science. *Current Protein and Peptide Science*, **9** (2008), 28–38.

[MF09] T. Milenković, I. Filippis, M. Lappe, N. Pržulj. Optimized Null Model for Protein Structure Networks. PLoS ONE, 4, (2009) DOI: 10.1371/journal.pone.0005967

Appendix A

Simple Dialogs

A.1 MOTIVATION

In the initial chapters, most of the exercise scripts require interaction with a user. For the most part, this has been accomplished by the user typing text into the Python Shell and then getting results back in the same window when the script executes a print statement. At times, interaction with the user has been done with the help of a dialog window provided by the `tkFileDialog` module as shown in Script 2.2 and Script 2.3.

The purpose of this appendix is to introduce the reader to scripting techniques that will give her program the ability to launch dialog windows that provide a graphical user interface (GUI). There are two possible scenarios (depending on how the GUI is designed):

1. A Python script is opened from the Python Shell and subsequently run. While this application is running, it may launch a dialog window that acts as a GUI for the user. Typically, such a GUI gives the user a richer interactive experience than simply interacting with the Python Shell and it can also be used to eliminate any unfortunate or confusing input coming from the user.

2. If the GUI dialogs (typically more than one) provide a *complete* interaction with the user, then execution of the script can be initiated by the **File... Open...** menu of the Chimera window. In this case, the Python Shell can be avoided if the script does not execute any print statements, all output from the script being displayed in the main Chimera window or in some dialog window.

In the subsequent sections, we consider the construction of some simple dialogs (implemented via Tkinter) that may be used to improve the more frequent user interactions that would otherwise be done by means of the Python Shell. It is not meant to be a complete lesson covering GUI design principles, and it certainly does not give a complete treatment of Tkinter. More extensive coverage of Tkinter can be found in various Internet sites or you could read Grayson's book [GR05].

A.2 MODAL AND MODELESS DIALOGS

There are two types of user interface dialogs: modal and modeless. A modal dialog typically requires user input and ideally, execution in the main window will not continue until that interaction has been completed. If the dialog window has been launched and the user does not wish to proceed with the interrogation, then a Cancel button is usually provided so that an appropriate resumption of execution can proceed without the information. A modeless dialog window allows the user to interact with both the dialog and the main Chimera window without first having to close or cancel the dialog.

The logic behind a modal dialog is normally simpler than that used for a modeless dialog: the modal dialog is displayed, the information is obtained from the user, the dialog is closed, and execution proceeds with the script that had launched the dialog. By contrast, all permitted interactions with a modeless dialog must have a meaningful relationship with the current state of the executing script associated with the main window. In an effort to properly manage this interaction, the modeless dialog will often use strategies that prohibit certain dialog activities that are not appropriate for the current state of the program's execution. This can be done by disabling (graying out) various controls (also called widgets) in the dialog window.

A.3 BASIC DESIGN GUIDELINES

Dialogs can be classified into various categories according to their functionality:

- Dialogs launched by the user:
    - Property dialogs use nouns in various titles so that the user can understand the significance of the information being displayed (e.g., "File size on disk:").
    - Function dialogs have titles involving a verb that describes the command that the user can invoke (e.g., "Select" or "Apply").

- Dialogs launched by the application:
  - Process dialogs give information about current status (e.g., time remaining in a lengthy file transfer or space remaining on a disk drive).
  - Bulletin dialogs provide transitory status or may provide information and then block further processing until the user acknowledges that information (an example that you have probably seen already when installing new software would be the "Accept" button to indicate acceptance of the terms in a license agreement appearing in a text box widget).

In all cases, the dialog will contain control and status widgets that should be accompanied by text that informs the user about the significance of the widget or group of widgets. The text may be a title that is in a Label widget located near a group of other widgets (e.g., a title explaining the significance of a group of buttons each having additional text on each button to explain what that particular button does).

The layout of widgets within a dialog should present a clear visual hierarchy typically defined by means of nested groupings. A grouping may be implicit or it may be visually explicit. For example, functionally related buttons may be grouped within a parent frame that provides a visual indication of this grouping (e.g., the `Frame` and `LabelFrame` widgets are visually represented by showing a raised border that surrounds the buttons in a group).

A.4 A MODELESS DIALOG WITH SEVERAL WIDGETS

We will start by discussing the widgets of the dialog window illustrated in Figure A.1. To see the script for this dialog along with a short mainline script that launches the dialog, find the file: `Script A_01 WidgetBuffet.py` in the Appendix A scripts.

The goal of this Appendix is to see how Python script snippets can be used to produce the various widgets in the modeless dialog of Figure A.1. Since the purpose of this dialog is simply to show a selection of widgets, the layout design of the dialog does not adhere to any particular guidelines derived from HCI (human–computer interaction) principles. Nonetheless, it does illustrate how we can achieve logical and visual nesting of widgets through the use of frame widgets (function calls `tk.LabelFrame` and `tk.Frame`).

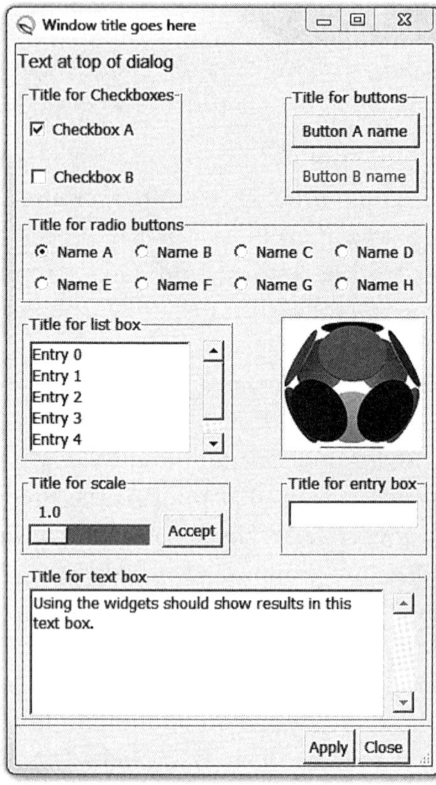

FIGURE A.1 The widget buffet modeless dialog.

After studying the script and understanding how it works, you should be able to copy and modify relevant parts of the `Script A_01 WidgetBuffet` script so that your own program can work with a simple GUI. The script contains most of the widgets that you might need in a GUI. Menu widgets are not covered because they are somewhat more complicated to design and implement.

The script for this dialog, contained in `Script A_01 WidgetBuffet`, covers about three pages and may be somewhat daunting for the programmer who is not familiar with Tkinter. However, it is quite modular in its construction and easily understandable if you approach it one section at a time.

The script can be separated into four parts:

1. A preamble (imports modules, provides the class definition header, and initiates the fill of the dialog window).
2. Definitions for the widgets (organized on a frame-by-frame basis).

3. Definitions for the callback functions (invoked when a widget is activated).
4. The statements for the mainline program.

We will first describe the preamble and then discuss the various widgets and their corresponding callback functions.

A.4.1 Preamble Statements

The script starts with the following statements:

```
import chimera, Tkinter as tk
from chimera.baseDialog import ModelessDialog

class WidgetBuffet(ModelessDialog):
    buttons = ("Apply", "Close")
    help = False    # Leave this out to get a Help button.
    title = "Window title goes here"

    def fillInUI(self, parent):
        # Text to be displayed in the dialog window:
        topText = tk.Label(parent, text = " Text at top of dialog",
                           font = ("Helvetica", 11, "normal"))
        topText.grid(row = 0, column = 0, columnspan = 2,
                     pady = 4, sticky = tk.W)
```

The first statement imports both `chimera` and `Tkinter`. Note that a `Tkinter` prefix will be designated by using `tk`. The next statement brings in the `ModelessDialog` class and the next statement defines the `WidgetBuffet` class that indicates inheritance from the `ModelessDialog` class. The next three statements specify the buttons that will appear (or not appear) at the bottom of the dialog and the title that will be placed in the window pane. Note that the help button does not appear because of the `help = False` statement.

The initialization function `fillInUI` sets up the widgets in the dialog. The first widget is an optional `Label` that provides some text to describe the purpose of the dialog. The arguments for the `Label` instantiation are reasonably self-explanatory. The statement after that uses the grid manager to specify the location of the `Label` widget within the dialog window. We next consider Tkinter strategies for placement of the other widgets.

A.4.2 Widget Placement

Tkinter has three approaches for the layout of widgets:

1. The Pack Manager is the easiest one to use, but it is somewhat limited in its capabilities.
2. The Grid Manager is more flexible than the pack manager and provides a versatile mechanism for widget layout. It requires more directives from the programmer. You are not allowed to have the Grid Manager and the Pack Manager working together within one dialog.
3. The Place Manager is the most capable but requires more work to get everything set up just right. For simple dialogs it is implementation overkill and is not recommended.

Considering the previous points, we will work with the Grid Manager. The Grid Manager considers a layout to be organized as a two-dimensional array of cells organized in rows and columns. You do not have to specify the number of rows and columns; this will be determined by the Grid Manager. A single cell is considered to be in a particular row and column. The width of a cell will be determined by the width of the column in which it lies, and the height of a cell will be determined by the height of the row in which it lies. The height of a row will be equal to the height of the widget (in that row) with the largest height. The width of a column will be equal to the width of the widget (in that column) with the largest width. As discussed below, these rules are modified if the widget spans more than one column or more than one row. Specification of column span is done using the `columnspan` keyword.

Normally, a cell will contain a single widget (e.g., a button, checkbox, scroll bar, etc.). In this appendix, we encourage the option of having a cell* contain a single frame widget instantiated by using `tk.Frame` or `tk.LabelFrame`. Each frame essentially establishes a lower level grid system with cells containing other widgets. This provides a nesting hierarchy for all the widgets and their placement in the dialog is fairly easy to reorganize if necessary. Figure A.2 shows the grid system that is being used for the dialog illustrated in Figure A.1. The high-level grid system is delineated using heavy dashed lines, while the various low-level grid systems are delineated using narrower dashed lines that are superimposed over the ridges that visually delineate a frame.

* We start with one frame occupying a single cell, but as we will see later, it is possible for a frame to span two or more adjacent cells.

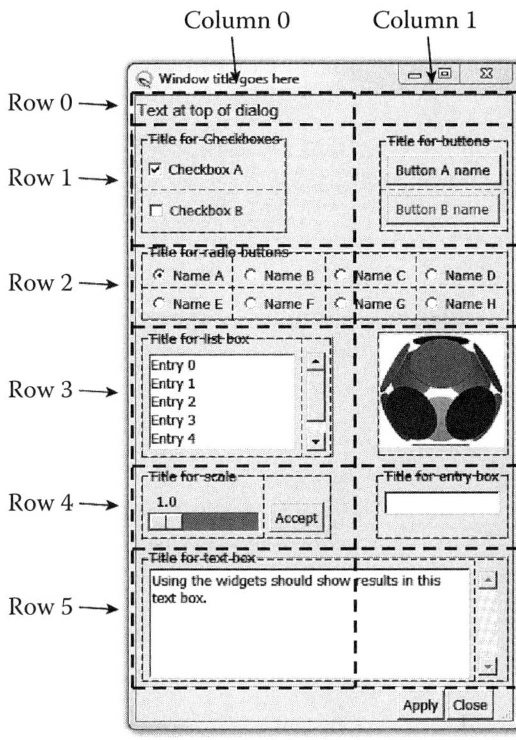

FIGURE A.2 Grid systems for the dialog of Figure A.1.

With some careful inspection you can verify that the high-level grid system has 12 cells organized in an array with 6 rows and 2 columns. These cells are used to organize nine "high-level" widgets as follows:

1. The `Label` widget referenced as `topText` in the script sets up a text string at the top of the dialog. The statement to do this relies on the Grid Manager and reads as follows:

```
topText.grid(row = 0, column = 0, columnspan = 2,
             pady = 4, sticky = tk.W)
```

   The row and column settings indicate that the label occupies all of row 0. It starts in column 0 and spans two columns (not really necessary for this short string of text but advisable if the text string is made longer). There is some vertical padding so that the text is not cramped by the edges of the cell and the "sticky" parameter specifies that the widget should hug the "West" side of the cell.

2. The next significant widget is a `LabelFrame` that contains a title label and two check buttons. In the high-level grid system, it occupies the cell in row 1 and column 0.

3. The third widget is a `LabelFrame` that contains a title label and two press buttons. In the high-level grid, it occupies the cell in row 1 and column 1.

4. The next `LabelFrame` contains a title label and eight radio buttons. In the high-level grid, it occupies both cells in row 2 because it starts in column 0 and spans two columns.

5. The fifth widget is a `LabelFrame` that contains a title label, a list box, and a scroll bar. In the high-level grid, it occupies the cell in row 3, column 0.

6. The sixth widget is a `Frame` that contains a `Canvas` widget. In the high-level grid, it occupies the cell in row 3, column 1.

7. The seventh widget is a `LabelFrame` that contains a title label, a `Scale` widget, and a button. High-level grid coordinates are row 4, column 0.

8. The eighth widget is a `LabelFrame` that contains a title label and an `Entry` box. High-level grid coordinates are row 4, column 1.

9. The last widget is a `LabelFrame` that contains a `Text` widget. In the high-level grid, it occupies both cells in row 5 because it starts in column 0 and spans two columns.

A.4.3 Widgets in the WidgetBuffet Dialog

We can now discuss the script snippets that produce the various widgets appearing in the eight frames.

A.4.3.1 The Checkbutton Class

The Checkbutton class can be used to generate check boxes in a GUI. Consider the following script:

```
# Setting up two check boxes:
# Checking the box will change the value
# of the IntVar and will initiate the execution
# of the function specified by the command option.
checkBoxFrame = tk.LabelFrame(parent, text = "Title for Checkboxes",
```

```
                    bd = 2, relief = tk.RIDGE)
checkBoxFrame.grid(row = 1, column = 0,
                    padx = 4, pady = 4, sticky = tk.W)

self.checkedBoxA = tk.IntVar()
cbA = tk.Checkbutton(checkBoxFrame, text = "Checkbox A ",
                    variable = self.checkedBoxA,
                    command = self.boxAfunc, indicatoron = 1)
cbA.grid(row = 0, column = 0, pady = 6, sticky = tk.E)
cbA.select()    # This button initialized with a check mark.

self.checkedBoxB = tk.IntVar()
cbB = tk.Checkbutton(checkBoxFrame, text = "Checkbox B ",
                    variable = self.checkedBoxB,
                    command = self.boxBfunc, indicatoron = 1)
cbB.grid(row = 1, column = 0, pady = 6, sticky = tk.E)
```

The script starts with the instantiation of the `LabelFrame` widget. Note that its utilization of the high-level grid will be stipulated by using the `parent` parameter. The border defining the visual extent of the frame is established by the option: `relief = tk.RIDGE`. It should be mentioned that this option has no visual effect if you leave out the option `bd = 2`, which specifies the width of this border. The subsequent statements instantiate the two check boxes, all within this frame. Note that they make use of the Grid Manager, but the grid system is relative to the frame (stipulated by the `checkBoxFrame` parameter in each instantiation). The cell coordinates for these two widgets are relative to this frame (see the narrow dashed lines within cell (1,0) of the high-level grid) in Figure A.2. For example, the first Checkbutton widget is in row 0, column 0 of this lower level grid. If you were to replace the `checkBoxFrame` parameter with `parent`, then the label widget would overwrite the text at the top of the dialog.

Each `Checkbutton` instantiation statement contains an option that specifies the callback function for that button. For example, a mouse click in checkbox A will initiate execution of the callback function `boxAfunc`:

```
# Callback functions needed for check boxes:
def boxAfunc(self):
    line = "Box A checked. Value = %1d" % self.checkedBoxA.get()
    self.textWidget.insert(tk.END, line + "\n")
    self.textWidget.yview(tk.END)
```

Normally, the statements in this function would perform an action related to this check box. In this dialog, the action is simply to have a text message displayed in the text box (referenced as `self.textWidget`) at the bottom of the dialog. Note that the indicator value (0 or 1) associated with the checkbox is available to the callback function by means of the `self.checkedBoxA.get()` call. The `get()` function is a method associated with `self.checkedBoxA`, which is designated as an IntVar() and used by the `variable` option in the `Checkbutton` instantiation statement. As a final observation, we see that the statement `cbA.select()` initializes checkbox A to be in its selected state.

A.4.3.2 Normal Buttons

The script:

```
# Setting up two buttons (one disabled for now):
buttonsFrame = tk.LabelFrame(parent, text = "Title for buttons",
                             bd = 2, relief = tk.RIDGE)
buttonsFrame.grid(row = 1, column = 1, padx = 4,
                  pady = 4, sticky = tk.E)

self.btnA = tk.Button(buttonsFrame, text = " Button A name ",
                      command = lambda: self.btnFunc("A"))
self.btnA.grid(row = 0, column = 0, padx = 4, pady = 4)

self.btnB = tk.Button(buttonsFrame, text = " Button B name ",
                      command = lambda: self.btnFunc("B"),
                      state = tk.DISABLED)
self.btnB.grid(row = 1, column = 0, padx = 4, pady = 4)
```

sets up two buttons in the frame residing in the cell at location (1, 1). If you understood the script for the checkboxes, then this script should be no trouble. Each button press initiates the execution of the callback function `self.btnFunc`. To make this script a bit more interesting, we are using the same callback function for both buttons. To identify the button that was pushed, we pass a parameter to the callback function by using a `lambda` expression. Using a `lambda` expression is necessary if this `command` option is to work properly after a mouse click on the button. Note that specifying the option as: `command = self.btnFunc("A")` will cause the `self.btnFunc` callback to be called during the instantiation of the button and this is not what we want to do! We want it to be called when the button is depressed using the mouse. To add more interest to this script, the option `state = tk.DISABLED` has been used in the instantiation for Button B. Consequently, when the dialog is launched, this button is initially in a "grayed-out" disabled state. To give Button A "something to do," we have included the statement `self.btnB.configure(state = tk.NORMAL)` in its callback function and this will put Button B into its normal (not disabled) state.

A.4.3.3 A Group of Radio Buttons

The script:

```
# Set up radio buttons as specified by a list:
rButtonsFrame = tk.LabelFrame(parent, text = "Title for radio buttons",
                              bd = 2, relief = tk.RIDGE)
rButtonsFrame.grid(row = 2, column = 0, columnspan = 2, padx = 4, pady = 4,
                   sticky = tk.W + tk.E)

# Use tk.IntVar() here for integer result values.
self.rBtnResult = tk.StringVar()
radioBtn_L = [("Name A", "Result A", 0, 0),
              ("Name B", "Result B", 0, 1),
              ("Name C", "Result C", 0, 2),
              ("Name D", "Result D", 0, 3),
              ("Name E", "Result E", 1, 0),
              ("Name F", "Result F", 1, 1),
              ("Name G", "Result G", 1, 2),
              ("Name H", "Result H", 1, 3)]

self.rBtnResult.set("Result A") # This button starts off selected.
for text, rslt, rw, cl in radioBtn_L:
    rBtn = tk.Radiobutton(rButtonsFrame, text = text,
                          variable = self.rBtnResult,
                          value = rslt, command = self.rBtnfunc)
    rBtn.grid(row = rw, column = cl, padx = 4, sticky = tk.W)
```

sets up a `Label` widget and eight radio buttons within the frame referenced as `rButtonsFrame`. Having learned about the `command` and `variable` options in the previous sections, this script should be easy to understand. The names of buttons and their associated result values have been specified in the `radioBtn_L` list. Tuples in the list also contain the low-level grid coordinates for each button. Note that the frame for the button has a column span of two cells and so the frame is correspondingly bigger than the previous frames. The locations of the button are in a low-level grid delineated by the narrow lines within cells (2, 0) and (2, 1) of the high-level grid in Figure A.2. For example, the radio button "Name G" can be found in cell (2,1) of this frame. The callback function `self.rBtnfunc` is the same for all radio buttons. The `value` option of the radio button instantiation statement specifies a result for each radio button. Each result is a string such as "`Result A`" and this is made available to the callback function when it executes the statement `self.rBtnResult.get()`. In this call, `self.rBtnResult` is a `StringVar`, which is designated in the radio button instantiation as the parameter for the `variable` option. Finally, note that `self.rBtnResult.set("Result A")` designates the radio button that is "preselected" when the dialog is launched.

A.4.3.4 A List Box Widget with a Scroll Bar Widget

The script:

```
listBoxFrame = tk.LabelFrame(parent, text = "Title for list box",
                             bd = 2, relief = tk.RIDGE)
listBoxFrame.grid(row = 3, column = 0, padx = 4, pady = 4,
                  sticky = tk.W + tk.NS)

self.lstBx = tk.Listbox(listBoxFrame, height = 5, width = 18,
                        selectmode = tk.SINGLE, bg = "white")
                        # Use MULTIPLE for checkbox like operation.

scroll = tk.Scrollbar(listBoxFrame, command = self.lstBx.yview)
self.lstBx.configure(yscrollcommand = scroll.set)
scroll.grid(row = 0, column = 1, padx = 5, pady = 4,
            sticky = tk.W +tk.N + tk.S)

lstBx_L = ["Entry 0", "Entry 1", "Entry 2",
           "Entry 3", "Entry 4", "Entry 5"]
for item in lstBx_L: self.lstBx.insert(tk.END, item)
self.lstBx.grid(row = 0, column = 0, padx = 4, pady = 4,
                sticky = tk.E)
self.lstBx.bind("<Double-Button-1>", self.Apply)
```

sets up a `Listbox` widget and a `Scrollbar` widget within the frame located in the cell at row 3 and column 0 of the high-level grid. The low-level grid for this frame has a grid system with one row and two columns. Row 0 contains the list box in column 0 along with the vertical scroll bar in column 1. Note that the height of row 0 is determined by the list box vertical height, which is specified by the option `height = 5`, where 5 is the number of visible lines in the list box. The width of the list box, here set to be 18, is also given in terms of a character count. This dialog has the option `selectmode = tk.SINGLE` meaning that one can only select a single item from the list box. Using `selectmode = tk.MULTIPLE` would allow multiple selections.

By using the function call: `self.lstBx.curselection()`, we can determine which list box entry was selected. For the sake of convenience, this statement has been put into the callback function for the Apply button. Note that selection of a list box line such as Entry 2 will produce a `curselection` return value that is the tuple ('2',). Since the selection mode is `SINGLE`, there is only one entry in this tuple. As you would expect, using a `MULTIPLE` selection mode produces tuples with multiple entries.

As mentioned earlier, column 1 of row 0 contains a scrollbar widget. Notice how the script facilitates communication between the listbox widget and the scrollbar widget. The instantiation of the `Scrollbar` widget includes the option: `command = self.lstBx.yview`, which references the listbox and sets the orientation of the scrollbar to be vertical (specified as `yview`). The next statement: `self.lstBx.configure(yscrollcommand = scroll.set)` configures the listbox so that it knows about the `scroll` widget. Note how the listbox and the scrollbar can be visually associated by having the listbox move to the "East side" of its low-level cell (`sticky = tk.E`), while the scrollbar snuggles up to the "West side" of its cell (`sticky = tk.W +tk.N + tk.S`). The `tk.N` and `tk.S` specifications ensure that the vertical height of the scrollbar match the height of the listbox. The final layout has the listbox adjacent to the scrollbar with a small space between the two, as dictated by the `padx` settings.

The listbox is populated with entries by using an `insert` function within a `for` loop that works through a list of character strings. Note that each entry is inserted at the end of the growing list in the listbox.

The last statement: `self.lstBx.bind("<Double-Button-1>", self.Apply)` is interesting. It produces a "binding" between an event and a callback function. More precisely, double clicking of an entry in the listbox will cause the callback function for the Apply button to be executed. This is a convenient way for the user to get immediate action on a listbox selection.

A.4.3.5 A Canvas Widget

The script:

```
# Set up a canvas to the right of the list box:
canvasFrame = tk.Frame(parent, bd = 2, relief = tk.RIDGE)
canvasFrame.grid(row = 3, column = 1, padx = 4, pady = 4,
                 sticky = tk.E)

# Title left out (a picture is worth a thousand words...)

picture = tk.Canvas(canvasFrame, width = 110, height = 107)
picture.grid(row = 0, column = 0)

gfImage = tk.PhotoImage(file = "canvasExample.gif")
picture.create_image(0, 0, image = gfImage, anchor = tk.NW)
# Store a reference to the image as an attribute of the canvas:
picture.img = gfImage  # Remove this and the image disappears.
```

puts a `Canvas` widget into the frame that is in cell (3,1) of the high-level grid. A canvas can be used to display graphics or images. There are several methods associated with the widget allowing the programmer to insert lines, circles, polygons, and so on. These capabilities will be ignored because they are beyond our needs. For the dialog being considered, the canvas widget simply displays the graphic `canvasExample.gif`. Note that the `Canvas` widget has a width and height that are specified as options in the `Canvas` instantiation.

The image is accessed by opening a file having a pathname that is specified by the `file` option in the `PhotoImage` function. The image is transferred to the `Canvas` widget by means of the function `create_image`. The pixel coordinates (0,0) and the `anchor` option are used to locate the image within the two-dimensional canvas.

Note that it is important for the image to have a reference that is an attribute of the canvas. This is accomplished with the statement: `picture.img = gfImage`. Without this statement, the image will not be retained in the dialog and the canvas will be empty.

A.4.3.6 A Scale Widget

A scale is useful if the user is required to input a number that is restricted to some prespecified range of possible values. An example would be the requirement to stipulate a threshold value prior to the execution of an algorithm.

The script:

```
# Setting up a scale:
scaleFrame = tk.LabelFrame(parent, text = "Title for scale",
                           bd = 2, relief = tk.RIDGE)
scaleFrame.grid(row = 4, column = 0, padx = 4, pady = 4,
                sticky = tk.W)

self.scaleVar = tk.DoubleVar()
scale = tk.Scale(scaleFrame, variable = self.scaleVar,
                 orient = tk.HORIZONTAL, length = 100,
                 from_ = 1.0, to = 12.0, resolution = 0.1,
                 troughcolor = "goldenrod1")
scale.grid(row = 0, column = 0, sticky = tk.W)

acceptButton = tk.Button(scaleFrame, text = "Accept",
                         command = self.acceptSelection)
acceptButton.grid(row = 0, column = 1, padx = 4, pady = 4,
                  sticky = tk.S + tk.W)
```

sets up a `Scale` widget and a button in the frame occupying cell (4, 0) of the high-level grid. Statements dealing with the `LabelFrame` and `Button` instantiations are similar to those covered in earlier sections. Note that the callback function for the Accept button is `acceptSelection()`.

Let us review some of the options that appear in the `Scale` instantiation statement. The option `orient = tk.HORIZONTAL` establishes a horizontal orientation for the scale while the option `length = 100` specifies its length in pixels. The options `from_` and `to` will dictate the range of values that the scale will put out. Since `from` is a Python keyword, Tkinter uses the underscore in the `from_` option to avoid a Python error. The `resolution = 0.1` option dictates that output from the scale will increment in steps of 0.1 as the scale slider is moved.

The value provided by the scale is retained in the `self.scaleVar`. This is specified by the option `variable = self.scaleVar`. This variable is initialized as a `DoubleVar()` and it is accessed in the `acceptSelection` callback function where it is displayed in the text box.

A.4.3.7 An Entry Box Widget

Entry boxes are useful for getting short character strings from a user, for example, a Protein Data Bank (PDB) identifier.

The script:

```
# Setting up a entry box:
entryFrame = tk.LabelFrame(parent, text = "Title for entry box",
                           bd = 2, relief = tk.RIDGE)
entryFrame.grid(row = 4, column = 1, padx = 4, pady = 4,
                sticky = tk.E + tk.NS)

self.entryVar = tk.StringVar()
entry = tk.Entry(entryFrame, textvariable = self.entryVar,
                 width = 14, bg = "white")
entry.grid(row = 0, column = 0, padx = 4, pady = 4,
           sticky = tk.E)
# Uncomment the next line if the entry box is to start DISABLED.
#entry.config(state = tk.DISABLED)
```

places an `Entry` widget into the frame within cell (4,1) of the high-level grid. If you understood all the Tkinter statements in the preceding sections, then there is nothing really new here. Just two small notes: the option `width = 14` is specifying the entrybox width in terms of character length and the `Entry` widget can be disabled by means of the `config` function.

A.4.3.8 A Text Box Widget

A text box can be used for entry of longer strings of text but, in this dialog, it is used to provide status output.

The script:

```
# Setting up a text box:
textFrame = tk.LabelFrame(parent, text = "Title for text box",
                          bd = 2, relief = tk.RIDGE)
textFrame.grid(row = 5, column = 0, columnspan = 2,
               padx = 4, pady = 4, sticky = tk.W + tk.E)

self.textWidget = tk.Text(textFrame, width = 40, height = 6,
                          wrap = tk.WORD, bg = "white")
self.textWidget.grid(row = 0, column = 0, padx = 4,
                     sticky = tk.W)

textScroll = tk.Scrollbar(textFrame,
                          command = self.textWidget.yview)
self.textWidget.configure(yscrollcommand = textScroll.set)
textScroll.grid(row = 0, column = 1, padx = 2, pady = 4,
               sticky = tk.W +tk.N + tk.S)
```

places a `Text` widget into the labeled frame that is in cells (5, 0) and (5, 1) of the high-level grid. Note that the `LabelFrame` widget spans two columns. Again, if you understood all the Tkinter statements in the preceding sections, then there is nothing really new here. The scrollbar is set up in a fashion that is similar to what we have seen for the list box. The `Text` widget has a width and height that is specified by a character count. Note that the textbox has a `wrap` option that determines how a line continues to the next line. By stipulating `wrap = tk.WORD`, we will not get lines ending in mid-word unless they are extremely long (longer than the width of the text box).

In our implementation of the dialog, character strings are put into the text box by the callback functions. In all such cases, a character string stored in the variable `line` is created and then inserted at the end of the text box by using the function call: `self.textWidget.insert(tk.END, line + "\n")`. Note the use of "`\n`" so that each callback addition to the text appears as a new line. The final statement: `self.textWidget.yview(tk.END)` is used to ensure that the text scrolls up as each new line is put into the text box. This avoids the need for the user to constantly scroll the text box to see the latest new lines.

A.4.4 Testing the WidgetBuffet Dialog

The `widgetBuffet.py` script contains a very simple script to test the dialog:

```
wb = WidgetBuffet()
out = "Using the widgets should show results in this text box.\n"
wb.textWidget.insert(tk.END, out)
```

The dialog class is instantiated and this is enough to launch it. The next statement shows how a mainline script can access a widget in the dialog. To illustrate this, a character string has been sent to the text box at the bottom of the dialog.

It should be noted that this dialog does not do any interactions with the Python Shell. Consequently, you could simply launch it as an application by using the **File... Open...** menu of the Chimera window and all further interactions would not involve the Python Shell.

A.5 BUILDING A MODELESS DIALOG

Understanding the underlying construction and functionality of the widgets in the `WidgetBuffet` dialog, we can construct other modeless dialogs by simply "stealing and modifying" its statements. One would go through the following steps:

1. **Preamble**: Copy over the preamble statements seen in Section A.4.1. You will want to rename the dialog and define the text to be placed in the window pane and at the top of the dialog. You can also bring in various standard buttons that are available, for example: "OK" and "Close." Any other buttons that you name will be set up at the bottom of the dialog but you will be responsible for providing the callback functions.
2. **Layout**: Copy over the needed frames along with the widgets that they contain. Modifications will likely include the following:
    a. Decide where the frames will appear relative to the high-level grid.
    b. Decide what goes into a frame (title label, number of buttons, etc.).

c. Determine the layout of these widgets relative to the low-level grid within the frame. This will involve appropriate use of the "`sticky`" options and may require some modifications of the width, height, and padding options. You will probably test the dialog several times, making several changes until the layout is both user-friendly and aesthetically pleasing.

d. Be sure that the communication between the widget and the callback is properly handled (creating the appropriate `IntVar()`, `StringVar()`, or `DoubleVar()` variables).

3. **Callback functions**: Implement the callback functions. This will include statements that deal with interactions between the dialog and the mainline program. You will also need to determine whether a callback function responds to one widget or to several (as in the Button example discussed in Section A.4.3.2).

4. **Interactions with the mainline**: Write the code for the mainline program that will instantiate the dialog and subsequently work with the widgets that it contains.

A.6 BUILDING A MODAL DIALOG

Most of the script snippets that we have seen in the previous sections can be applied to modal dialogs. Let us now consider how a modal dialog differs from a modeless dialog by reviewing the statements in the file: `Script A_02 Sample ModalDialog.py`. It is a small dialog so be careful not to lose it among any other windows that may be open on your desktop.

Here is the script:

```
import chimera, Tkinter as tk
from chimera.baseDialog import ModalDialog

class userInput(ModalDialog):
    title = "User Input"
    buttons = ('OK', 'Cancel')
    # Leave out next statement to get a Help button.
    help = False

    def __init__(self, requestToUser):
        self.requestToUser = requestToUser
        ModalDialog.__init__(self)
```

```
    def fillInUI(self, parent):
        topText = tk.Label(parent,
                           text = self.requestToUser)
        topText.grid(row = 0, column = 0)

        # Next line is needed in case the user does a Cancel.
        self.result = None
        self.entry = tk.Entry(parent, width = 8)
        self.entry.grid(row = 0, column = 1)

    #=====================================================
    # CALLBACK functions:

    def OK(self):
        self.result = self.entry.get()
        ModalDialog.Cancel(self, value=None)

#============================================================
# Mainline:

md = userInput("  Type in a PDB ID:  ")
md.run(chimera.tkgui.app)
if md.result: print md.result
```

Notice that inheritance is from `ModalDialog` and not `ModelessDialog`. Object initialization is done by the `__init__` call which takes in a string parameter that will be used to define `topText`, which is a `Label` widget at the top of the dialog. The `__init__` function also does initialization of the `ModalDialog` base class: `ModalDialog.__init__(self)`. Functionality of statements in the `fillInUI` function is essentially the same as what we saw earlier for the modeless dialog. Care must be taken to ensure that a final result is available to the mainline program when the user clicks on the OK button. In the given script, the widgets are not enclosed in one or more frames because the layout is very simple. The `Label` and `Entry` widgets are positioned using a grid with one row and two columns.

The callback function for the OK button gets a string result from the `Entry` widget and sets it up in the `self.result` variable, which can be accessed by the mainline code immediately after the dialog disappears due to the execution of: `ModalDialog.Cancel(self, value=None)`. Note that there is no callback function for the Cancel button because this functionality is handled by a base class method.

The mainline is fairly simple. It instantiates the dialog but then another statement is required to run it. Since it is a modal dialog, the statement following `md.run(chimera.tkgui.app)` will not run until the user dismisses the dialog by clicking on either the OK button or the Cancel button.

REFERENCE

[GR05] J. E. Grayson. *Python and Tkinter Programming*. Manning Publications, Greenwich, CT, 2005.

Appendix B

Scenographics

B.1 MOTIVATION AND INTRODUCTION

There are many occasions when it is desirable to put more graphical content into a molecular display. We might wish to use an arrow to show a possible direction of movement for an atom or a cylinder-like object to represent the inertial axis of a helix. In various mathematical models, an amino acid side chain may be approximated by an ellipsoid with a corresponding visual component in the display. The `Scenographics`* package in the `StructBio` folder provides various methods that can be used to place three-dimensional (3D) objects and surfaces into the current display of a molecule. All objects can be positioned in 3D display space by specifying coordinates and, in some cases, the directional orientation of the object. Object size and color can also be specified.

The `Scenographics` package has several classes that allow a programmer to place scenographic elements into a Chimera scene. These fall into the following six categories:

1. Solids
2. Surfaces
3. Lines

* As noted in Wikipedia (http://en.wikipedia.org/wiki/Scenography), the term *scenography* goes back to 1862 and was used by Antonio Caimi who excelled in *Arte scenografica*, which is the study and practice of design for stage settings. These endeavors "required ingenious engineering to create movable sets, or create illusions of environments." In our case, the ingenuity will require the application of linear algebra and graphics algorithms to create extra elements in the molecular display.

4. Parametric surfaces
5. Ellipsoids
6. Labels

These classes are not completely independent. For example, the class that generates parametric surfaces will use the classes that generate surfaces and lines. Considering these dependencies, we will cover these classes in the order just specified.

B.2 USING THE `Solids` CLASS

The `Solids` class has methods to draw various 3D objects including struts, tubes, spheres, cubes, boxes, disks, spindles, arrows, and the Platonic solids (tetrahedrons, hexahedrons, octahedrons, dodecahedrons, and icosahedrons). Figure B.1 shows some typical examples, whereas Figure B.2 presents a whimsical view of the five regular convex polyhedra referred to as the Platonic solids. Figure B.1 was produced by the script `Script B_01 Display SolidsTest.py` and Figure B.2 was produced by the script `Script B_02 Platonic Garden.py`.

Using the class as shown in these scripts requires the following import statements:

```
from StructBio.Scenographics.solids import Solids
from StructBio.Utilities.miscFunctions import colorByName
```

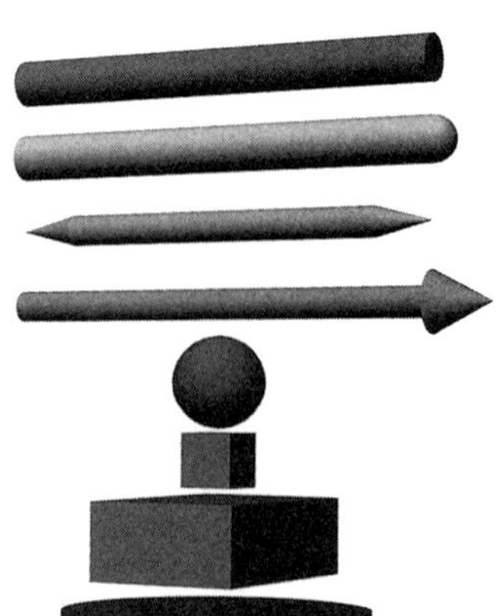

FIGURE B.1 **(See color insert.)** Some examples of the available three-dimensional (3D) objects from the `Solids` class.

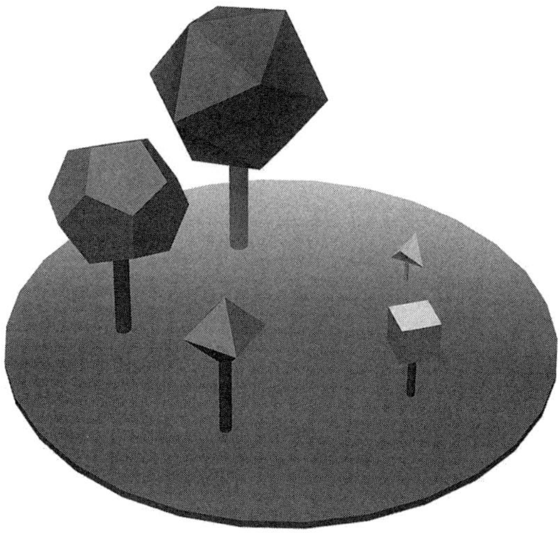

FIGURE B.2 (**See color insert.**) A Platonic garden.

The first statement brings in the `Solids` class definition. The importation of the `colorByName` function found in the `StructBio.Utilities.miscFunctions` package is required if you wish to specify a color by using a character string such as "`light sea green`" which is one of the Chimera built-in colors.* Typical use of `Solids` will also require the importation of `numpy` so that the script can compute 3D coordinates using data representations provided by `array()`. After the import statements, the script can instantiate `Solids` objects, for example:

```
solidsDisp = Solids()
```

The default model name is provided by the Python string `"Solids graphics"`, but this can be changed by giving the class initialization function a character string argument, for example:

```
solidsDisp = Solids("Platonic solids")
```

as seen in `Script B_02 Platonic Garden.py`.

After the object instantiation, the script can add various objects to the display, for example:

```
solidDisp.addSphere(center, radius, rgb)
```

* http://www.cgl.ucsf.edu/chimera/docs/UsersGuide/colortables.html.

where center, radius, and rgb specify the obvious attributes for the sphere. Once all the required objects have been added to the scene, we can initiate their display as follows:

```
solidDisp.display()
```

Here is a list of the function headers in the Solids class:

```
def addStrut(self, p1, p2, radius, rgb):
def addTube(self, p1, p2, radius, rgb):
def addSphere(self, center, radius, rgb):
def addCube(self, center, side, rgb):
def addBox(self, center, sides_T, rgb):
def addDisk(self, center, normal, thickness, radius, rgb):
def addSpindle(self, p1, p2, radius, rgb):
def addArrow(self, p1, p2, radius, rgb):
def addTetrahedron(self, center, size, rgb, transparency = 1.0):
def addHexahedron(self, center, size, rgb, transparency = 1.0):
def addOctahedron(self, center, size, rgb, transparency = 1.0):
def addDodecahedron(self, center, size, rgb, transparency = 1.0):
def addIcosahedron(self, center, size, rgb, transparency = 1.0):
def addPolygon(self, polyPoints_L, rgb, transparency = 1.0):
```

Most of these methods have an obvious functionality. The addStrut function puts a cylinder (capped at both ends) into the display, whereas the addTube function generates a *test-tube*-like object (a cylinder that is capped at one end and has a half sphere at the other end). Figure B.1 shows a strut, tube, spindle, arrow, sphere, cube, box, and disk.

The parameters for the function calls are fairly straightforward. The requirements are as follows:

p1, p2, center are arrays that hold coordinates for 3D points.

radius, side, thickness are floating point positive numbers.

rgb designates a color (a 3-tuple of floating point numbers each in the range [0.0, 1.0]).

transparency is a number in the range [0.0, 1.0].

normal is an array that holds a three-component vector that designates a direction.

sides_T is a tuple that designates the lengths of the sides of a box (three positive floating point numbers).

Figure B.2 was generated by using Script B_02 Platonic Garden. The function calls are quite straightforward. There is no setting of the parameters that will rotate these objects.

B.3 USING THE `Surfaces` CLASS

The `Surfaces` class has methods that generate a triangulated surface. The import statement to access the class is as follows:

```
from StructBio.Scenographics.surfaces import Surfaces
```

A sample script using the class can be found in `Script B_03 Display Polygons test.py`. As expected, generating a surface starts with a class instantiation such as

```
pd = Surfaces()
```

After this, you can add surface elements by doing function calls to `addPolygon` or `addFace`. The parameter sequence is the same for both function invocations. The first argument is a list of vertices (each being a `numpy` array) and the next two arguments define the color and transparency of the surface. After all polygonal additions are completed, the surface can be displayed by executing the `display()` function call.

If you run the script, you will notice that rotation of the scene will cause the polygons defined by `addFace` to disappear when you attempt to look at them from *behind the surface*. Why is this useful? If the surface pieces are enclosing a 3D object, then there will be pieces on the far side of the object that cannot be seen because they are concealed by the part of the surface that is closer to the observation point. If the surface comprises pieces generated by `addPolygon`, then they will be handled by the graphics engine even though they will eventually be obscured from view. If the pieces are generated by `addFace`, then they are simply omitted when seen from behind. This *back-face culling* in the graphics pipeline provides a performance advantage because surface pieces that are not seen will not be processed for display.

A cautionary note: One must be careful to distinguish the front face from the back face of a polygon. This orientation is defined by the ordering of the vertices. The front (visible) face will have its vertices listed in counter-clockwise fashion in the `addFace` invocation. So if you are building a surface by using `addFace`, you will have to ensure that the vertices are listed in the required order. This will avoid unintended *surface holes*.

Note that `Script B_03 Display Polygons test.py` also has a user interaction via the Python Shell. This interaction shows how the function calls `hide()` and `show()` can be used to control the appearance of the surfaces in the display.

In Section B.5 we will discuss classes that provide more sophisticated surface constructions that are defined by means of parameterized equations.

B.4 USING THE `Lines` CLASS

The `Lines` class has methods that generate an unbroken line that is made from piecewise linear line segments. The import statement to access the class is as follows:

```
from StructBio.Scenographics.lines import Lines
```

A sample script using the class can be found in `Script B_04 Display Lines test.py`. As expected, generating a line starts with a class instantiation such as

```
pd = Lines()
```

After this you can add line elements by doing function calls to `addLine`. The first argument is a list of vertices (each being a `numpy` array) and the next argument defines the color of the line. After all line additions are completed, the lines can be displayed by executing the `display()` function call. The `Lines` class has `hide()` and `show()` methods with the same functionality as in the `Surfaces` class.

B.5 INTRODUCING THE `ParametricSurfaces` CLASS

The `ParametricSurfaces` class is a base class that provides the data structures and methods that are to be used by other derived classes. The `ParametricSurfaces` class will allow you to construct surfaces that are much more complex than what we have seen so far. The class probably has more functionality than you will need but it should hopefully meet any extra future requirements that are not currently foreseen. To understand how this class is used to build a surface, we consider the following steps:

- Step 1: Start with a curve in the x–y plane. This curve is defined as a sequence of points held in the list `baseCurvePts _ L`. Each entry in the list is a two component `numpy` array. The number of points is `numBasePts`.

- Step 2: Use an algorithm that will replicate the base curve to form a *cross section* with some particular 3D position and orientation. This is done repeatedly to generate a list of cross sections stored in `crossSections_L`.
- Step 3: The succession of cross sections can be used to define either a mesh or a surface.

If you inspect the `__init__` function of the `ParametricSurfaces` class,* you will see the declarations for these three variables. Figure B.3 gives an elementary example of a base curve with its nine points represented by small spheres. The details for Steps 2 and 3 will be covered later. For now, we describe these steps in more general terms.

Figure B.4 shows the base curve replicated to generate four cross sections (as in Step 2). Again, for emphasis, each of the points in a cross section is represented by a small sphere.

We can now ask for a mesh that will connect the corresponding points in the successive cross sections. This is Step 3 and is illustrated in Figure B.5.

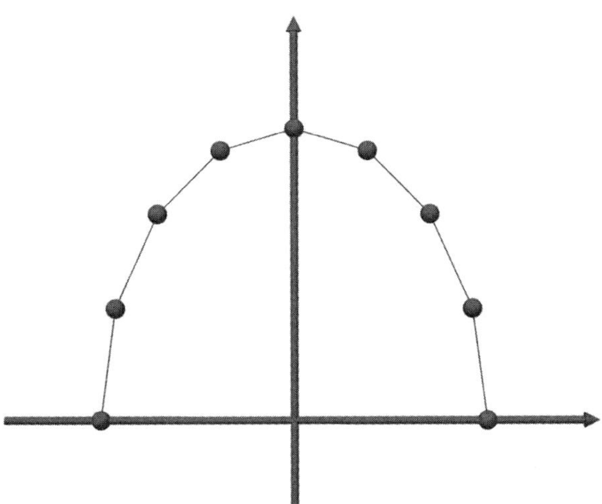

FIGURE B.3 Base curve points.

* See the package component `StructBio.Scenographics.parametricSurfaces`, available from the book's website.

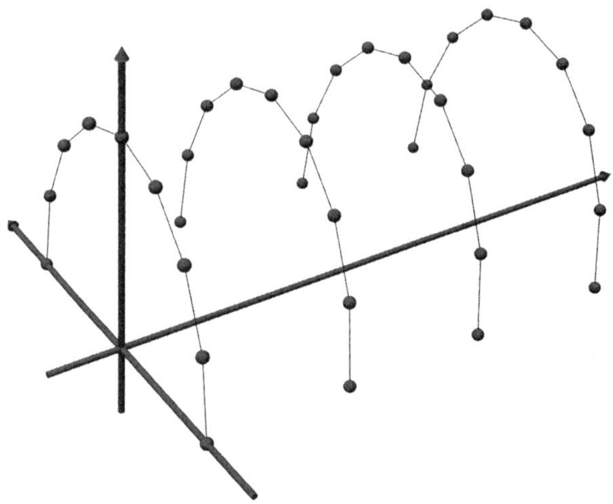

FIGURE B.4 Cross sections.

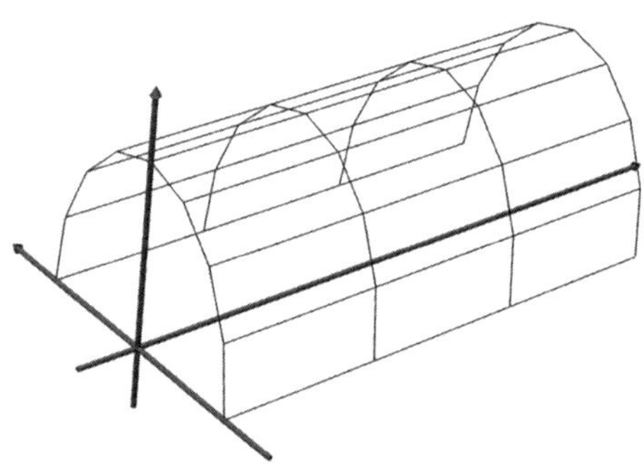

FIGURE B.5 Generation of a mesh.

The surface can be defined by filling in the mesh with polygonal surface elements as in Figure B.6. Note that the surface has been given a transparency of 0.8 and so the *z* axis can be seen beneath the surface.

Our mesh and surface constructions will allow some extra variations. Setting `closedBase = True` causes the mesh generation or surface generation methods to treat the base curve points as defining a closed curve. The successive cross sections then define a *cylinder* type of surface as illustrated in Figure B.7a. When the surface has a closed base curve we can *cap* the ends. If we set `startCap = True`, then the first cross section defines a polygon that will be filled in with a surface as illustrated in Figure B.7b. A similar setting can be used for `endCap`.

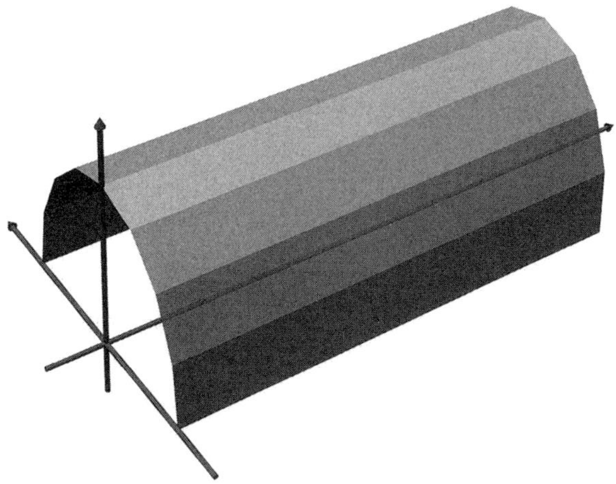

FIGURE B.6 Generation of a surface.

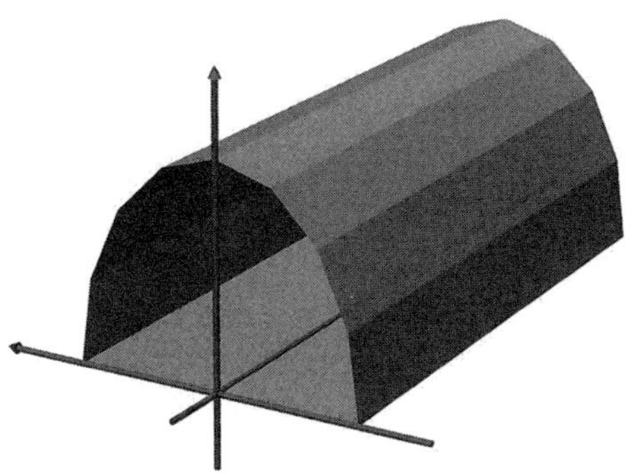

(a)

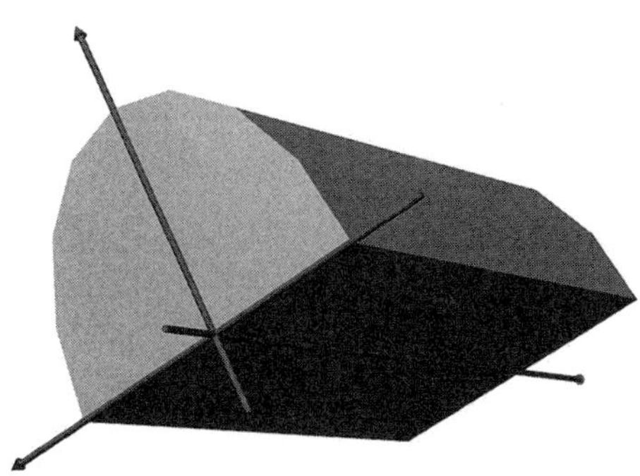

(b)

FIGURE B.7 (a) Surface with closed base. (b) Surface with start cap.

If you inspect the script that defines the ParametricSurfaces base class, you will find various methods that are related to Step 3 discussed earlier. A quick overview of these functions will be described next:

```
__init__(self, baseCurvePts_L, modelName = "")
```

This is the initialization method for the base class instantiation. It accepts the list of points of the base curve and sets up the required data structures needed for surface or mesh generation. Setting modelName can be used to give the model a name that will be seen in the Model Panel dialog (accessed using the menu invocation: **Favorites... Model Panel**).

```
rotateAndMove(self, rotMat, translationVector)
```

This method is used by the derived classes to reposition and possibly rotate the vertices stored in the list of cross sections.

```
removeMesh(self)
```

This is used to eliminate the Chimera model holding the mesh lines.

```
removeSurface(self)
```

This is used to eliminate the Chimera model holding the surface.

```
displayMesh(self, color, closedBase = False,
            startCap = False, endCap = False, closedSpine = False)
```

This method accesses the cross-section list and uses that data to generate the mesh lines. Notice the default settings for closedBase, startCap, and endCap. The closedSpine parameter will be covered later in this section when discussing Figure B.10.

```
displaySurface(self, color, transparency = 1.0,
               closedBase = False, startCap = False,
               endCap = False, closedSpine = False)
```

The displaySurface method will display a surface with the color and transparency specified. Notice the default settings for closedBase, startCap, and endCap. It is possible to use both displaySurface and displayMesh to get the display of a surface with an embedded mesh.

```
showSurface(self), hideSurface(self), showMesh(self), and
hideMesh(self)
```

These functions have the same functionality as the `show` and `hide` methods found in the `Surfaces` class.

As mentioned earlier, all these methods are related to Step 3. The functionality needed to set up the `crossSections_L` list (done in Step 2) is accomplished by classes that are derived from the `ParametricSurfaces` class. Each of these classes has its own strategy for the positioning and orienting of the cross sections. These classes can be listed as follows:

- `FrenetFrameSurface`
- `ExtrusionSurface`
- `RevolutionSurface`
- `RuledSurface`

B.5.1 Using the `FrenetFrameSurface` Class

The `FrenetFrameSurface` class is the most complicated class to use, but we cover it first because the `ExtrusionSurface` and `RevolutionSurface` classes are actually derived from this class and give the programmer an easier approach to working with this class.

The class can be imported by using

```
from StructBio.Scenographics.parametricSurfaces \
   import FrenetFrameSurface
```

The object instantiation is done via the `__init__` method that starts with the following header:

```
__init__(self, baseCurvePts_L, spinePts_L,
         rotation3Dmats_L, xScale_L, yScale_L,
         modelName = "")
```

As described earlier, we assume a base curve is defined in the *x*, *y* plane by specifying a list of two-dimensional (2D) points: `baseCurvePts_L`. When a `FrenetFrameSurface` object is instantiated, it will be given a list of points in 3D space defining a *spine curve*. In the `__init__` function, these points are provided by `spinePts_L`. Three more lists `rotation3Dmats _L`, `xScale _L`, and `yScale _L` define operations for 3D rotation and scaling.

We describe the functionality of this class by providing more details for Step 2 introduced at the beginning of Section B.5. Recall the purpose of Step 2: We use an algorithm that will replicate the base curve to form a *cross section* with some particular 3D position and orientation. This is done repeatedly to generate a list of cross sections stored in `crossSections_L`. We can now describe the operations required to generate the ith cross section, that is stored in `crossSections_L[i]`:

- Step 2(a): Copy the points in the `baseCurvePts_L` list to `crossSections _ L[i]`.
- Step 2(b): The x coordinate for every point in `crossSections _ L[i]` is scaled by a factor stored in `xScale_L[i]`.
- Step 2(c): The y coordinate for every point in `crossSections _ L[i]` is scaled by a factor stored in `yScale_L[i]`.
- Step 2(d): Every point in `crossSections_L[i]` is subjected to a 3D rotation operation specified by the rotation matrix stored in `rotation3Dmats_L[i]`.
- Step 2(e): Translate the points in `crossSections_L[i]` by using the vector stored in `spinePts_L[i]`.

Figure B.8 gives an example demonstrating these operations. The base curve is shown on the lower left. Three arrows have been set up to show the x, y, and z axes. In this case only four cross sections have been created. No scaling was done for the x coordinate, but the y coordinate of each base curve point was subject to scaling using the four scaling factors [1.0, 0.8, 0.6, 0.4]. Four rotation matrices were generated in ad hoc manner and used to rotate the scaled base curve prior to translation:

$$\begin{bmatrix} \cos(i\pi/12) & 0 & \sin(i\pi/12) \\ \sin(i\pi/12) & 0 & -\cos(i\pi/12) \\ 0 & 1 & 0 \end{bmatrix} \quad i = 0,1,2,3$$

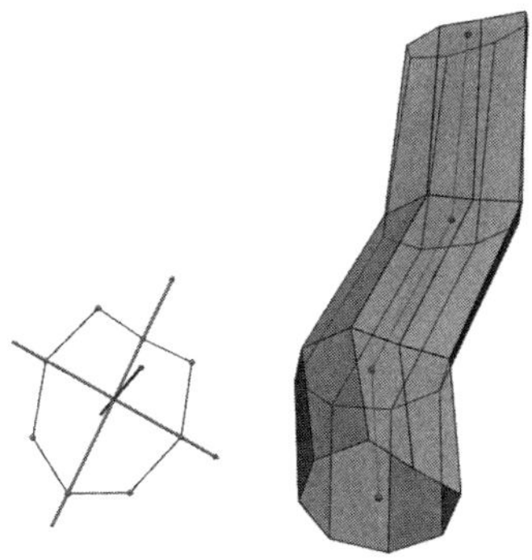

FIGURE B.8 Generation of a simple Frenet frame surface.

The four spine points stored in `spinePts_L` were [[5, 1, 1], [4, 2, 2], [4, 4, 3], [3, 6, 4]]. These points have been emphasized in Figure B.8 as small spheres.

After instantiation of the class object, `displayMesh` was used to generate the mesh and `displaySurface` (color gray with a transparency of 0.5) was used to show the surface. The script for Figure B.8 can be found in the file `Script B_05 Frenet frame surface Example 1.py`. This example has only four cross sections, but it has some visual complexity because both the *y* scaling factor and the rotation matrix change as we go from one cross section to the next. Normally, a Frenet frame surface would have many more cross sections and they would be closer together. Furthermore, the rotations and scaling factors would vary in small steps to facilitate a smooth surface continuity that is lacking in Figure B.8.

It should be noted that typical use of the Frenet Frame approach involves moving the base *profile* along the spine curve in the generation of the successive cross sections. Each cross section corresponds to the base curve being given an orientation and position that is established by rotating and translating its local frame of reference now called the Frenet frame. As depicted in Figure B.9, this is done in such a way that the *z* axis of the local frame moves along the spine curve so that it is tangential to the spine curve.

This places a constraint on the generation of the rotation matrices. So, when using the `FrenetFrameSurface` class you should take care to ensure that the rotation matrices force the *z* axis to be parallel with a tangent to the spine curve. There are typically two ways to do this: The rotation matrices can be formulated by taking into account the global topology of the surface (as in the next example) or rotation matrix `rotation3Dmats_L[i]` can be built so that the *z* axis of the local frame is parallel to a vector that

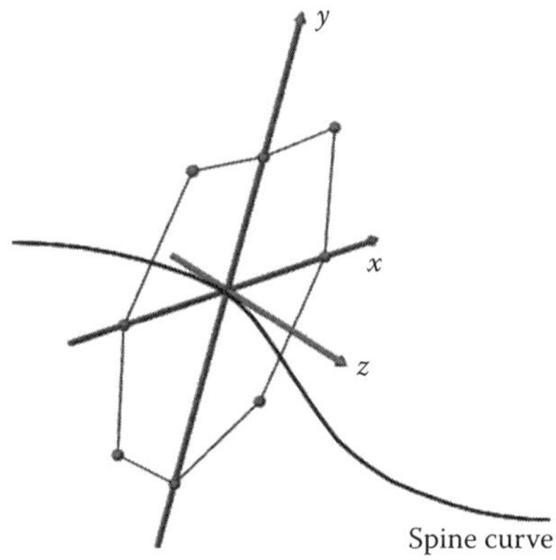

FIGURE B.9 Base curve in Frenet local frame moving along the spine curve.

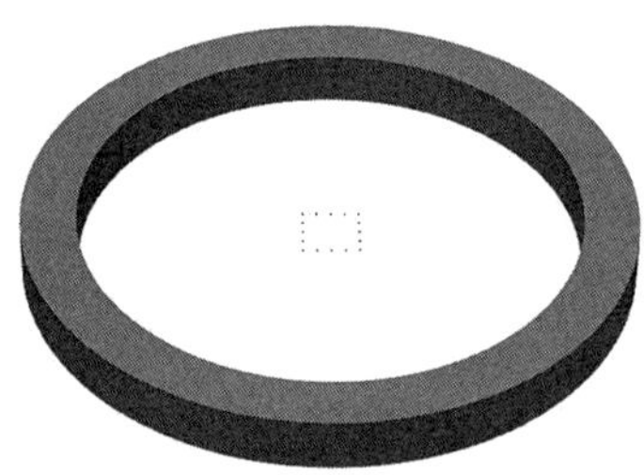

FIGURE B.10 Base curve in Frenet local frame moving along a simple circle.

goes from `spinePts_L[i-1]` to `spinePts_ L[i+1]` (with special attention being given to the end points of the spine to avoid an out-of-bounds error when accessing the `spinePts_L` list). Note that the generation of Figure B.8 involved a somewhat arbitrary choice of rotation matrices and so this constraint was not put in place.

The script in the file `Script B_06 Frenet frame surface Example 2.py` provides another example of Frenet surface generation with the rotation matrices carefully designed so that the *z* axis of the local frame is tangent to a spine that is a simple circle.

Here are some relevant notes for Figure B.10:

- The base curve is defined by a set of 16 points that form a square. They have been shown in the figure as a set of small spheres at the center of the figure.
- The surface involves 256 cross sections and so we have a very smooth continuity in areas that show surface curvature.

- The rotation matrices have been generated so that the z axis of the moving local Frenet frame is always parallel to the x, y plane but its x, y components (relative to the global frame) slowly change to give a constant rotation that keeps this z axis tangent to a circle.
- Note the parameter setting `closedSpine = True`. This closes the surface so that it *comes back on itself*, a useful feature when drawing surfaces of a periodic nature, for example, a toroid.

We finish this section with a final example of a Frenet frame surface that has some utility for the display of pie chart data. In a recent paper by Kim et al. [KN11] they use a pie chart to report on the classification of proteins with respect to biological function. The style of the pie chart is quite compelling because it is represented as a set of colored 3D pie sections that have been moved away from the pie center by using a small translation as in Figure B.11.

We now describe the generation of a single pie piece. We start with a base curve in the x, y plane that is a *half ellipse* as shown in Figure B.12. Arrows have been put in place to show the x, y, and z axes. The semi-minor axis has length of 1 unit and the semi-major axis has a length of 10 units. There are 64 points in the base curve. Note that the ellipse has been shifted along the x axis in the negative direction so that the profile edge goes through the origin.

We will need a spine curve for each pie piece. We start with a circle of radius that is the *same as the length of the semimajor axis* of the pie profile

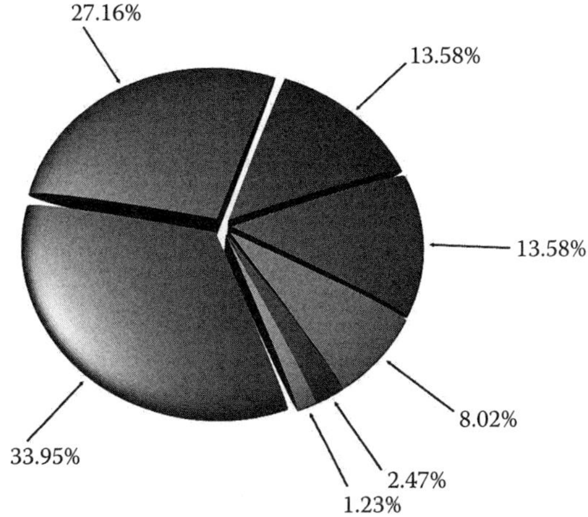

FIGURE B.11 **(See color insert.)** Pie base curves.

(10 units) and consider the perimeter of this circle to be a collection of arcs. The length of the various arcs will be in proportion to the given data values. Each arc is then translated away from the origin along a radial vector that goes from the origin to the center of the arc. Points on each arc will define the `spinePts_L` lists, one list needed for each pie piece. Figure B.13 illustrates this construction. In this example, five data values were used: [22., 44., 55., 13., 22.]. The data were sorted and the inner circle was divided into five arcs with lengths in proportion to these data values. In Figure B.13 five small spheres are used to mark the break points separating the arcs. Each arc was moved away from the center along a radial line going from the origin to the center point on the arc.

The script to generate the pie surfaces shown in Figure B.13 can be found in the file named `Script B_07 Pie a la Frenet.py`. For this figure, the data values are [22., 44., 55., 2., 4., 13., 22.]. This is the same as the previous list with two extra small numbers put in to give a more interesting final result. The script uses only 70 lines of code and this includes code for the arrows and labels* that associate a pie piece with its data value. Note that rotation matrices have been constructed so that the z axis of the local Frenet frame is tangent to an arc acting as a spine for a pie piece.

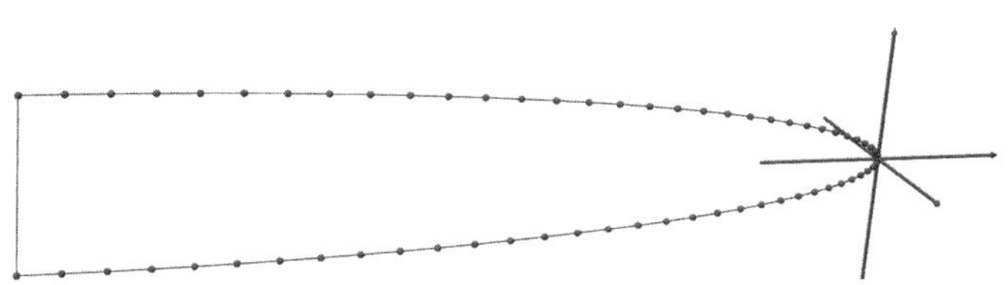

FIGURE B.12 Pie chart.

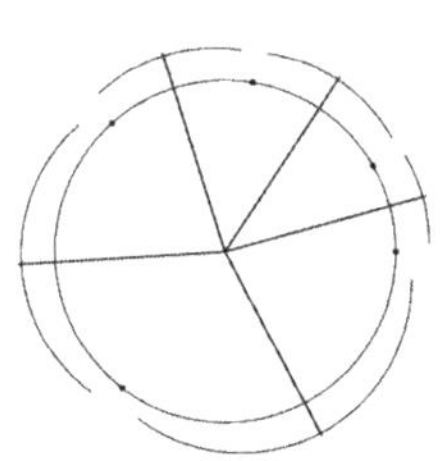

FIGURE B.13 Pie base curve profile.

(Incidentally, Figure B.13 was generated using the `Lines` and `Solids` classes.)

* Labels will be covered in Section B.7 of this appendix.

A GUI (graphical user interface) version of this script has been put into the Appendix D directory. The extra code to support the GUI will make the script somewhat longer.

The `FrenetFrameSurface` class gives you considerable power in the display of various sophisticated surfaces. For example, if you have a sudden irresistible desire to draw a Klein bottle,* then this is the class to use! Unfortunately, these capabilities come with a cost: You will need to spend some extra effort to generate the lists used in the `__init__` parameters of the class instantiation. This can be somewhat onerous especially when building the rotation matrices that are required. But it is often the case that the surface requires a spine and set of associated rotation matrices that are easier to generate because the surface is simpler. Thus we introduce two classes that are derived from the `FrenetFrameSurface` class:

1. The `ExtrusionSurface` class: The spine is a simple straight line. The rotation matrices perform 2D rotations of the base curve prior to the translation operations.
2. The `RevolutionSurface` class: The spine is a circle that goes around the y axis. The rotation matrices are used to *revolve* the base curve around the y axis.

By using these classes we can avoid the complications typically associated with the instantiation of a `FrenetFrameSurface` object.

B.5.2 Using the `ExtrusionSurface` Class

The `ExtrusionSurface` class generates a sequence of cross sections by extruding† a base curve along a straight line. The class can be imported by using

```
from StructBio.Scenographics.parametricSurfaces \
   import ExtrusionSurface
```

The object instantiation is done via the `__init__` method that starts with the following header:

```
__init__(self, baseCurvePts_L, zMin, zMax,
         numCrossSections, rotation2Dmats_L,
         xScale_L, yScale_L, modelName = "")
```

* See http://en.wikipedia.org/wiki/Klein_bottle. Note that the surface is nonorientable, so you would *not* be able to fill in the mesh using calls to `addFace`.

† The term extrusion is often used when describing the manufacturing of aluminum materials. See http://en.wikipedia.org/wiki/Extrusion.

As described earlier, we assume a base curve is defined in the *x,y* plane by specifying a list of 2D points: `baseCurvePts_L`. Initialization of the object is essentially a special case of the initialization done for the `FrenetFrameSurface` class. Prior to calling the __ init __ function for the `FrenetFrameSurface` class the following lists are created:

- The `spinePts_L` list is defined by generating `numCrossSections` points along the *z* axis ranging from `zMin` to `zMax`.
- The `rotation3Dmats_L` list is built from the `rotation2Dmats_L` list.
- The `xScale_L` and `yScale_L` lists are simply passed along unmodified.

The `rotation3Dmats_L` list allows you to do a 2D rotation relative to the *x*, *y* plane while creating the successive cross sections. Bear in mind that an extrusion along the *z* axis does not signify a restriction because we can use the `rotateAndMove` function of the `ParametricSurface` class to reorient and reposition the cross sections prior to the generation of a mesh or surface.

The script in the file `Script B_08 Rotini extrusion.py` provides an interesting example of an extrusion that involves a list of 2D rotation matrices (Figure B.14). It is colored to look like rotini pasta.* If you run the script, you can see that the 64 points in the base curve have been visually emphasized using small red spheres.

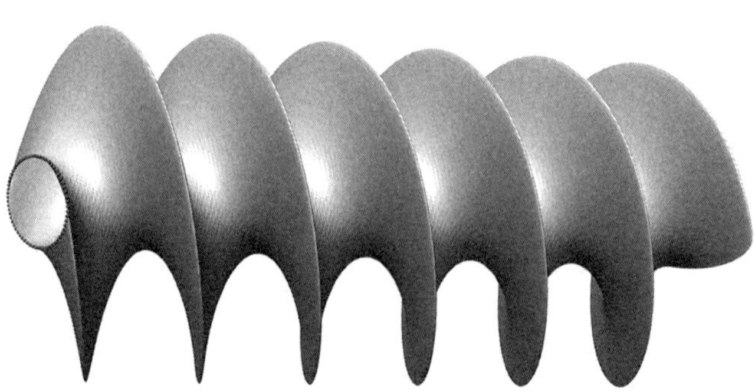

FIGURE B.14 Rotini extrusion.

* See http://en.wikipedia.org/wiki/Rotini.

If the elliptical base curve was extruded along the z axis, we would get a simple cylinder with elliptical cross section. To make the script more interesting, the extrusion works with a list of 2D rotation matrices that rotate the cross sections right after they are copied from the base curve. The script is relatively easy to read and is only 30 lines long (not counting the seven lines of code for the small spheres).

B.5.3 Using the `RevolutionSurface` Class

The `RevolutionSurface` class generates a sequence of cross sections by revolving a base curve in the x, y plane around the y axis. The class can be imported by using

```
from StructBio.Scenographics.parametricSurfaces \
   import RevolutionSurface
```

The object instantiation is done via the `__init__` method that starts with the following header:

```
__init__(self,baseCurvePts_L,thetaMin, thetaMax,
        numCrossSections, xScale_L, yScale_L,
        modelName = "")
```

In this case, we assume a base curve is defined in the right half of the x, y plane. The curve is specified by a list of 2D points: `baseCurvePts _ L`. Initialization of the object is essentially a special case of the initialization done for the `FrenetFrameSurface` class. Prior to calling the __init__ function for the `FrenetFrameSurface` class the following lists are created:

- The `spinePts_L` list is simply a list of zeros that are `numCrossSections` in number.
- The `rotation3Dmats_L` list is designed to take a copy of the base curve and rotate it over the x, z plane (i.e., around the y axis) to get a succession of cross sections. The amount of rotation is bounded below by `thetaMin` and bounded above by `thetaMax`.
- The `xScale_L` and `yScale_L` lists are simply passed along unmodified.

Note that the rotation around the *y* axis does not signify a restriction in placement of the final display object because we can use the `rotateAndMove` function of the `ParametricSurface` class to reorient and reposition the cross sections prior to the generation of a mesh or surface.

The script in the file `Script B_09 Vase revolution surface.py` gives an interesting example of a revolution surface. Like the rotini model it is not something that you are likely to put in a molecular display, but it is a convenient structure to illustrate the use of the class. Figure B.15 shows the base curve. Points defining the curve are emphasized with small spheres. The surface defined by this profile is shown in Figure B.16. To get the neck shadow to be cast on the lower part of the vase you can experiment with the menu invocation **Tools... Viewing Controls... Effects**.

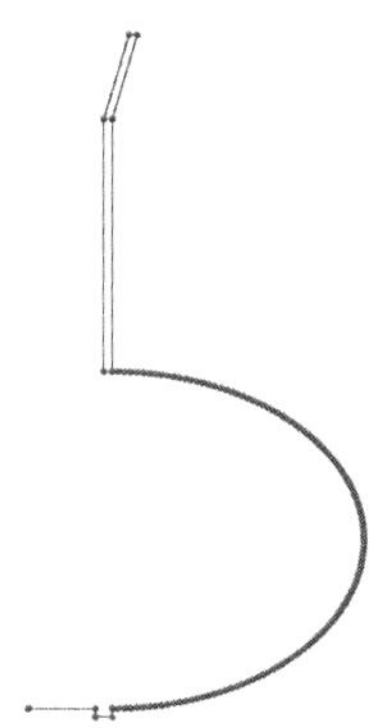

FIGURE B.15 Vase base curve points.

FIGURE B.16 Vase revolution surface.

B.5.4 Using the RuledSurface Class

The RuledSurface class generates a sequence of cross sections by moving a straight line through space to develop the surface. The ends of the line are constrained to move along two curves situated in the 3D space. So, in some sense we are using two spine curves instead of one. Consequently, we cannot consider the class to be a derivation of the FrenetFrameSurface class, but it can be treated as a derivation of the ParametricSurfaces class. The class can be imported by using

```
from StructBio.Scenographics.parametricSurfaces import RuledSurface
```

The object instantiation is done via the __init__ method that starts with the following header:

```
__init__(self, firstCurvePts_L, finalCurvePts_L,
         numCrossSections, modelName = "")
```

Instead of visualizing a ruled surface as a straight line moving through space with end points constrained to lie on two curves, we can consider the surface to be a set of cross sections. We start with a *first curve* defined by points stored in the list firstCurvePts _L. These points are shown in Figure B.17 as a set of small cubes. Next we specify a *final curve* defined by points stored in the list finalCurvePts_L. These points are shown in Figure B.17 as a set of small spheres. The number of points in each curve must be the same. Construction of the surface proceeds by running straight lines between the corresponding points in the two lists. In Figure B.17 we see that each cube is joined to a sphere by means of a straight line. Each straight line is then subjected to a linear interpolation procedure that places extra uniformly spaced points on that line. The corresponding interpolation points are then linked to form cross sections.

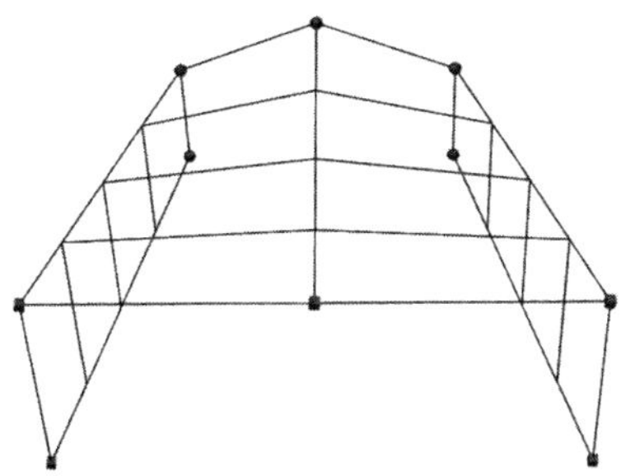

FIGURE B.17 Simple example of a ruled surface.

In Figure B.17, we are assuming that `numCrossSections` is 5. The first curve and final curve both count as a cross section, so we need three extra interpolation points on each straight line to derive the points needed to define the three additional cross sections.

Usually, the number of cross sections can be a small number because there is no significant curvature for a rectangle bounded by two consecutive straight lines. There can be a small amount of twist because the opposite sides of a rectangle are not necessarily parallel. Although the number of cross sections may be small, the number of points used on the first and final curves may be considerably larger so that these curves appear to be smoothly rendered. This is demonstrated in Figure B.18 that shows a ruled surface that has a circle as a first curve and a square as a final curve. The number of points in each curve is 512. This is somewhat high, but the surfaces are very smooth. Figure B.18 shows the surfaces with ends capped and not capped.

To demonstrate that ruled surfaces can be more complicated, we end this section with Figure B.19 that illustrates a neck ruff (popular during Elizabethan times*). In this case, the first curve is a simple circle, whereas

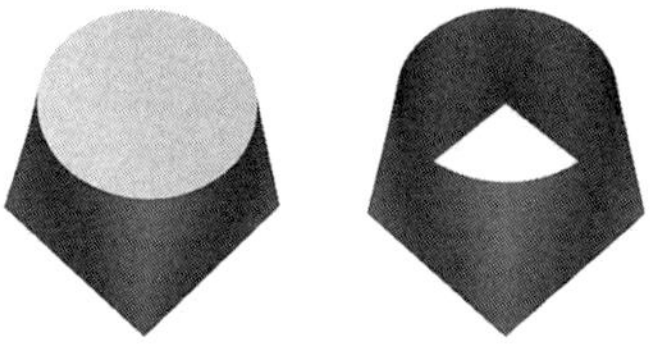

FIGURE B.18 Circle to square ruled surfaces (capped and not capped).

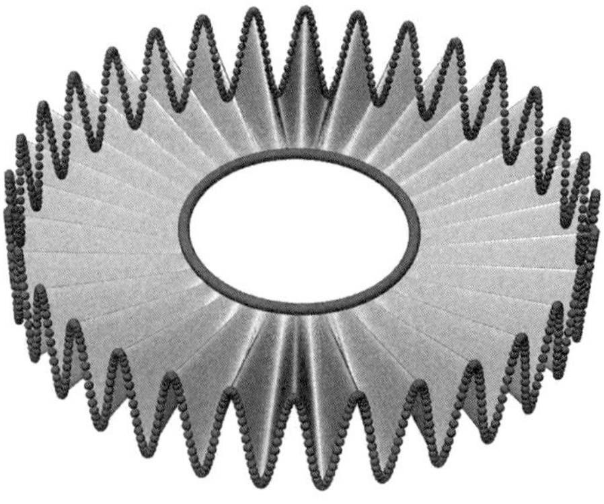

FIGURE B.19 Ruffle ruled surface.

* http://en.wikipedia.org/wiki/Ruff_(clothing).

the final curve is a circular sinusoidal wave form. Small spheres have been placed on the curves to designate the end points of the straight lines and to give the illustration an interesting visual effect.

B.6 USING THE `Ellipsoid` CLASS

As mentioned in the introduction, there are situations when you may wish to place an ellipsoid structure into a scene. For example, in the study of protein domain movements (see [BS08] and [BF92]) the authors work with the apo and holo conformations of a protein. They use an ellipsoid to enclose a domain in both forms and use this visual aid to show the amount of shift and rotation that a domain experiences on ligand binding.

The `Ellipsoid` class can be used for such an application. The class can be imported by using

```
from StructBio.Scenographics.ellipsoid import Ellipsoid
```

The object instantiation is done via the `__init__` method that starts with the following header:

```
__init__(self, length, width, height, center, axes_L,
         numCrossSections, numBasePts,
         ellipsoidName = "Ellipsoid")
```

Display of an ellipsoid is often preceded by calculations that derive a set of inertial axes for some given set of atoms, such as a protein domain, a secondary structure, or a side chain. We will assume that such a calculation provides a list of orthonormal axes that are passed to the __init__ function via `axes_L`. The list should be ordered so that `axes_L[0]` is the primary axis corresponding to the largest eigenvalue, whereas `axes_L[2]` is the axis corresponding to the smallest eigenvalue. The inertial axes will be used to establish the orientation of the ellipsoid, whereas the `center` parameter specifies the location of the ellipsoid center in the display space. The `length` parameter determines where the ellipsoid intersects the primary axis. In a similar fashion the `width` and `height` parameters correspond to `axes_L[1]` and `axes_L[2]`, respectively. These parameters define the lengths of the *semiaxes*. In other words, if the inertial axes were simply the standard *x*, *y*, and *z* axes (primary axis being the *x* axis, etc.) and the center was the origin, then the equation of the ellipsoid surface would be

$$\left(\frac{x}{\text{length}}\right)^2 + \left(\frac{y}{\text{width}}\right)^2 + \left(\frac{z}{\text{height}}\right)^2 = 1$$

The `Ellipsoid` class is derived from the `RevolutionSurface` class. There are two parameters that control the fineness of the mesh and hence the smoothness of the surface. The variable `numCrossSections` controls the number of cross sections that are generated when the base curve is revolved around its axis. This essentially determines the number of meridian lines that run in the same direction as the major axis. The variable `numBasePts` controls the number of points on the base curve. This essentially determines the number of latitude lines that are orthogonal to the meridian lines (similar to the parallels of latitude on a world globe).

The script in the file `Script B_12 Ellipsoid surface example.py` contains a small test program for this class. It displays an ellipse with length = 8, width = 4, height = 2, center having coordinates [3, 4, 5], and orientation given by an axes list defined via

```
u = np.pi/8; v = np.pi/4; w = -np.pi/3
cu = np.cos (u); cv = np.cos (v); cw = np.cos (w)
su = np.sin (u); sv = np.sin (v); sw = np.sin (w)


axes_L = [np.array([cw*cv, -sw*cu + cw*sv*su, sw*su + cw*sv*cu]),
          np.array([sw*cv, cw*cu + sw*sv*su, -cw*su + sw*sv*cu]),
          np.array([-sv, cv*su, cv*cu])]
```

Although this definition seems a bit daunting, it is really a rotation matrix that can change the orientation of an object in 3D space (see p. 231 of [Bu09]). It gives us an easy strategy to create orthonormal vectors. Apart from this, the script is quite easy to understand. Extra code has been put in place to show the inertial axes (scaled with respect to axes lengths) within the ellipsoid. They appear as spindles within the ellipsoid. The results are shown in Figure B.20. In the lower left corner you can see three arrows to indicate the orthonormal x, y, and z axes of the display space. These axes have been put in place so that you can more easily assess the position and orientation of the ellipsoid. The ellipsoid has been covered with both a mesh and a surface (transparency set to 0.3 so that the ellipsoid axes are easily visible).

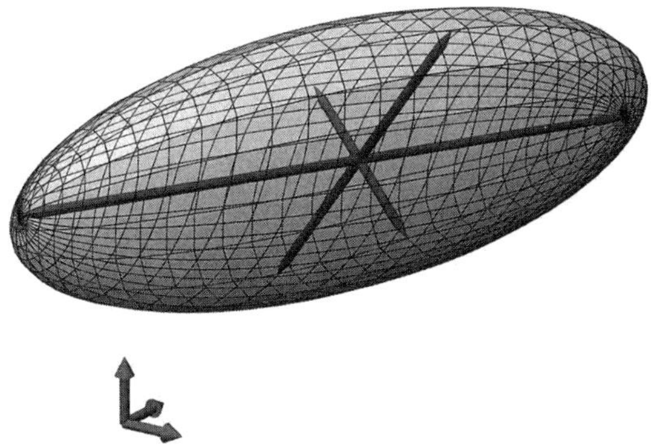

FIGURE B.20 Rotated ellipsoid with surface and mesh.

B.7 USING THE LabelGroups CLASS

As seen in Figure B.11, it is often very useful to add informative labels to scenographic objects. The LabelGroups class does this by using the molecule construction capabilities of Chimera and subsequently using the atom labels to decorate the scene. The class instantiation and its methods isolate the user from these implementation details. One can specify the position and color of the labels without any regard for this "under the hood" activity. Using Chimera in this way does represent a type of *kludge*, but it is quite effective and allows us to build a set of labels with an organization that is hierarchical in nature: the class instantiation produces a LabelGroups object that contains various label groups each containing an ordered set of labels (see Figure B.21).

The class can be imported by using

```
from StructBio.Scenographics.labels import LabelGroups
```

The object instantiation is done via the __ init __ method that starts with the following header:

```
__init__(self, labelGroupsName = "Labels")
```

This function will instantiate a LabelGroups object. The labelGroupsName can be used to give a name to the object and this will appear in the Model Panel after the showLabels method is executed. Other functions in the class can be described as follows:

```
addLabelGroup(self, groupName)
```

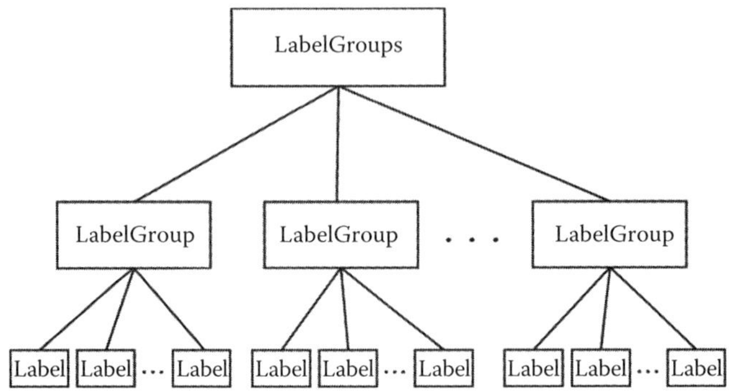

FIGURE B.21 Label hierarchy.

This function adds a label group to the previously instantiated `LabelGroups` object. The `groupName` parameter is a character string that will be used later to identify this particular label group.

```
addLabel(self, groupName, label, coords_T)
```

This function places a label into the scene. It also associates the label with a particular group by using the `groupName` specification. The `label` parameter holds the character string that will appear in the scene. The `coords _ T` parameter is a `numpy` array holding the 3D coordinates of the point where the label will appear. Labels have an ordering in the group that is determined by their order of addition to the group.

```
setLabels(self, groupName, colorName, stride = 1)
```

This function can be used to change the colors of labels in a particular group specified by the `groupName` parameter. The `colorName` parameter holds a character string specifying any one of the standard Chimera colors (as seen in the Color Actions dialog that you would usually access by the menu invocation: **Actions... Color... all options...** followed by checking Show all colors). The `stride` parameter allows you to apply this action to a subset of the labels in the group, for example, `stride = 3` will affect labels with indexes 0, 3, 6, 9, and so on. This is useful when labels are associated with the ticks situated on the axis for a plot (as in Appendix D).

```
showLabels(self)
```

This is used to pass the underlying molecular model to Chimera so that the labels will appear in the scene.

```
clearLabels(self, groupName)
```

This function will replace every label in the specified group with an empty string.

```
destroyLabels(self)
```

This is used to eliminate the underlying molecular model.

You can experiment with the `LabelGroups` class by running and modifying the script in the file `Script B_13 LabelGroups test.py`. Note that the script expects you to pay attention to the Python Shell window, hitting the Enter key when you wish to go onto the next step.

REFERENCES

[Bu09] F. Burkowski. *Structural Bioinformatics: An Algorithmic Approach.* Chapman & Hall/CRC, London, 2009.

[BS08] M. Brylinski and J. Skolnick. What is the relationship between the global structures of apo and holo proteins? *Proteins*, **70** (2008), 363–377.

[BF92] M. F. Browner, E. B. Fauman, and R. J. Fletterick. Tracking conformational states in allosteric transitions of phosphorylase. *Biochemistry*, **31** (1992), 11297–11304.

[KN11] D.-N. Kim, C.-T. Nguyen, and M. Bathe. Conformational dynamics of supramolecular protein assemblies. *Journal of Structural Biology*, **173** (2011), 261–270.

Appendix C

The Graph Class

C.1 MOTIVATION

In its simplest terms, a graph is defined by specifying a set of vertices and a set of arcs, each arc joining two vertices. As such, it provides an excellent mathematical model to describe molecular structure. In the most recognizable application of the idea, the set of vertices acts as an abstraction for the atoms within the molecule, whereas the arcs going between vertices correspond to the bonds between the atoms. In more sophisticated models, a vertex may represent a larger part of the molecule, such as a protein side chain, whereas an arc between vertices may represent some type of relationship between two side chains, for example, an indication that they are in contact with each other (see [XX00]). Typically, residues are considered to be in contact if their distance of separation is less than some threshold (say, 10 Å). Depending on the application, separation may be calculated as the distance between alpha carbon atoms or the distance between the centroids of the side chain atoms.

The motivation for the use of graphs arises from the possibility that some algorithm applied to the graph has a corresponding biological significance. For example, using graphs to model residue interactions has played a significant role in the study of allosteric interactions (see the study done by Pandini and others in [PF12]). In [PF12], the authors generated a graph network that models local correlated motions of residues and graph algorithms were used to discover various paths within that network.

There are several studies that use a Delaunay tessellation to define the notion of *neighboring residues*. As described in [ST96], the Delaunay tessellation of a protein structure leads to a set of space-filling irregular tetrahedrons, or Delaunay simplexes, each of the four vertices representing a protein side chain. Researchers have done various statistical analyses, and it has been discovered that the residue composition of these simplexes is not random. Instead, there are nonrandom preferences in the makeup of these simplexes. This has led to several consequential studies dealing with Delaunay tessellations and their relationships to topics such as the following:

- Four-body statistical potentials [CL01], [GT01], [KT03], [MS97]
- Protein structure similarity [BS04], [BV03]
- Protein shape, pockets, and cavities [LE98a], [LE98b], [LE98c], [XB07]
- Mechanical properties [KY97]
- Structure motifs [WY98]
- Domain partitions [WH99]

The paper by Huan and others [HB05] studies packing motifs and compares the use of contact distance with that of Delaunay tessellation. Finally, Delaunay tessellation applied to atoms, rather than residues, can be found in [KG97] and [MS02].

C.2 INTRODUCING THE `GraphBase` CLASS

Formally, we define a graph as a five-element tuple $G=(V,E,W_V,W_E,\Sigma_V)$ where V is a set of vertices and E is a set of undirected arcs* and $E \subseteq V \times V$. Each of the next three members in the tuple is a function:

1. W_V defines a weight for each vertex: $V \to \mathbb{R}$.
2. W_E defines a weight for each arc: $E \to \mathbb{R}$.
3. Σ_V is a bijection that defines a unique label for each vertex: $V \to L$, where L is a set of labels such that $|V|=|L|$.

We will also need the following definition for an *induced* subgraph: given the graph $G=(V,E,W_V,W_E,\Sigma_V)$ and the vertex set $\overline{V} \subseteq V$, the subgraph

* Although it is common to refer to the set E as having *edges*, we will reserve that term for a somewhat different use (this will be discussed in Section C.3).

induced by $\overline{V}$ is $\overline{G} = \left(\overline{V}, \overline{E}, W_{\overline{V}}, W_{\overline{E}}, \Sigma_{\overline{V}}\right)$ where $\overline{E} = \left\{(u,v) \in E \mid u \in \overline{V}, v \in \overline{V}\right\}$. As expected, the weight and label functions for $\overline{G}$ will be the same as those used in G (with domains restricted to the subgraph).

The script `graph.py` in the `StructBio` directory contains various class definitions that should be useful when you need to generate a graph object (including some frequently used graph algorithms). We will sometimes use the phrase *abstract graph* to emphasize that the graph exists solely as a Python object derived from the `GraphBase` class and it does not produce any display elements. In Section C.3, we introduce the `Graph` class, which generates display elements.

To generate an abstract graph, you would go through the following steps:

1. Import the required classes using the following import statement:

```
from StructBio.Graphs.graph import Vertex, Arc, GraphBase
```

2. Instantiate a graph object using the `GraphBase` class.

3. For each vertex of the graph, instantiate a vertex object using the `Vertex` class. The instantiation function allows you to specify the vertex weight (a floating point number with a default value of 0.0). Follow the instantiation with a call to the `addVertex` function of the graph object. This accomplishes two things: the vertex object is associated with its vertex label (a character string) and the vertex is added to the graph.

4. For each arc in the graph, instantiate an arc object using the `Arc` class. The instantiation function allows you to specify the arc weight (a floating point number with a default value of 0.0). Follow the instantiation with a call to the `addArc` function of the graph object. The `addArc` function has three parameters: the first two parameters specify the labels of the vertices defining the arc, and the last parameter is the arc object.*

* It should be mentioned that the `addVertex` and `addArc` functions are not constrained to work with only `Vertex` and `Arc` objects. They can add to the graph any objects that have the required attributes: `vertexWeight` and `arcWeight`, respectively. In particular, `addVertex` can add to the graph any object that is instantiated from a class that is *derived* from the `Vertex` class. This is why we will refer to the `Vertex` class as a *base* class. Note that both `Arc` and `GraphBase` are also considered to be base classes.

If you need a list of arcs, then the function call `getArcTupleList()` will generate a list of tuples, each tuple holding the two vertex labels that defined the arc.

Once the graph object is built, you can use any of the following functions to generate an induced subgraph:

- `getSubgraph_VertexWeightsWithinRange(self, minVal, maxVal)`

  This function returns the subgraph that is induced by vertices with weights that are within the closed interval specified by the two floating point values `minVal` and `maxVal`.

- `getSubgraph_VertexDegreeWithinRange(self, minVal, maxVal)`

  This function returns the subgraph induced by vertices with degrees* that are within the closed interval specified by the two floating point values `minVal` and `maxVal`.

- `getSubgraph_ArcsRemoved(self, arcTuples_L)`

  This function returns a graph that is the same as the self object but with the arcs listed in `arcTuples_L` removed.

- `getShortestPath(self, startingNodeLabel, endingNodeLabel)`

  This function considers the arc weights to be arc lengths. It returns the subgraph induced by the vertices that are on the shortest path running from the vertex designated by `startingNodeLabel` to the vertex designated by `endingNodeLabel`.

- `computeConnectedComponents(self, minSize)`

  This function uses a depth first search algorithm to return a list of subgraphs, each corresponding to a connected component of the graph. The parameter `minSize` specifies the minimum size of a subgraph. In other words, any connected component with fewer than `minSize` vertices will be ignored and does not produce a list entry.

As an example of a simple weighted graph, consider Figure C.1.

* The *degree* of a vertex is the number of arcs connected to the vertex (with loops counted twice—not relevant in this context).

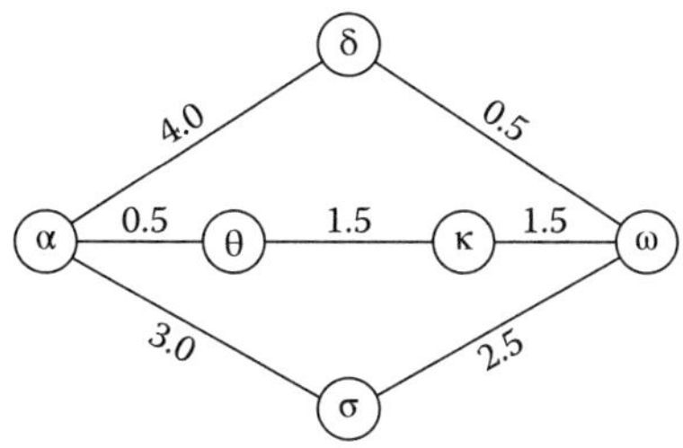

FIGURE C.1 A simple weighted abstract graph.

The following script* can be used to build the graph:
Script C.1: Simple `GraphBase` Example

```
import sets
from sets import Set
from StructBio.Graphs.graph import Vertex, Arc, GraphBase

g = GraphBase()
g.addVertex("alpha", Vertex())
g.addVertex("delta", Vertex())
g.addVertex("theta", Vertex())
g.addVertex("kappa", Vertex())
g.addVertex("sigma", Vertex())
g.addVertex("omega", Vertex())
g.addArc("alpha", "delta", Arc(4.0))
g.addArc("delta", "omega", Arc(0.5))
g.addArc("alpha", "theta", Arc(0.5))
g.addArc("theta", "kappa", Arc(1.5))
g.addArc("kappa", "omega", Arc(1.5))
g.addArc("alpha", "sigma", Arc(3.0))
g.addArc("sigma", "omega", Arc(2.5))

sp = g.getShortestPath("alpha", "omega")
print sp.orderedPathLabels_L
```

Note the following points:

- We access the needed classes by doing the following import:
  `from StructBio.Graphs.graph import Vertex, Arc, GraphBase`
- The abstract graph object `g` is instantiated by executing `g = GraphBase()`.

* See the Appendix C directory available from the book's website: `Script C_01 Simple GraphBase example.py`

- A statement such as `g.addVertex("alpha", Vertex())` is used to instantiate a vertex object (default weight 0.0), assign it a label, and add it to the graph `g`.
- A statement such as `g.addArc("alpha", "delta", Arc(4.0))` is used to instantiate an arc object with specified weight and subsequently add it to the graph `g`. The arc will join the vertices with labels specified by the first two parameters.
- The script illustrates the use of the function `getShortestPath` to calculate the shortest path between the vertex labeled `alpha` and the vertex labeled `omega`. This produces a subgraph object referenced by the variable `sp`.
- The subgraph `sp` has all the attributes of a graph within the `graphBase` class. For example, we could get the labels of the vertices in this subgraph by accessing the keys of the vertex dictionary, `sp.vertices_D.keys()`. Information about the arcs could be accessed by using `sp.arcs_D`. The `graph.py` script gives more information about these attributes.
- Although it is possible to trace the nodes along the shortest path by working with the usual attributes of a subgraph (`vertices_D` and `arcs_D`), this is somewhat inconvenient. Consequently, when a subgraph has been created by using the `getShortestPath` method, the algorithm will also define the attribute `orderedPathLabels_L`, which contains a list of the labels of the nodes in the shortest path with an ordering that goes from the starting vertex to the ending vertex. This list has been accessed by the print statement `print sp.orderedPathLabels_L` to print the labels of the vertices in the shortest path. The printed output will be `['alpha', 'theta', 'kappa', 'omega']`.

C.3 INTRODUCING THE Graph CLASS

The `GraphBase` class is useful for various applications that need an abstract graph to be constructed with a subsequent analysis that can be accomplished by the extraction of subgraphs. Although this functionality may sometimes be useful, a more significant role of `GraphBase`, `Vertex`, and `Arc` will be their utilization as base classes for derived classes that provide extra capabilities for the display of graphs in a scene that may also contain molecular structures. Specifically, we will work with the following derived classes:

- `Graph` derived from `GraphBase`
- `Node` derived from `Vertex`
- `Edge` derived from `Arc`

All these derived classes work with the `Solids` class to provide display elements that are typically related to other structures in the display (e.g., a molecule). Consider a short program that produces a very simple graph: two nodes linked by a single edge (see Figure C.2):

Script C.2: A Simple Graph

```
import numpy
from numpy import array as ar
from StructBio.Graphs.graph import Node, Edge, Graph

g = Graph("Simple graph")
# Red node:
n1 = Node(0., ar((1., 0., 0.)), 0.7,(1., 0., 0.))
# Blue node:
n2 = Node(0., ar((9., 0., 0.)), 0.7,(0., 0., 1.))
e12 = Edge(0., 0.3,(0., 1., 0.))
g.addNode("Node_1", n1)
g.addNode("Node_2", n2)
g.addEdge("Node_1", "Node_2", e12)

g.solids.display()
print g.vertices_D.keys()
```

The following are some notes explaining the code:*

- The script starts with importing the needed classes (`Graph`, `Node`, and `Edge`).

FIGURE C.2 A very simple graph.

* See the Appendix C directory available from the book's website: `Script C_02 A simple graph.py`

- Instantiation of a `Graph` object includes an initialization statement that instantiates a `Solids` object. Since the `Graph` class inherits from the `GraphBase` class, the initialization will also create the data structures that will track vertices and arcs. Note that the graph name "Simple graph" will appear in the Model Panel viewable if you do the menu invocation **Favorites... Model Panel**.
- Syntax for the instantiation of a `Node` object is

```
Node(nodeWeight, coords, radius, color).
```

In our script, all node weights are 0.0 and the coordinates are specified by `numpy` arrays. The specified radius will be the radius of the sphere that is generated in the display when the `addNode` function is executed. The last parameter is a 3-tuple that specifies the RGB components of the sphere color.

- Syntax for the instantiation of an `Arc` object is

```
Arc(edgeWeight, radius, color).
```

In the script, edge weight is 0 and the radius and color are 0.3 and `(0.0, 1.0, 0.0)` corresponding to green.

- Syntax for the addition of a `Node` object is

```
addNode(nodeLabel, node).
```

This will assign the specified node label to the node and will add the node as a vertex to the underlying abstract graph. The function will also display a sphere with the radius and color dictated by the attributes of the node object.

- Syntax for the addition of an `Edge` object is

```
addEdge(nodeAlabel, nodeBlabel, edge).
```

This will add a corresponding arc to the underlying abstract graph and will display a spindle with radius and color determined by the attributes of the `Edge` object. In the display, the location of the edge ends will be determined by the nodes specified by the node labels.

- After all the nodes and edges have been established, the display is made visible by using the function call `g.solids.display()`.
- The script ends with the extra statement `print g.vertices_D.keys()`. This generates the output `['Node_2', 'Node_1']` in the Python shell window. It was added to the end of the script as a demonstration that attributes of the underlying abstract graph can be accessed if needed.

C.3.1 Reasons for Using Derived Classes

It is worthwhile to emphasize the design aspects of the classes introduced in Sections C.2 and C.3. The base classes primarily deal with abstract graphs as defined by the mathematical definition $G=(V,E,W_V,W_E,\Sigma_V)$. All the algorithms strictly deal with vertices, arcs, weights, and labels. There is no dependency on display attributes or coordinates, and so the algorithms are forced to focus on abstract attributes while having the freedom to ignore attributes associated with the display.

Meanwhile, users who wish to design algorithms related to molecular structure can build the display objects: the spheres and spindles related to their application. While this construction is being done, the underlying abstract graph is also created without the user having to deal with the data structures that manage vertices and arcs.

The net result is a clear separation of concerns: the `GraphBase`, `Vertex`, and `Arc` classes relate to mathematically oriented graph algorithms while the `Graph`, `Node`, and `Edge` classes relate to the display and constructions that correspond to the biological structures within the application.

C.4 METHODS WITHIN THE `Graph` CLASS

Now that we know how a graph can be created, we can discuss the methods that can be applied to these graphs. All these methods deal with the extraction of subgraphs with the exception of Delaunay tessellation, which is a function that can be used to construct the edges of a graph after the nodes are established.

C.4.1 Subgraph Extraction Determined by Node Weights

Consider the following script:

Script C.3: Example for Extraction of a Subgraph

```
import chimera, numpy
from chimera import runCommand
from numpy import array as ar, random as rn

from StructBio.Utilities.miscFunctions import colorByName
from StructBio.Graphs.graph import Graph, Node, Edge

runCommand("set bg_color white")

g = Graph()

g.addNode("A", Node(1., ar((1., 0., 0.)), 0.7,
          colorByName("dark red")))
g.addNode("C", Node(1., ar((9., 1., 0.)), 0.7,
          colorByName("red")))
g.addNode("E", Node(1., ar((1., 0., 9.)), 0.7,
          colorByName("orange red")))
g.addNode("G", Node(1., ar((2., 7., 9.)), 0.7,
          colorByName("violet red")))
g.addNode("B", Node(3.2, ar((0., 8., 2.)), 0.7,
          colorByName("dark green")))
g.addNode("D", Node(3.4, ar((9., 8., 1.)), 0.7,
          colorByName("spring green")))
g.addNode("F", Node(3.5, ar((8., 2., 7.)), 0.7,
          colorByName("lime green")))
g.addNode("H", Node(3.7, ar((7., 6., 8.)), 0.7,
          colorByName("light green")))

for n in g.vertices_D.keys():
     for m in g.vertices_D.keys():
         if n < = m: continue
         if rn.random() < 0.5:
             spnd_color = tuple((ar(g.vertices_D[n].color) +
                                 ar(g.vertices_D[m].color))/2.0)
             g.addEdge(n, m, Edge(0.0, 0.3, spnd_color))

g.solids.display()

redG = g.getSubgraph_NodeWeightsWithinRange(1.0, 1.0,
                                          "Reddish subgraph")
redG.solids.display()

greenG = g.getSubgraph_NodeWeightsWithinRange(3.0, 4.0,
                                            "Greenish subgraph")
greenG.solids.display()
```

The script uses the derived classes to construct a graph with nodes that are approximately positioned near the corners of a cube. Figure C.3 is in gray scale, but if you run the script, you will see that nodes with weight 1.0 have colors with a reddish tone, whereas nodes with weights between 3.0 and 4.0 have colors with a greenish tone. The specification of different colors is meant to illustrate the use of the `colorByName` function that is available in the `miscFunctions` module. To avoid an extensive sequence of statements that generate edges, the script uses two nested loops that randomly drop edges into the display. The color of an edge is the *average* of the colors of the end nodes.

The script is given to illustrate the use of the `getSubgraph_NodeWeightsWithinRange` method. It will extract a subgraph that is induced by selecting nodes with weights falling within a specified range. Execution of the script produced the display shown in Figure C.3. As the edges are determined randomly, you will likely get a different graph if you run the script. To verify that the function is actually returning subgraphs each contained in its own Chimera model, we can consult the Model Panel. By setting the checkboxes in the Model Panel dialog box, we can show or hide any of the three graphs in the display. Figure C.4 shows the resulting display when we choose to display both subgraphs simultaneously. The figure includes part of the Model Panel so that you can see how the checkboxes were set. Of course, in a more sophisticated program we would expect the script to selectively show or hide the models without any need for the user to interact with the Model Panel.

The function `getSubgraph_VertexDegreeWithinRange` behaves in a very similar fashion except that it extracts the vertices having degrees within the specified range. These vertices are used to generate an induced subgraph.

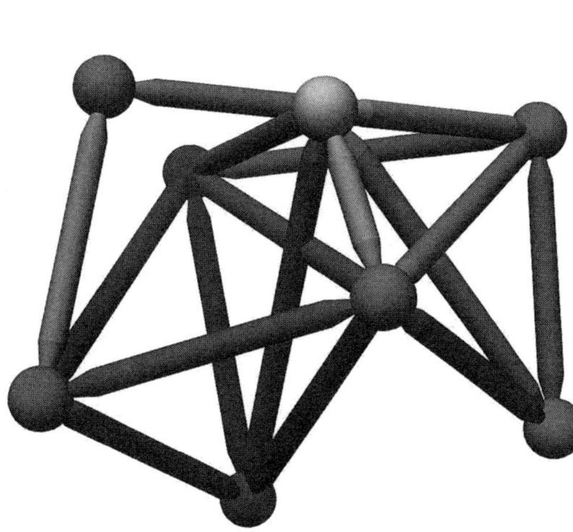

FIGURE C.3 Full graph produced by Script C.3.

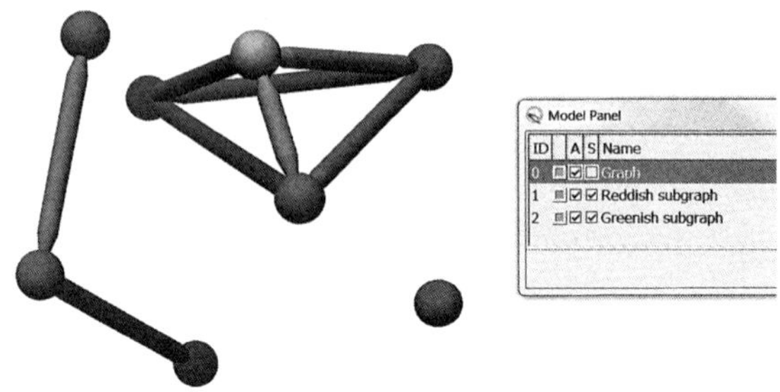

FIGURE C.4 Subgraphs produced by Script C.3.

C.4.2 A Function to Produce Delaunay Tessellations

Section C.1 stressed the importance of Delaunay tessellations for various research studies. The `graph.py` module includes a function that will produce a Delaunay tessellation when given the three-dimensional coordinates of the nodes of a graph. The algorithm that generates the tessellation must deal with node coordinates, and so this function is more naturally given as a method within the `Graph` class. In practice, it will determine all the necessary edges to provide the tessellation and these new edges and existing nodes will form the arcs and vertices of an underlying abstract graph that can be the input for other graph algorithms. Using the function is reasonably straightforward as exemplified by the following script:

Script C.4: Using the Function for Delaunay Tessellation

```
import chimera, numpy
from chimera import runCommand
from numpy import array, random as rn
from StructBio.Graphs.graph import Graph, Node, Edge

runCommand("set bg_color white")

maxX = 32.; maxY = 16.; maxZ = 16.
numNodes = 12

g = Graph("Delaunay Tessellation")

for ix in range(numNodes):
    c = array((rn.random()*maxX,
               rn.random()*maxY,
               rn.random()*maxZ))
    nd = Node(0.0, c, 0.7, tuple((rn.random(),
                                   rn.random(),
                                   rn.random())))
    g.addNode("V" + "%03d"%ix, nd)
```

```
g.buildDelaunayTessellation(15.0)
g.solids.display()
```

The script instantiates a graph that is referenced by the variable g. The graph name is "Delaunay tessellation," and this name will appear in the Model Panel. The script starts by creating 12 nodes randomly placed in a box with sides having lengths 32, 16, and 16 Angstroms. The nodes have randomly generated colors. The function call g.buildDelaunayTessellation(15.0) will generate the tessellation. The parameter in the function call (15.0 in the script) sets an upper limit on the length of an edge in the tessellation. If any edge in a tetrahedron has a length greater than this threshold, then the tetrahedron is rejected and does not contribute to the graph. Figure C.5 illustrates a typical display generated by the script. Note that it is possible for a node to be excluded from the tessellation if it does not have three neighboring nodes within threshold distance.

C.4.3 Subgraphs Corresponding to Connected Components

Computing the connected components of an undirected graph has many applications. The Graph class has a method to do this. Consider the following script:

Script C.5: Connected Components Generated from a Delaunay Tessellation

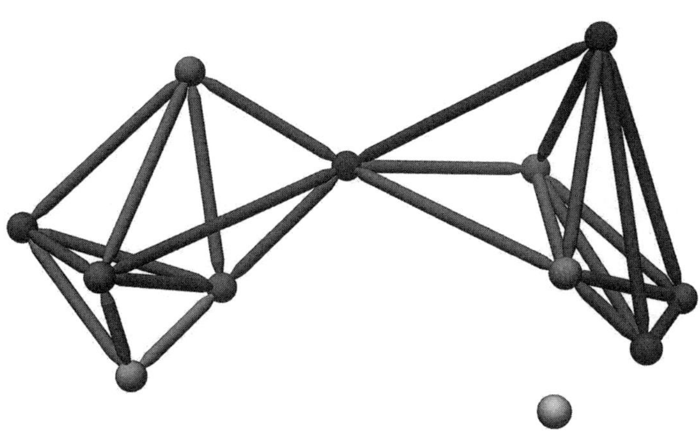

FIGURE C.5 Subgraphs produced by Script C.3.

```
import chimera, numpy
from chimera import runCommand
from numpy import array, random as rn
from StructBio.Graphs.graph import Graph, Node

runCommand("set bg_color white")

maxX = 64.; maxY = 32.; maxZ = 16.
numNodes = 32

g = Graph("Delaunay Tessellation")

for ix in range(numNodes):
    c = array((rn.random()*maxX,
               rn.random()*maxY,
               rn.random()*maxZ))
    g.addNode("V" + "%03d"%ix,
              Node(0.0, c, 0.7, tuple((rn.random(),
                                       rn.random(),
                                       rn.random())))))

g.buildDelaunayTessellation(15.0)
g.solids.display()

connectedComponents_L = g.getConnectedComponents \
                        (2, "Connected component")
for cc in connectedComponents_L:
    cc.solids.display()
```

The function `getConnectedComponents` will generate a list of subgraphs each of which is a connected component within the given graph. To test this function, it is convenient to start with a random graph that has edges generated from a Delaunay tessellation. The distance threshold for the tessellation has been set somewhat low, thus increasing the chances that the tessellation will produce more than one connected component. A typical result of the script execution is presented in Figure C.6.

The parameters for the function call include a minimum node count for any connected component. In this case, the minimum is set to 2 and, consequently, the isolated node near the bottom of the display will not be considered as one of the connected components to be entered into the return list. Note that the function uses the given string "`Connected component`" to generate the names of the models in the Model Panel. The function has arranged for various models to have distinct names by appending their list index to the name string.

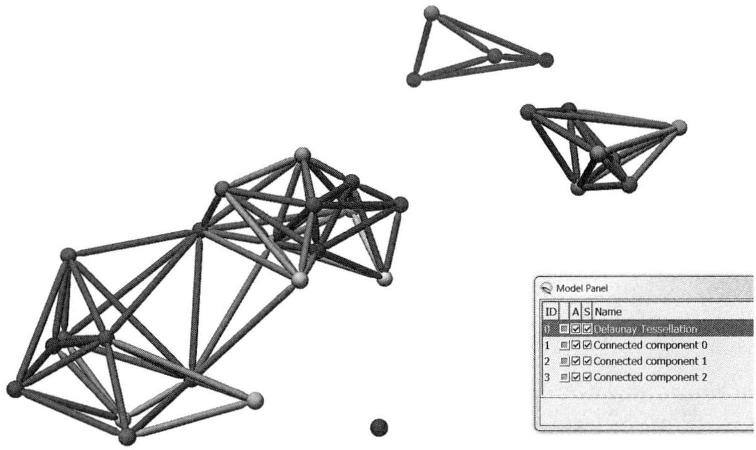

FIGURE C.6 **(See color insert.)** Connected components produced by Script C.5.

C.4.4 A Subgraph That Is a Shortest Path

The calculation of a shortest (minimum-weight) path has several uses, for example, the demonstration of signal propagation in the modeling of allosteric behavior in a protein. The `Graph` class provides the function `getShortestPathSubgraph` to calculate the shortest path between two nodes. In the function call, the first two parameters specify these nodes by means of their labels. The third parameter is a character string that provides the name of the model holding the display of the subgraph. Again, it is convenient to use a Delaunay tessellation on random nodes to test the function. Consider the following code:

Script C.6: Shortest Path through a Delaunay Tessellation

```
import chimera, numpy
from chimera import runCommand
from numpy import array, random as rn
from StructBio.Graphs.graph import Graph, Node

runCommand("set bg_color white")

maxX = 64.; maxY = 32.; maxZ = 16.
numNodes = 24

g = Graph("Delaunay Tessellation")

c = array((0.0, 0.0, 0.0))
g.addNode("V001", Node(0.0, c, 2.0,
                       tuple((0.0, 1.0, 0.0))))
c = array((maxX, maxY, maxZ))
g.addNode("V" + "%03d"%numNodes, Node(0.0, c, 2.0,
                                      tuple((1.0, 0.0, 0.0))))
```

```
for ix in range(2, numNodes):
    c = array((rn.random()*maxX,
               rn.random()*maxY,
               rn.random()*maxZ))
    g.addNode("V" + "%03d"%ix,
              Node(0.0, c, 0.7, tuple((rn.random(),
                                       rn.random(),
                                       rn.random()))))

g.buildDelaunayTessellation(24.0)
g.solids.display()

shortestPath = g.getShortestPathSubgraph \
               ("V001", "V" + "%03d"%numNodes,
                "Shortest Path")
shortestPath.solids.display()

for nodeLabel in shortestPath.orderedPathLabels_L:
    c = shortestPath.vertices_D[nodeLabel].coords
    print "%4s  [%6.2f,%6.2f,%6.2f]" % \
          (nodeLabel, c[0], c[1], c[2])
```

The statements that generate the random nodes and subsequent Delaunay tessellation are very similar to that seen in the previous examples except for the addition of two larger spheres to represent the starting node (colored green) and the ending node (colored red). These colors can be seen if you run the script. The coordinates of these nodes place them at the extreme diagonally opposite corners of the bounding box. Using the attributes inherited from the `GraphBase` class, the script has access to a list of labels of the nodes in the shortest path. To demonstrate access to these attributes, the script prints out the node labels and the node locations. Figure C.7 illustrates the graph that is processed by the shortest path function, whereas Figure C.8 shows the result when the Model Panel is used to display the path apart from the full graph.

Now that we have reviewed all the functionality of the `Graph` class, we should be ready to apply this module to problems that require a graph representation.

C.4.5 Subgraph Extraction Determined by Node Degree

The function `getSubgraph_NodeDegreeWithinRange` extracts a set of nodes each of which has a degree that is within a specified range. It then generates the induced subgraph using an approach that is very similar to that used by the `getSubgraph_NodeWeightsWithinRange` function.

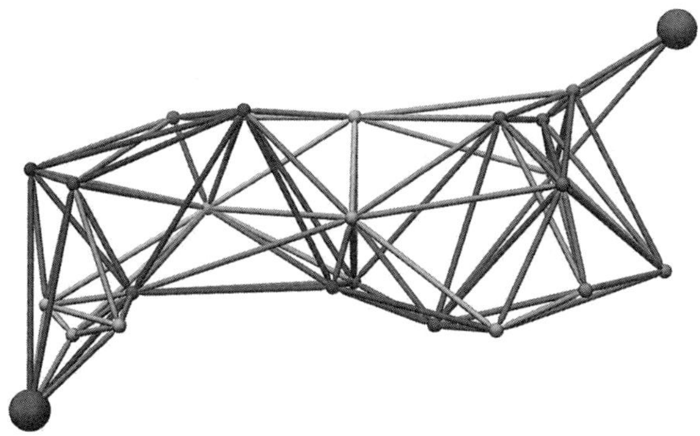

FIGURE C.7 Start graph for the shortest path test (Script C.6).

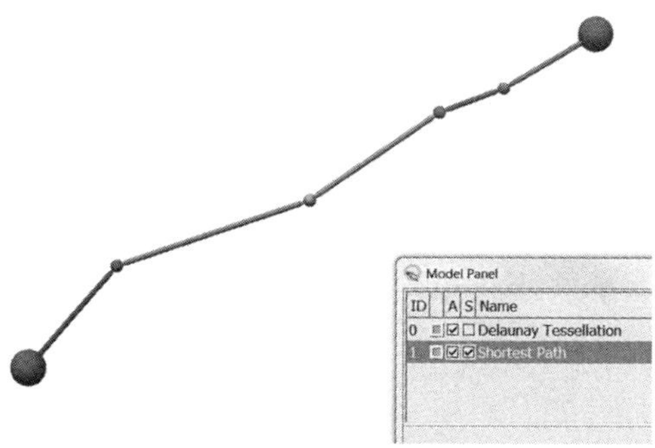

FIGURE C.8 Shortest path produced by Script C.6.

Note that the resulting subgraph may have nodes with degrees outside the specified range. Script C.07 (available from the book's website) can be used to test this function. The functionality is quite similar to the function `getSubgraph_NodeWeightsWithinRange`, so we will not consider the script any further.

C.4.6 Subgraph Generation by Removal of Edges

The function `getSubgraph_EdgesRemoved` creates a subgraph with the same vertices and all edges except for those listed in `edgeTuples_L`. A script to test this function can be found in the file `Script C_08 Delaunay Tessallation with edge removal test.py`.

See Exercise 6 of Chapter 8 for a typical application of the `Graph` class.

REFERENCES

[BS04] D. L. Bostick, M. Shen, and I. I. Vaisman. A simple topological representation of protein structure: Implications for new, fast, and robust structural classification. *Proteins*, **56** (2004), 487–501.

[BV03] D. L. Bostick and I. I. Vaisman. A new topological method to measure protein structure similarity. *Biochemical and Biophysical Research Communications*, **304** (2003), 320–325.

[CL01] C. W. Carter Jr, B. C. LeFebvre, S. A. Cammer, A. Tropsha, and M. H. Edgell. Four-body potentials reveal protein-specific correlations to stability changes caused by hydrophobic core mutations. *Journal of Molecular Biology*, **311** (2001), 625–638.

[GT01] H. H. Gan, A. Tropsha, and T. Schlick. Lattice protein folding with two- and four-body statistical potentials. *Proteins*, **43** (2001), 161–174.

[HB05] J. Huan, D. Bandyopadhyay, W. Wang, J. Snoeyink, J. Prins, and A. Tropsha. Comparing graph representations of protein structure for mining family-specific residue-based packing motifs. *Journal of Computational Biology*, **12** (2005), 657–671.

[KT03] B. Krishnamoorthy and A. Tropsha. Development of a four-body statistical pseudo-potential to discriminate native from non-native protein conformations. *Bioinformatics*, **19** (2003), 1540–1548.

[KG97] N. Kobayashi and N. Go. A method to search for similar protein local structures at ligand binding sites and its application to adenine recognition. *European Biophysics Journal*, **26** (1997), 135–144.

[KY97] N. Kobayashi, T. Yamato, and N. Go. Mechanical property of a TIM-barrel protein. *Proteins*, **28** (1997), 109–116.

[LE98a] J. Liang, H. Edelsbrunner, P. Fu, P. V. Sudhakar, and S. Subramaniam. Analytical shape computation of macromolecules: I. Molecular area and volume through alpha shape. *Proteins*, **33** (1998), 1–17.

[LE98b] J. Liang, H. Edelsbrunner, P. Fu, P. V. Sudhakar, and S. Subramaniam. Analytical shape computation of macromolecules: II. Inaccessible cavities in proteins. *Proteins*, **33** (1998), 18–29.

[LE98c] J. Liang, H. Edelsbrunner, and C. Woodward. Anatomy of protein pockets and cavities: Measurement of binding site geometry and implications for ligand design. *Protein Science*, **7** (1998), 1884–1897.

[MS02] B. J. McConkey, V. Sobolev, and M. Edelman. Quantification of protein surfaces, volumes and atom-atom contacts using a constrained Voronoi procedure. *Bioinformatics*, **18** (2002), 1365–1373.

[MS97] P. J. Munson and R. K. Singh. Statistical significance of hierarchical multi-body potentials based on Delaunay tessellation and their application in sequence-structure alignment. *Protein Science*, **6** (1997), 1467–1481.

[PF12] A. Pandini, A. Fornili, F. Fraternali, and J. Kleinjung. Detection of allosteric signal transmission by information-theoretic analysis of protein dynamics. *FASEB Journal*, **26** (2012), 868–881.

[ST96] R. K. Singh, A. Tropsha, and I. I. Vaisman. Delaunay tessellation of proteins: Four body nearest neighbor propensities of amino acid residues. *Journal of Computational Biology*, **3** (1996), 213–221.

[WH99] L. Wernisch, M. Hunting, and S. J. Wodak. Identification of structural domains in proteins by a graph heuristic. *Proteins*, **35** (1999), 338–352.

[WY98] H. Wako and T. Yamato. Novel method to detect a motif of local structures in different protein conformations. *Protein Engineering*, **11** (1998), 981–990.

[XB07] L. Xie and P. E. Bourne. A robust and efficient algorithm for the shape description of protein structures and its application in predicting ligand binding sites. *BMC Bioinformatics*, **8** (Suppl 4) (2007), S9.

[XX00] Y. Xu, D. Xu, and H. N. Gabow. Protein domain decomposition using a graph-theoretic approach. *Bioinformatics*, **16** (2000), 1091–1104.

Appendix D

2D and 3D Plots

D.1 MOTIVATION

The *informatics* component of *bioinformatics* deals with the retrieval, management, processing, and display of information. To quote from Wikipedia:*

Informatics is the science of information, the practice of information processing, and the engineering of information systems. Informatics studies the structure, algorithms, behavior, and interactions of natural and artificial systems that store, process, access, and communicate information. In addition, it studies human–computer interaction and how an interface can be built to maximize user efficiency.

In many situations, the communication of information will involve the presentation of data output in the form of a plot. We have already seen examples of this in earlier chapters.

The dimensionality of a plot may be one of the following:

- Two dimension (2D): The data appear in a plane. For example, a scatter plot with point markers appearing in a plane or a curve $g(x, y) = 0$ represented by a line in the x, y plane.
- Three dimension (3D): The data appear in 3D space. The representation may illustrate points in a 3D space (a scatter plot), a line usually defined parametrically, or in a surface; for example: $z = f(x, y)$.

* http://en.wikipedia.org/wiki/Informatics_(academic_field).

D.2 PLOTS CONSIDERED IN THIS APPENDIX

Using Chimera, a plot can be displayed in a variety of settings:

- The script can work with the `matplotlib` module to generate a plot that will be displayed in a separate window that is produced by routines within `matplotlib`.
- The script can work with the `matplotlib` module to generate a plot that will be displayed within a canvas widget that is part of a modeless dialog.
- The script can use the `Solids` and `Surfaces` classes (see Appendix B) to generate a plot that is shown within the Chimera window. This last possibility is particularly suitable for 3D plots because we can use the display facilities of Chimera to rotate and position a complex plot when we want to draw attention to some specific features of the plot. Another advantage to this approach is the lighting effects that Chimera provides. This can give the plot an enhanced impression of its 3D verisimilitude.

More details about matplotlib facilities [Hu07] can be found at http://matplotlib.sourceforge.net/.

D.3 BAR CHARTS

Consider the following script, which builds a bar chart that compares the values stored in two arrays `data_A` and `data_B`. Note that the data are passed to the array as a single list of 2-tuples. Using a single list will force the user to package `data_A` and `data_B` using arrays of equal length. You can adapt this script to make it more suitable for your own needs, for example, by changing the colors and widths of the bars. Figure D.1 shows the plot that is displayed.

```
import matplotlib.pyplot as plot
import numpy as np

def barChart(xAxisLabel, yAxisLabel, title, data_L):

    # The variable data_L is a list of 2-tuples
    # holding y values to be displayed in the bar chart.
    unzippedData = zip(*data_L)
    dataA = unzippedData[0]
    dataB = unzippedData[1]
```

```
    width = 0.4
    xVals = np.arange(1,len(data_L) + 1)

    plot.ion()
    plot.subplot(1,1,1)
    plot.bar(xVals, data_A, width, color = 'g')
    plot.bar(xVals + width, data_B, width, color = 'r')

    plot.xticks(arange(0, len(data_L) + 1, 1))

    plot.xlabel(xAxisLabel)
    plot.ylabel(yAxisLabel)
    plot.title(title)

    raw_input("Hit Enter to dismiss the plot.")
    plot.close()

#=====================================================
# Mainline for testing:

dataVals_L = [[2,3], [4,5], [6,8], [7,6], [3,1]]
barChart("x-axis label goes here.",
         "Y-axis label goes here.",
         "Title", dataVals_L)
```

Script D.1: Bar Chart Example

Note the use of the `raw_input("Hit Enter to dismiss the plot.")` statement that is followed by the `plot.close()` function call. This allows your script to pass control back to the Python Shell window when you are done with the plot window. The mainline statements set up a short input list of data tuples and then initiate a call to the `barChart` function.

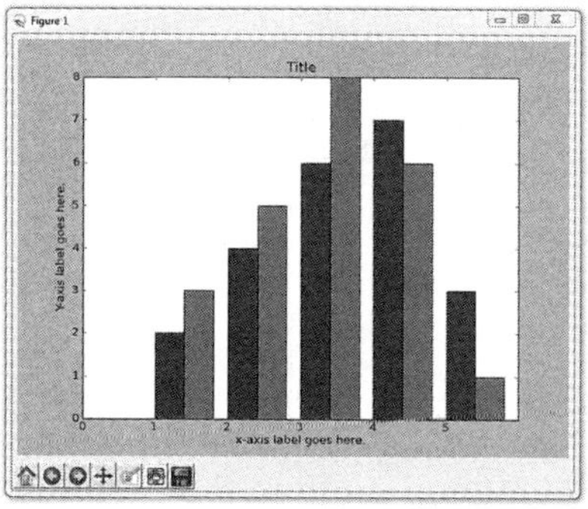

FIGURE D.1 **(See color insert.)** Bar chart from Script D.01.

D.4 2D DENSITY PLOT

A density plot is similar to the contact maps seen in Chapter 3. Formally, the plot gives a visual representation of a mapping that one can write as $z = f(x, y)$. Each x, y pair in the domain of the function will correspond to a pixel position in the 2D plot, and the z value corresponding to this x, y pair will be recorded as a particular color value for the pixel. More generally, the plot routine can be given the data stored in a 2D array. The script is as follows:

```
import numpy as np
import matplotlib.pyplot as plot
from matplotlib.colors import ListedColormap


def generate2Dplot(zG_A, plotTitle):
    # zG_A should be a square array holding values
    # in the range [0.0, 1.0].

    # [0,.5) black; [.5, 1.] white:
    #cmap = ListedColormap(['#000000', '#FFFFFF'])

    # [0., 0.3333] black; [.3333, .6666] gray;
    # [.6666, 1.0] white:
    #cmap = ListedColormap(['#000000', '#888888',
    #                        '#FFFFFF'])

    # [0., 0.2] black; [.2, .4] dark gray;
    # [.4, .6] gray; [.6, .8] light gray;
    # [.8, 1.] white
    #cmap = ListedColormap(['#000000', '#444444',
    #                        '#888888', '#CCCCCC',
    #                        '#FFFFFF'])

    # Higher resolution color map:
    cmap = ListedColormap(['#000000', '#222222',
                           '#444444', '#666666',
                           '#888888', '#AAAAAA',
                           '#CCCCCC', '#EEEEEE',
                           '#FFFFFF'])

    n = int(zG_A.shape[0])
    xSeq = np.arange(0, n+1, 1)
    ySeq = np.arange(0, n+1, 1)

    # Use the next statement to have the main
    # diagonal head to the "north west".
    # ySeq = np.arange(n, -1, -1)

    xG, yG = np.meshgrid(xSeq, ySeq)
```

```
    plot.ion()
    plot.pcolor(xG, yG, zG_A, cmap = cmap)
    #plot.axis([0, n-1, n-1, 0])
    plot.axis([0, n, n, 0])
    plot.axes().set_aspect('equal')

    plot.title(plotTitle)

    raw_input("Hit Enter to dismiss the plot.")
    plot.close()

#=======================================================
# Mainline for testing:

n = 200
zVals_A = np.zeros((n,n), float)
tp = 2.0*np.pi

for ix in range(n):
    for jx in range(n):
        zVals_A[ix, jx] = 0.25*(np.sin(tp*ix/n) +
                                np.cos(3.0*tp*jx/n))**2

generate2Dplot(zVals_A, "0.25*(sin(2pi*x) + cos(6pi*y))**2")
```

Script D.2: Density Plot Example

Here are some relevant points related to Script D.2:

- Note that the script assumes that the data given to it in the `zVals_A` array have been scaled so that each value lies in the range [0.0, 1.0].
- The statement defining the variable `cmap` requires a list of colors specified in hexadecimal format. In Script D.2, gray levels are used. The number of gray levels will depend on the number of entries in this list. The script provides four different possibilities. You can modify this statement to use more gray levels or use color ranges as required.
- Note that the labeling on the vertical axis is consistent with row numbering that you would have in a data matrix. If you wish the order of the labels to be consistent with the usual Euclidean x, y plane, then choose `ySeq = np.arange(n, -1, -1)` and use the statement `plot.axis([0, n, 0, n])` instead of `plot.axis([0, n, n, 0])`.
- The mainline uses two nested loops and a trigonometric function to define the contents of the data array. This was simply done to get data values into the 200 by 200 array. Ignoring this expediency, we see that the mainline is just a simple invocation of the plotting function: `generate2Dplot(zVals_A, "title")`.

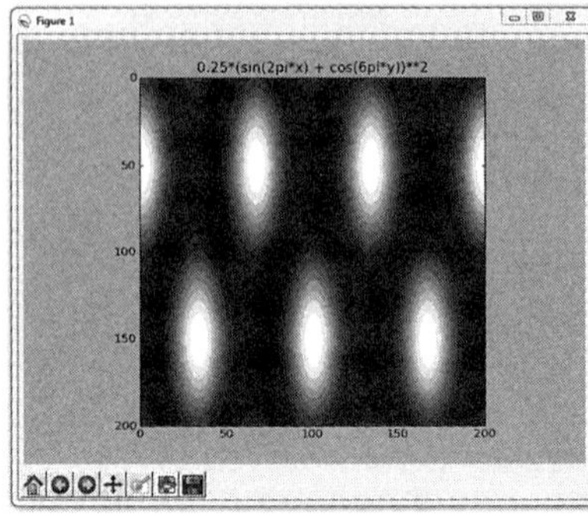

FIGURE D.2 Density plot produced by Script D.02.

Figure D.2 shows the plot that is produced by the script.

D.5 2D SCATTER PLOT

The next script (`Script D_03 Scatter plot example.py`) gives you a template for a generic scatter plot:

```
import matplotlib.pyplot as plt
plt.ion()
fig = plt.figure(figsize = (8,8))

ax = fig.add_subplot(111, aspect = 'equal')
ax.set_title("Your title here", fontsize = 14)
ax.set_xlabel("X axis title", fontsize = 12)
ax.set_ylabel("Y axis title", fontsize = 12)
ax.set_xlim([-10, 10])
ax.set_ylim([-10, 10])
ax.set_xticks([-8, -6, -4, -2, 0, 2, 4, 6, 8])
ax.set_yticks([-8, -6, -4, -2, 0, 2, 4, 6, 8])
ax.grid(True, linestyle = '-', color = '0.75')
ax.hlines(0, -10, 10, colors='k', linestyles='dashed')
ax.vlines(0, -10, 10, colors='k', linestyles='dashed')

ax.scatter([3., 5.], [-8., -3.],
           s = 20, color = 'r', marker = 's')
ax.scatter([1., 2., -3., 6.], [5.3, 7.4, 2., -0.2],
           s = 20, color = 'g', marker = 'd')
ax.scatter([2., 4., -7.], [9.3, 9.8, -2.],
           s = 20, color = 'b', marker = 'o')
fig.savefig("demoPlot.png", dpi = fig.dpi)

raw_input("Hit Enter to dismiss the plot.")
plt.close()
```

Script D.3: 2D Scatter Plot Demo

If you run the script, you should get a colored version of the plot in Figure D.3 in a separate window.

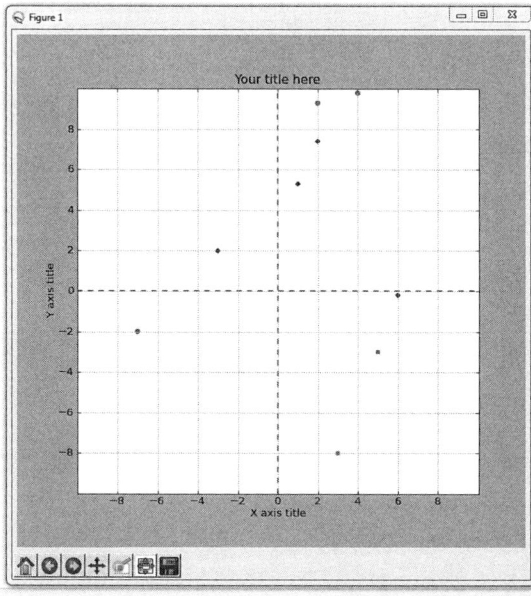

FIGURE D.3 Two-dimensional(2D) scatter plot demonstration.

The script starts with the importing of `matplotlib.pyplot`. We set interactive mode to "on" using the function call: `plt.ion()`. This causes an immediate change to the plot display whenever a plotting statement is executed. The next statement declares a `pyplot` object and establishes the size of the figure. The next block of statements defines the appearance of the plot including the titles and disposition of the axes. The significance of each statement should be reasonably obvious when you inspect Figure D.3. Execution of the `scatter()` functions displays the data points. For example, the first of these statements puts two red (`color = 'r'`) markers into the figure. Each is a small square as requested by the parameter setting: `marker = 's'`. The parameter setting `s = 20` establishes the size of the marker. The locations of the two points are defined by two lists, the first list giving the *x* coordinates and the second list giving the *y* coordinates. After the execution of the first `scatter()` statement, the red markers will be placed at positions with coordinates (3.0, –8.0) and (5.0, –3.0) in the *xy* plane. The next statement puts in four green markers each having a diamond shape, and the last statement puts in three blue markers each having a circular shape.

The `savefig()` function can be used to save the plot image as a `.png` file in the local directory. You can make the line into a comment if this is not desired. The final two lines allow you to dismiss the plot with control returning to the Python shell window.

D.6 2D PLOTTING PARAMETRIC CURVES

The next script taken from the file `Script D_04 Plotting parametric curves.py` can be used as a template for plotting parametric curves in the Euclidean plane.

```
import numpy as np, matplotlib.pyplot as plt

def parametricPlot(x_L, y_L, color = "black",
                   plotTitle = "", aspect = 1.0,
                   xAxisLabel = "", yAxisLabel = "",
                   gridFlag = False):
    plt.ion()

    fig = plt.figure()
    fig.canvas.set_window_title("Parametric Plot")
    subplt = fig.add_subplot(111, aspect = aspect)

#    subplt.set_xlim([10., 28.])
#    subplt.set_ylim([-4., 12.])
#    subplt.set_xticks(np.array([10., 14., 18.,
#                      22., 26.,28.]))
#    subplt.set_yticks(np.array([-4., 0., 4.,
#                      8., 12.]))

    subplt.grid(gridFlag)
    plt.setp(subplt.get_yticklabels(), fontsize = 10)
    plt.setp(subplt.get_xticklabels(), fontsize = 10)

    subplt.set_title(plotTitle, fontsize = 14)
    subplt.set_xlabel(xAxisLabel, fontsize = 12,
                      style = 'italic')
    subplt.set_ylabel(yAxisLabel, fontsize = 12,
                      style = 'italic')

    subplt.plot(x_L, y_L, color = color)

    raw_input("Hit Enter to dismiss the plot.")
    plt.close()

#=======================================================
# Mainline for testing:

x_L = []
y_L = []
for t in np.arange(0.0, 201.0, 1.0):
```

```
    theta = 2.0*np.pi*t/100.
    x_L.append(6.0*np.pi + 3.0*(2.0*np.sin(theta) -
               np.sin(2.0*theta)))
    y_L.append(6.0 + 3.0*(2.0*np.cos(theta) -
               np.cos(2.0*theta)))


parametricPlot(x_L, y_L, plotTitle = "Cardioid",
               xAxisLabel = "x", yAxisLabel = "y")
```

Script D.4: Plotting a Parametric Curve

Here are some relevant notes for the script:

- Matplotlib will automatically generate the labels for tick points on both the x and y axes. If you wish to put in your own labels, perhaps with a different spacing, then you can use the function calls to set limits and ticks. These are seen in the script as comments but can easily be modified and made executable.
- The script insures that `aspect` is set to 1.0 as a default value. If you ignore this constraint, then the plot will not preserve the aspect ratio if the plot window is resized in a way that distorts the aspect ratio.
- Note that the title for the plot window is specified by calling the function `fig.canvas.set_window_title`, whereas the title for the plot is handled by calling the function `subplt.set_title`.

The mainline uses `parametricPlot` to plot a particular cardioid function (Figure D.4). A cardioid curve is often seen in a coffee cup when light is reflected from the interior rim of the cup. The accumulated light rays form an envelope that is seen as a cardioid curve. The parametric equations for a cardioid are the following:

$$x = a(2\sin t - \sin 2t)$$
$$y = a(2\cos t - \cos 2t)$$

The script uses a similar definition but with the curve shifted in both x and y directions so that the example can illustrate the automatic generation of the labels on the axes.

We can, of course, consider the plotting of the explicit equation $y = f(x)$ to be a parametric plot. It would simply be treated parametrically as: $x = t$, $y = f(t)$ and the `parametricPlot` function would be given the usual `x_L` and `y_L` lists.

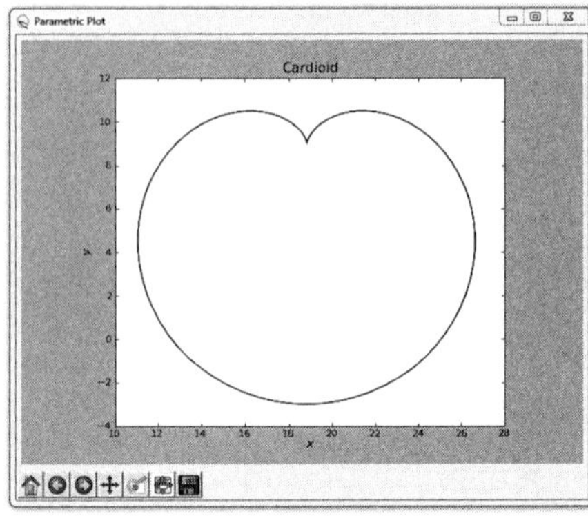

FIGURE D.4 Parametric plot of a cardioid curve.

D.7 3D SURFACE PLOTS

The next script taken from the file `Script D_05 Surface plot example.py` can be used as a template for plotting 3D surfaces with matplotlib.

```
import numpy as np
import matplotlib.pyplot as plt
from mpl_toolkits.mplot3d import axes3d, Axes3D
from matplotlib import cm

#=======================================================
def matplotlibSurface(x_A, y_A, z_A,
                      windowTitle = "Surface Plot",
                      plotTitle = ""):
    plt.ion()

    fig = plt.figure()
    fig.canvas.set_window_title(windowTitle)
    ax = fig.gca(projection='3d', aspect = 1.0)

    surf = ax.plot_surface(x_A, y_A, z_A,
                           rstride=1, cstride=1,
                           cmap=cm.cool,
                           linewidth=0,
                           antialiased=False)

    ax.set_title(plotTitle, fontsize = 14)

    fig.colorbar(surf, shrink=0.5, aspect=5)

    raw_input("Hit Enter to dismiss the plot.")
    plt.close()


#=======================================================
# Mainline:
```

```
m = 20
n = 40

xVals_L = np.linspace(-1.0, 1.0, num = n,
                      endpoint = True)
yVals_L = np.linspace(-1.0, 1.0, num = m,
                      endpoint = True)

zVals_A = np.zeros((m, n), float)
# Saddle surface:
for i in range(m):
    for j in range(n):
        zVals_A[i, j] = np.sin(xVals_L[j]*yVals_L[i])

# Note both x_A and y_A have shape (m, n):
x_A, y_A = np.meshgrid(xVals_L, yVals_L)


matplotlibSurface(x_A, y_A, zVals_A,
                  plotTitle = "Saddle Surface")
```

Script D.5: 3D Surface Plot

The statements in the `matplotlibSurface` function are usually of matplotlib scripts that plot a 3D surface. Since matplotlib typically deals with matrix data, you will have to take a bit of extra care if you wish to plot a function $z = z(x, y)$ and get axes labeling that is consistent with the usual Euclidean frame of reference. In Script D.5, note that the array arguments `x_A`, `y_A`, and `z_A` all have the same shape, which is (`m`, `n`). This consistency will be assured by using the `np.meshgrid` function but only if the number of entries in the x list (stored in `xVals_L`) is the same as the number of columns in the `zVals_A` array and if the number of entries in the y list (stored in `yVals_L`) is the same as the number of rows in the `zVals_A` array. The result of the surface plot is shown in Figure D.5.

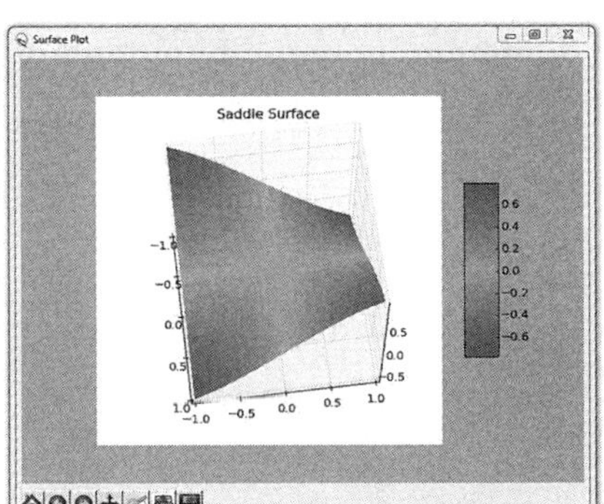

FIGURE D.5 Three-dimensional(3D) surface plot.

We have now covered bar charts, density plots, scatter plots, parametric plots, and surface plots. These can be used as templates and will likely be useful for many of the applications that require plotting facilities. In the subsequent sections of this appendix, we cover additional topics that are somewhat more sophisticated. We start with techniques for combining matplotlib with Tkinter and then move onto Python scripts that do strip charts and 3D surface plots in the Chimera display.

D.8 USING MATPLOTLIB WITH A MODELESS DIALOG

Section 4.4.1 contains a sample script to compute a 2D scatter plot, which is displayed in a separate matplotlib window. It would be convenient if such a plot could be displayed in a modeless dialog window, which contained other graphical user interface (GUI) elements, such as entry and button widgets, that would define the requirements of the plot or provide any other user interaction required by the application. Figure D.6 illustrates a very simple dialog containing a plot and the script that generates the dialog can be used as a starting point to implement more complicated dialogs such as those described in Exercises 12 and 13 of Chapter 4.

The dialog in Figure D.6 was generated by the script in the file `Script D_06 Tk & simple plot.py`.

```
import chimera, Tkinter as tk
from chimera.baseDialog import ModelessDialog
import matplotlib.pyplot as plt
from matplotlib.backends.backend_tkagg import FigureCanvasTkAgg
from matplotlib.figure import Figure

class TkPlusPlot(ModelessDialog):
    buttons = ("Close")
    help = False    # Leave this out to get a Help button.
    title = "Window title goes here"

    def fillInUI(self, parent):
        # Text to be displayed in the dialog window:
        topText = tk.Label(parent,
                           text = " Text at top of dialog",
                           font = ("Helvetica", 11, "normal"))
        topText.grid(row = 0, column = 0, columnspan = 2,
                     pady = 4, sticky = tk.W)

        #-------------------------------------------
        fig = Figure(figsize = (6,6), dpi = 100)

        ax = fig.add_subplot(111, aspect = 'equal')
        ax.set_title("Your plot title goes here", fontsize = 14)
```

```
        ax.set_xlabel("X axis title", fontsize = 12)
        ax.set_ylabel("Y axis title", fontsize = 12)
        ax.set_xlim([-10, 10])
        ax.set_ylim([-10, 10])
        ax.set_xticks([-8, -6, -4, -2, 0, 2, 4, 6, 8])
        ax.set_yticks([-8, -6, -4, -2, 0, 2, 4, 6, 8])
        ax.grid(True, linestyle = '-', color = '0.75')
        ax.hlines(0, -10, 10, colors='k', linestyles='dashed')
        ax.vlines(0, -10, 10, colors='k', linestyles='dashed')

        ax.scatter([3., 5.], [-8., -3.],
                   s = 20, color = 'r', marker = 's')
        ax.scatter([1., 2., -3., 6.], [5.3, 7.4, 2., -0.2],
                   s = 20, color = 'g', marker = 'd')
        ax.scatter([2., 4., -7.], [9.3, 9.8, -2.],
                   s = 20, color = 'b', marker = 'o')
        #---------------------------------------------

        # Set up a canvas and put the Figure into it.
        canvasFrame = tk.Frame(parent, bd = 2, relief = tk.RIDGE)
        canvasFrame.grid(row = 2, column = 0,
                         padx = 4, pady = 4, sticky = tk.E)

        canvas = FigureCanvasTkAgg(fig, master = canvasFrame)

        canvas.get_tk_widget().grid(row = 0, column = 0)

#=====================================================
# Mainline for testing:
wb = TkPlusPlot()
```

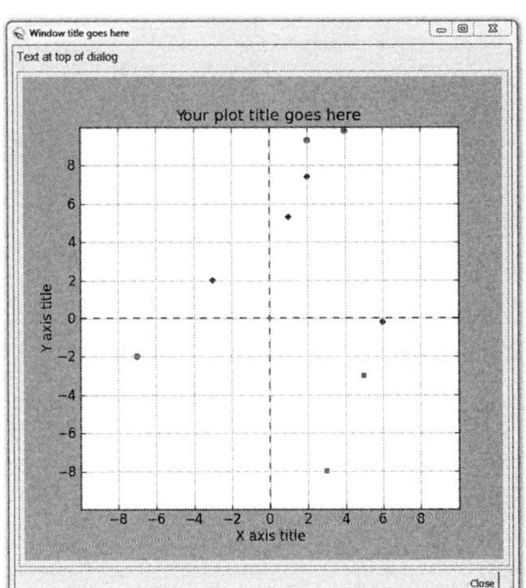

FIGURE D.6 Modeless dialog displaying a plot.

This script contains a set of statements situated between two dashed lines. These statements are essentially the same as those seen in Script 4.1 with a few unnecessary statements removed. The statement that defines `fig` involves the instantiation of a `Figure` object. The subsequent statements set up the plot elements. When this is completed, the script executes:

```
canvasFrame = tk.Frame(parent, bd = 2, relief = tk.RIDGE)
canvasFrame.grid(row = 2, column = 0, padx = 4,
                  pady = 4, sticky = tk.E)

canvas = FigureCanvasTkAgg(fig, master = canvasFrame)

canvas.get_tk_widget().grid(row = 0, column = 0)
```

The first two statements set up the frame. The next statement establishes the relationship between a canvas object and `fig`. The last statement positions the canvas within the frame and initiates the display of the plot within the dialog.

Other scripts that embed plots into dialogs can be found in the Chapter 4 directory of the software available for this book.

D.9 USING TKINTER TO IMPLEMENT A STRIP CHART

There are applications that require a plot of $y = f(x)$ that is difficult to put on the screen because the variable x ranges over a very long interval. In such a situation, it is convenient to use a strip chart equipped with a horizontal scroll bar that allows the viewer to access earlier portions of the plot (see Figure D.7). The script in the file `Script D_07 Strip chart.py` can be modified and used to record and display data that are best shown in a strip chart format. The entire script will not be discussed because the core techniques used by the script are dedicated to the functionality of a strip chart and likely not useable in a different application.

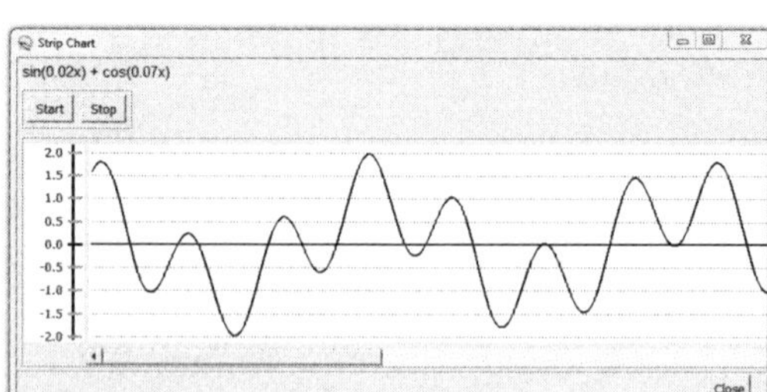

FIGURE D.7 Graphical user interface (GUI) for a strip chart.

To use the script, you should modify the opening statements of the program:

```
#-----------------------------------------------------
# Before using this strip chart,
# the following gloabal variables and
# functions have to be defined:

# Dimensions of canvas and limits on y-values:
CANVAS_H = 200       # Canvas plot height
CANVAS_W = 800       # Canvas plot width
Y_MIN = -2.1     # Minimum expected y value.
Y_MAX = 2.2      # Maximum expected y value.
# Lists for horizontal grid lines
GRID_DARK_L = [0.0]
GRID_GRAY_L = [2.0, 1.5, 1.0, 0.5,
               -0.5, -1.0, -1.5, -2.0]

PLOT_TITLE = " sin(0.02x) + cos(0.07x)"

# Maximum number of evaluations of F(x):
EVALUATIONS_MAX = 1000

def F(x):
    return np.sin(x*.02) + np.cos(x*0.07)
```

The semantics of the statements should be clear from the comments. Note that one should have an estimate by the minimum and maximum values of the data.

The remainder of the script deals with the setup of the modeless dialog. This includes the start and stop buttons and more importantly the canvas widget that is used to record the strip chart curve.

D.10 GENERATING A TICK LIST

A plot axis should have a sequence of ticks (e.g., small line segments) that indicate significant positions on the axis. Ticks should have labels that give the reader the ability to associate a tick with a numerical value specifying its position on the axis. To avoid visual clutter, we can provide labels for a subset of the ticks with the intention that the unlabeled tick positions can be inferred. The number of ticks typically varies from one application to the next and should be determined by the user.

The function `tickList(minVal, maxVal, numTicks)` will compute a list of ticks for graphs that have plotted values in the range [`minVal`, maxVal]. The number of ticks is specified by numTicks. The function can be imported by using the statement:

```
from StructBio.Utilities.miscFunctions import tickList
```

The objectives of the function can be stated as follows:

1. The position of the first tick should be less than or equal to minVal.
2. The position of the last tick should be greater than or equal to maxVal. This may require a slight increase in the number of ticks.
3. The interval length between two consecutive ticks should be specified by some simple number having, for example, one or two significant digits.
4. The first tick should have a rounded precision that is the same as the interval length.

The test program in the file Script D_08 TickList test.py gives a printout for various test cases that illustrate these objectives. Here is a copy of the tickList script:

```
import numpy as np

# Function to provide a list of tick values:
def tickList(minVal, maxVal, numTicks):
    incr = (maxVal - minVal)/numTicks
    incrRounded = round(incr, -int(np.floor(np.log10(incr))))

    # The first tick is equal to minVal and is rounded with
    # the same precision as incr.
    # If this rounding causes firstTick to be more than the
    # minVal, then we decrease firstTick by an amount equal
    # to the rounded version of incr.
    firstTick = round(minVal, -int(np.floor(np.log10(incr))))
    if firstTick > minVal:
        firstTick = round(minVal - incrRounded,
                    -int(np.floor(np.log10(incr))))

    t_L = []
    nextTick = firstTick
    while nextTick < maxVal + incrRounded:
        # Zero if sufficiently close to 0.0:
        if abs(nextTick) < incr * 0.00001:
           nextTick = 0.0
        t_L.append(nextTick)
        nextTick += incrRounded
```

```
    # We now have the correct tick values but the internal
    # representation may introduce trailing 9's or a result
    # that is just slightly too large.
    # So, we use formatting to do "final roundup".

    # Calculate number of digits in a large number.
    sig = int(np.ceil(np.log10(max(abs(minVal), abs(maxVal)))))
    frmt = '%.3g'
    # Display 3 significant digits,
    # but if a number is larger than 999 display them all.
    if sig > 3: frmt = '%.' + str(sig) +'g'
    t_L = ['%s' % float(frmt % x) for x in t_L]
    # Back to floats:
    t_L = [float(c) for c in t_L]
    return t_L
```

The first two lines of the script calculate a rounded version of `incr`, which is `(maxVal - minVal)/numTicks` the exact value for the interval between consecutive ticks. Note that rounding is done using `round (incr, -int(floor(log10(incr))))` so that the third point in the objective list is satisfied. The next three statements deal with the first point. The ensuing loop will generate the list of ticks as required, but they may not be suitable for display. One must realize that further computation using results of the round function does not necessarily produce a final result that has an exact internal floating point representation. To appreciate this, consider the following interactions with the Python Shell:

```
>>> round (5.616, 2)
5.62
>>> 1.1 + round (5.616, 2)
6.720000000000001 # The result is not 6.72!

>>> round (5.616, 2)
5.62
>>> 6.72 - round (5.616, 2)
1.0999999999999996 # The result is not 1.1.
```

To get around this small dilemma, we put the tick list through a final processing step that involves the conversion of the tick values to output strings that provide a more concise version of the tick labels.

If you run the `Script D_08 TickList test.py` program, you can see how it computes tick lists for various number ranges that have end points of varying magnitudes.

As a final note, you might wish to look at `Script D_09 x y plot with GUI.py`. Running the script will produce the dialog shown in Figure D.8. The

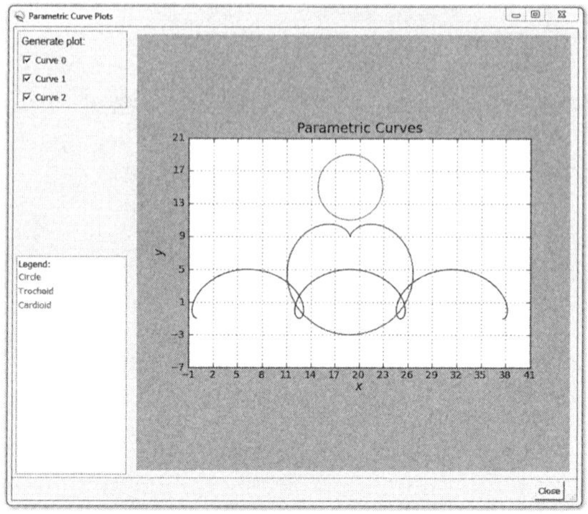

FIGURE D.8 GUI for multiple curve plots.

script was written to demonstrate a Tkinter plus matplotlib cooperation that is more complicated than that seen in Script D_06. The script is 161 lines long and will not be reproduced on these pages. Some points of interest:

- Note the use of a `CurvePlotBundle` class that is used to carry the data values along with the plot color and legend entry.
- The script shows a mainline test program that generates data for three curves but the script has been written to work with a larger number of curve plot bundles.
- The computation of ticks for any combination of curves to be displayed is done by computing the overall maximum and minimum values for the data followed by a call to the `tickList` function.

D.11 DATA STRUCTURE FOR COLORED SURFACES

Python scripts that work with 3D surface plots may generate the data for these plots, which are subsequently displayed within a GUI. To facilitate the transfer of surface information between scripts that generate surface data and scripts that display surface data, it would be expedient to have a self-contained data structure capable of storing the coordinates of points on the surface and the associated color information. Storage of this data will be done using the `SurfacePlotBundle` class that can be imported using the statement:

```
from StructBio.Plots.surfacePlotBundle import SurfacePlotBundle
```

In addition to color and transparency information, the class has attributes that track the six minimum and maximum values for the x, y, and z components of the coordinates of the points in the surface. These are useful as input for various calculations done by plot routines. Here is the class definition, initialization, and subsequent comments:

```
class SurfacePlotBundle(object):
  def __init__(self, x_A, y_A, z_A,
               color = np.array ((.5, .5, .5)),
               transparency = 1.0,
               surfaceName = "Surface Plot"):
```

The first three parameters are for three arrays each with the same shape. The entries defined by `x_A[i,j]`, `y_A[i,j]`, and `z_A[i,j]` define a 3D point $p\ (i, j) = (x, y, z)$ on the surface, where x = `x_A[i,j]`, y = `y_A[i,j]`, and z = `z_A[i,j]`. The colors and transparency parameters have default values as shown. If the default values are not used, here are the rules:

- Color:
  - The variable color can be a three component array (this color applied to the entire surface) or it can be an array with shape $(m, n, 3)$, where `z_A.shape` is (m, n).
  - The entry `color[i,j,:]` defines the rgb surface color at point $p(i, j)$.
- Transparency:
  - The variable transparency can be a simple float in the interval [0.0, 1.0], or it can be an array with the same shape as `z_A`.

Note that the `x_A`, `y_A`, and `z_A` arrays have the same meaning as the arrays used as input arguments for the `plot_surface` function in matplotlib. The color and transparency parameters allow a simple choice: use one value for the entire surface or use array structures that define the color and transparency for each discrete point defining the surface.

The collection of points defined by the `x_A`, `y_A`, and `z_A` arrays will be used to construct the surface for subsequent display. For us, the surface will be constructed by using the `addPolygon` function in the `Solids` class. The typical polygon has four points: $p(i, j)$, $p(i, j + 1)$, $p(i + 1, j + 1)$, and $p(i + 1, j)$.

To facilitate the transfer of a `SurfacePlotBundle` object from one application to the next, we make use of Python's ability to "pickle" a class object. Technically, this is Python's mechanism for object serialization.* It gives us an easy strategy for saving and recovering a class object without the need for carefully designed I/O (input/output) formats that would otherwise be required to save all the different components of the surface plot bundle.

The script in the file `Script D_10 Saddle surface plot bundle build.py` gives an example of code that is used to build a surface plot bundle. In the most general case, the values for the `x_A`, `y_A`, and `z_A` arrays are defined *parametrically*. Since we are defining a surface (characterized as a continuum having two degrees of freedom), the parametric definition of the surface will require two independent variables, call them u and v. Consequently, to define a surface plot bundle, we go through the following steps:

- Define `m` and `n` specifying the resolution of the surface (the higher, the smoother, but at the expense of more execution time and a larger pickle file).
- Define `uMin`, `uMax`, `vMin`, and `vMax` specifying the rectangular domain of the independent variables `u` and `v`.
- Define the surface name (this should be a character string that is reasonably short in case the string is used within a `Label` widget of a GUI).
- Define the default color as an RGB numpy array of shape (3,) with each array entry a floating point value in the range [0.0, 1.0] or build a more complicated array to specify color for all the points defining the surface.
- Define the transparency as a floating point number in the range [0.0, 1.0] or build an array to specify transparency for each point defining the surface.
- Define the functions $x(u, v)$, $y(u, v)$, and $z(u, v)$ with values stored in the `x_A`, `y_A`, and `z_A` arrays. By doing this, we are really defining a mapping function that takes (u, v) in a 2D space to a point (x, y, z)

* See: http://docs.python.org/3/library/pickle.html. Be sure to read the cautionary note about unpickling data from an unknown source.

in 3D space. These points will describe a surface because the (u, v) pairs are taken from a 2D continuum. If necessary, you may get z as a function of u and v by defining z in terms of x and y after they have been defined in terms of u and v.

Note: An application using the `SurfacePlotBundle` class can redefine the `surfaceName` attribute of an object.

This provides all the data needed to build the surface plot bundle, which can then be passed to the plotting routines. Here is a description of the various attributes stored in the class:

`x_A:`	An array holding the x components of coordinates of points defining the surface.
`y_A:`	An array holding the y components of coordinates of points defining the surface.
`z_A:`	An array holding the z components of coordinates of points defining the surface.
`color:`	An array with shape (3,) if the color argument designates a single color.
`color_A:`	An array with shape $(m, n, 3)$ if the color argument designates a color for each point defining the surface.
`transparency:`	A single float value if the transparency argument specifies a float value.
`transparency_A:`	An array with shape (m, n) if the transparency argument designates a transparency value for each point defining the surface.
`xMin:`	The minimum value of `x_A`.
`xMax:`	The maximum value of `x_A`.
`yMin:`	The minimum value of `y_A`.
`yMax:`	The maximum value of `y_A`.
`zMin:`	The minimum value of `z_A`.
`zMax:`	The maximum value of `z_A`.

D.12 USING CHIMERA TO GENERATE SURFACE PLOTS

We now consider a more ambitious strategy for generating 3D surface plots. This section presents various scripts that generate plots that will be shown in the Chimera window (another reason to have a tick list function).

Why use Chimera to display surface plots?

- Mouse drags give you rapid zoom and rotation operations allowing the user to position the surface in the most favorable position.
- You can exercise precise control of the scene. By using the menu invocation **Tools... Viewing Controls... Effects**, you can launch the Viewing dialog with the Effects tab selected. This allows control of depth cueing, transparency, and graphics quality. Incidentally, the quality of graphs is greatly improved by using a high subdivision value (say 18.0) and be sure to have the multisample checkbox checked to get better results with antialiasing. Viewing effects such as silhouettes and shadows are possible but of more questionable value for this application.
- The Lighting tab in the Viewing dialog gives full control over the light sources for the scene including control over sharpness, reflectivity. Note the selection for Intensity and Shininess.
- By using the Camera tab of the Viewing dialog, you can control the projection functionality. It is likely that you will have the projection choice set to perspective. This gives you the most realistic view of the surface set in a 3D scene. Setting the projection choice to orthographic eliminates perspective but it is a useful projection when you need to line up the surface with grid lines.

More information about the viewing tool can be found at http://www.cgl.ucsf.edu/chimera/docs/UsersGuide/sideview.html.

Running the script in the file `Script D_10 Saddle surface plot bundle build.py` will demonstrate these points (see Figure D.9). The goals of the script are as follows:

1. Demonstrate the advantages of generating plots in the Chimera display.
2. Give an example of how the surface plot kit can be used to construct plots.
3. Provide a useful function for the programmer: The script can be used to generate data input for other programs: You can save the surface data as a *surface plot bundle* for display in other applications covered in this appendix.

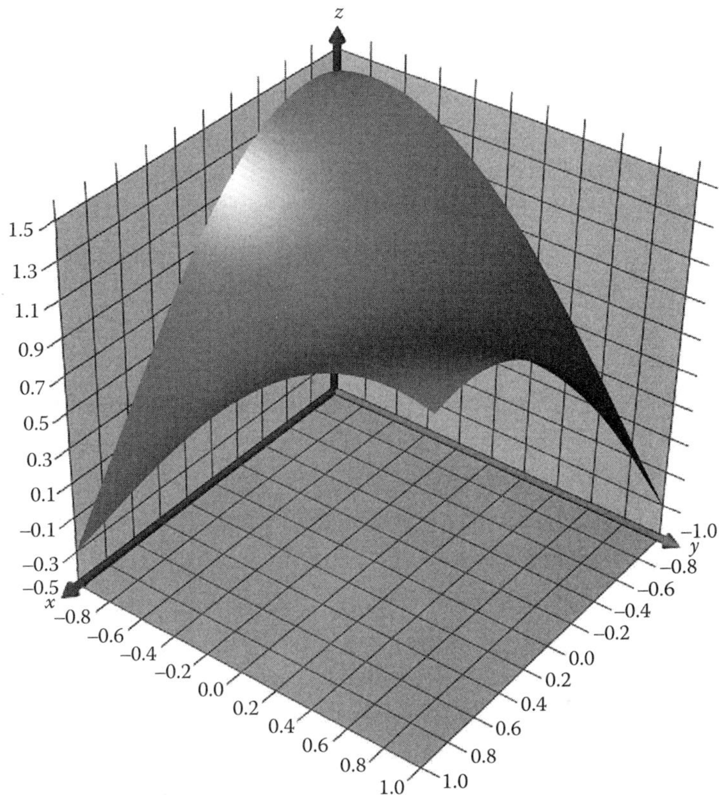

FIGURE D.9 (**See color insert.**) Surface plot in the Chimera display.

The part of the script that generates the plot is quite brief (just the 13 statements between the dashed lines) because it makes extensive use of classes that have been imported from the `StructBio.Plots.surfacePlotKit` package. The kit works with surface data that has been set up as a surface plot bundle.

The 3D plot shown in Figure D.9 has several components in addition to the surface being plotted. The graphics components seen in the plot as planes, grid lines, labels, and axes correspond to the surface kit classes and have been implemented by using the scenographics elements introduced in Appendix B. Let us now consider how classes in the surface plot kit are used to build the plot. The classes can be imported with the statement:

```
from StructBio.Plots.surfacePlotKit import\
   (TickLists, SurfacePlot, Axes, QuadrantPlanes,
    GridLines, TickLabels, plotMultipleSurfaces)
```

We will now describe the functionality of these classes:

```
class TickLists(object):
    __init__(self, surfacePlotBundles_L, numTicks)
```

The `TickLists` class works through the surface plot bundles specified in the `surfacePlotBundles_L` list and extracts the information necessary to create tick lists, for all three axes, that will be suitable for a plot that displays all the surfaces. The `numTicks` value is the approximate number of ticks that will be used for the axis corresponding to the dimension with the largest data variation. If there is a need to change the set of tick lists, (for example, there may be a change in `numTicks`) then the function `updateTickLists()` can be used to do such an update.

```
class SurfacePlot(Surfaces):
    __init__(self, spbObj, spbColoration = True,
             lowestZcolor = (0.,0.,1.),
             highestZcolor = (1.,0.,0.))
```

Instantiation of this class will generate a plot display of *one* surface specified by the surface plot bundle `spbObj`. There are two options for coloring the surface: If `spbColoration` is `True`, then the color is taken from the color attribute contained in the surface plot bundle object. If `spbColoration` is `False`, then the coloration of the surface at any point (x, y, z) depends on the value of z by computing where it lies on the interval specified by [`zMin`, `zMax`]. These extreme values are accessed from the surface plot bundle object. The color will be computed as a linear interpolation between the two color values specified by `lowestZcolor` and `highestZcolor`.

```
class Axes(Solids):
    __init__(self, ticks,
             xAxisColor = (.8, 0., 0.),
             yAxisColor = (0., .8, 0.),
             zAxisColor = (0., 0., .8))
```

Instantiation of this class creates a set of axes for the plot. The axes are aligned with the Chimera frame of reference. Consequently, when the plot is first seen in the display, the z axis will be pointing out toward the viewer, the x axis will be horizontal on the screen pointing to the right, and the y axis will be vertical pointing in the upward direction. The default colors of the axes will be red, green, and blue but can be changed by setting the color parameters in the `__init__` function. The length of each axis is determined by the highest value in the corresponding tick list. The `ticks` variable should be a reference to a `TickLists` object that was instantiated earlier. Note that the `modelName` will be "Axes", and this will be seen as an entry in Chimera's Model Panel.

```
class QuadrantPlanes(Surfaces):
    __init__(self, ticks,
              yzPlaneColor = (.8, 0., 0.),
              zxPlaneColor = (0., .8, 0.),
              xyPlaneColor = (0., 0., .8),
              transparency = 0.2)
```

Instantiation of the class produces a set of quadrant planes for the plot. The planes have positions and orientations that are consistent with the axes produced by the `Axes` instantiation (see the description of the `Axes` class). The `ticks` object will determine the length dimensions of the plane just as it did for the axes. The colors of the planes have default values that correspond to the color of the axis normal to that plane. They can be changed by setting the color parameters in the `__init__` invocation. The `modelName` will be "Quadrant Planes", and this will be seen as an entry in Chimera's Model Panel.

```
class GridLines(Solids):
    __init__(self, ticks)
```

Instantiation of the class generates a set of grid lines for the graph. The lengths and positions of the grid lines are determined by the `ticks` object instantiated earlier. The `modelName` will be "Grid Lines", and this will be seen as an entry in Chimera's Model Panel.

```
class TickLabels(LabelGroups):
    __init__(self, ticks)
```

Instantiation of the class generates a set of labels for the grid lines in the graph. The positions of the labels are determined by the `ticks` object instantiated earlier. The `modelName` will be "Labels for Grids & Axes", and this will be seen as an entry in Chimera's Model Panel. Since Chimera uses labels that are associated with atoms that are typically in a very complicated scene, Chimera will display labels "on top" of the scene so that they are not occluded by molecular structure. In the case of a plot, this might be annoying depending on the requirements of your visualization. For example, in Figure D.10 you can see the *z* label for the *z* axis shown on

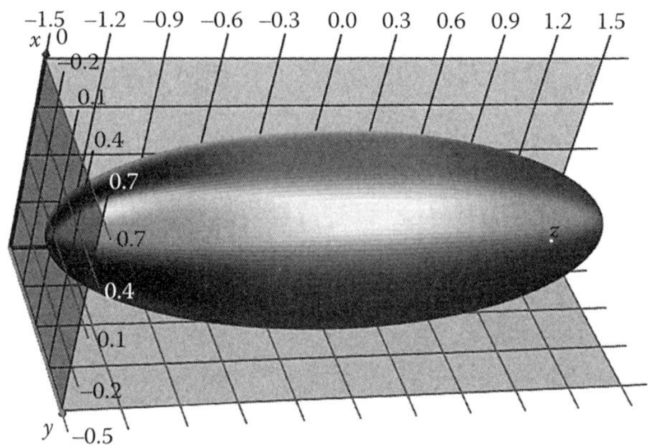

FIGURE D.10 **(See color insert.)** Surface plot of an ellipsoid in the Chimera display.

top of the ellipsoid surface. To turn off this behavior, you can do the menu invocation Actions... Label... options.... This will bring up a dialog window. You can then toggle the Labels on top: button to make it False.

```
plotMultipleSurfaces(spb_L, tickList,
                     spbColoration = True,
                     lowZendColor = (0., 0., 1.0),
                     highZendColor = (1.0, 0., 0.))
```

This function will display all surfaces designated by the surface plot bundle list `spb_L`. The semantics for the coloration parameters are essentially the same as those used for the `SurfacePlot` class except that `zMin` and `zMax` are replaced by the minimum and maximum values appropriate for the entire set of surfaces in the list. These values are extracted from the `tickList` object instantiated earlier.

```
fetchSPB()
```

Invocation of this function brings up a `tkFileDialog`, which will ask the user to specify the pathname for a file holding the pickled representation of a surface plot bundle object that will be passed back as the return value.

Here is a listing of the statements in Script D.10 that are used to construct the plot:

```
ticks = TickLists([spb], 10)
surfaceModels_L = plotMultipleSurfaces\
          ([spb],ticks, spbColoration = True)
```

```
axes = Axes(ticks)
axes.display()

qPs = QuadrantPlanes(ticks)
qPs.display()

gLs = GridLines(ticks)
gLs.display()

tLabels = TickLabels(ticks)
tLabels.setLabels("axesLabels", "black")
tLabels.setLabels("xT_yMax", "red",
                  stride = 1)
tLabels.setLabels("yT_xMax", "forest green",
                  stride = 1)
tLabels.setLabels("zT_xMax", "blue",
                  stride = 1)
```

Note that a unit of distance in the plot corresponds to a unit distance in the Chimera display space (1 Å). The script in file `Script D_11 Ellipsoid surface plot bundle build.py` is essentially the same at the previous script with an ellipsoid being generated (see Figure D.10) instead of a saddle surface. This example has been included because it demonstrates two aspects of plot generation not seen in the previous example. First, the surface is not generated by a simple $z = z(x, y)$ mapping. It requires a parametric definition of the surface. Second, the coloration of the surface is specified by an array of colors held in `colors_A` that is passed to the plot function via the surface plot bundle `spb`.

The script in the file `Script D_12 matplotlib for SPBs.py` has been included to demonstrate that it is fairly easy to write a script that can fetch a surface plot bundle and have it plotted using the `plot_surface` function in matplotlib.

As described earlier, the function `plotMultipleSurfaces` is able to plot more than one surface when it is given a list of surface plot bundles as done in Script D_13. Figure D.11 shows an example of such a plot. These surfaces have no biological relevance, and they are only used to illustrate the capabilities of the function. The gray surface is a type of saddle surface defined as

$$z(x, y) = \sin(xy) + \cos(xy)$$

while the colored surface is defined parametrically as

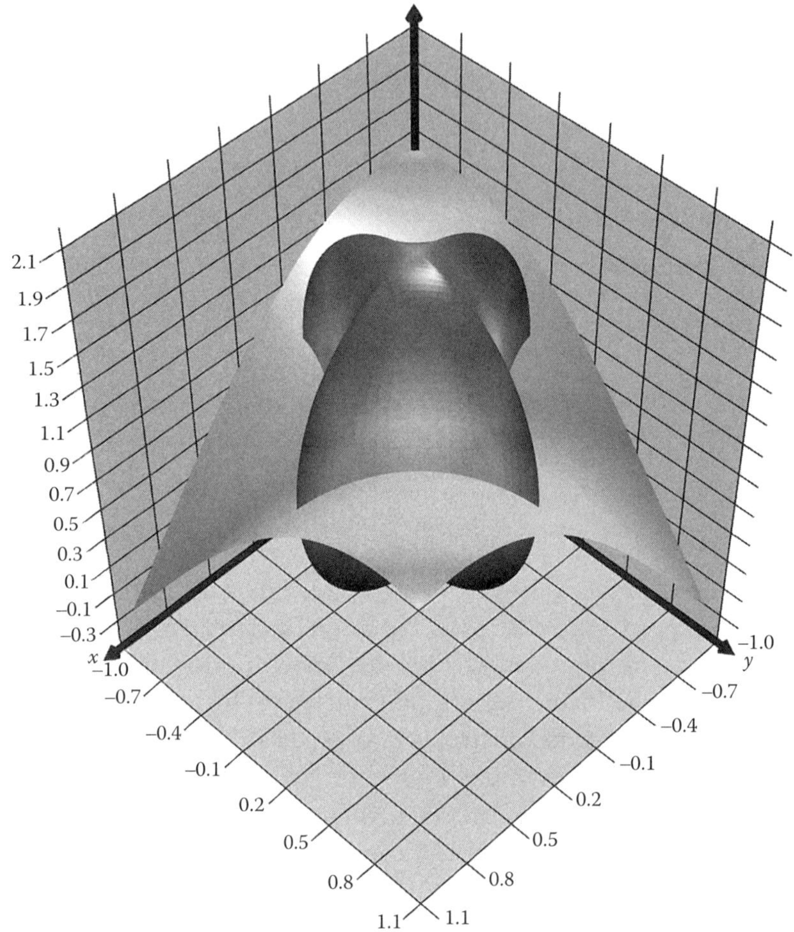

FIGURE D.11 Two intersecting surfaces.

$$x = 0.75\cos(v)\cos^3(u)$$
$$y = 0.75\cos(v)\sin^3(u)$$
$$z = 1.0 + \sin(v)$$

D.13 USER INTERFACE FOR THE SURFACE PLOTTING SCRIPT

Using the previous scripts for plotting a surface means that you will be accepting the presentation style unless you are willing to modify the code. Script D_14 is being offered as a tool that can display the surface along with support for surface color changes and various modifications of the quadrant planes, grid system, and axes. The GUI is shown is Figure D.12.

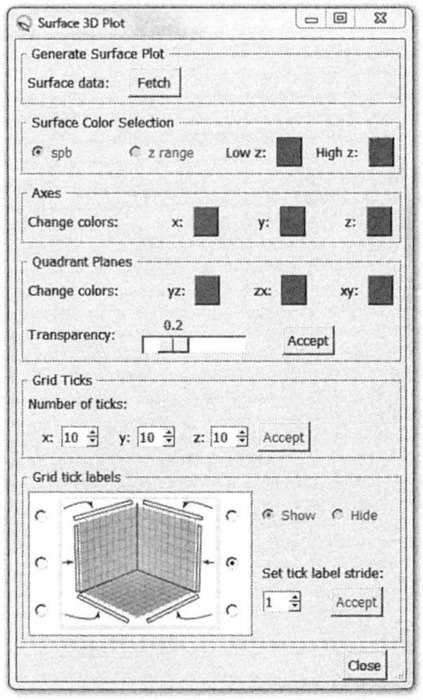

FIGURE D.12 GUI for a surface plot.

The Fetch button in the top frame can be used to fetch a surface plot bundle (this interface will only handle a single surface). After the script has generated the surface, its color may be changed. Other widgets in the GUI allow modification of the axes, quadrant planes, and grids.

The script has been written with modularity in mind and hopefully it can be adapted to your needs as required.

D.14 USER INTERFACE FOR THE PIE CHART SCRIPT

We conclude this appendix with a presentation of a user interface to create and modify pie charts. The interface relies on the pie chart Script B.7 that was introduced in Appendix B. The script is in the file `Script D_15 Pie plot & GUI.py`. To use the script, you will have to set up a small text file containing the data.

As an example, let us construct a fictitious data set that has six floating point numbers that are measuring some property that is related to secondary structure pairs. The text file would have some descriptive information in the first line followed by six lines each containing a floating point number and an associated label separated from the number by a blank. For example:

```
Foo index for secondary structure pairs
23.4 C-C
48.2 C-H
```

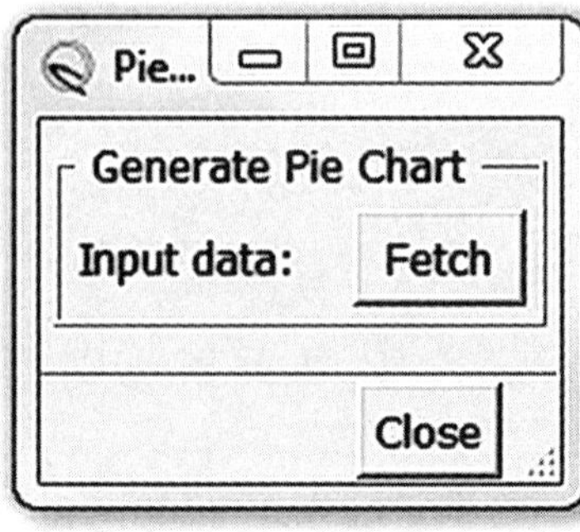

FIGURE D.13 Fetch dialog for pie chart.

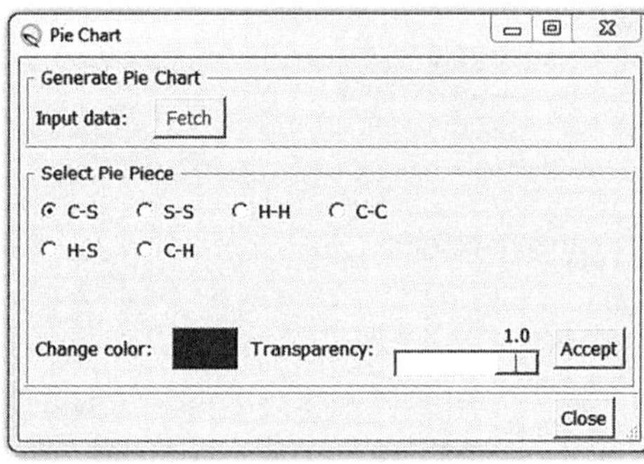

FIGURE D.14 Full dialog for pie chart.

```
3.5 C-S
16.2 H-H
33.1 H-S
11.7 S-S
```

Running Script D.14 produces the file fetch dialog (Figure D.13):

Clicking on the Fetch button will launch the Open dialog window allowing you to select the text file containing the pie chart data, for example: `Foo data for pie chart app.txt`. The pie chart script will read the data file, calculate the appropriate percentages, and then display the pie chart. This processing will also get the labels for the pie pieces, which will be set up in an expanded version of the dialog alongside radio buttons (see Figure D.14). The color and transparency of a pie piece can be changed by selecting a piece via the radio buttons.

REFERENCE

[Hu07] J. D. Hunter. Matplotlib: a 2D graphics environment. *Computing in Science and Engineering*, **9** (2007), 90–95.

Appendix E

Dynamic Programming

E.1 MOTIVATION

Dynamic programming is an algorithm originally developed by Richard Bellman in the early 1950s. At that time, he was working on applied problems dealing with multistage decision processes. He decided to use the term *Dynamic Programming* instead of *multistage decision processes* as a name for the algorithm that we are about to describe in this chapter. The *multistage decision* nature of the algorithm will eventually become clear.

For bioinformatics in general, dynamic programming is undoubtedly the most useful paradigm for the design of algorithms. Most students will likely have covered this topic in an earlier year of their undergraduate program, but it is so important that we cover it once again in detail. Dynamic programming is presented at an abstract level to present the general steps involved with the strategy.

E.2 INTRODUCTION

Dynamic programs solve optimization problems. The overall strategy underlying the algorithm is that the given problem can be solved by first solving smaller subproblems that have the same structure. The following two features characterize these subproblems:

1. Unlike the divide-and-conquer algorithm, which relies on subproblems being independent, dynamic programming handles problems that have overlapping subproblems.

2. We can apply a recursive strategy: We find an optimal solution of each subproblem by first finding optimal solutions for its contained subproblems.

E.3 DYNAMIC PROGRAMMING EXAMPLE: THE AL GORE RHYTHM FOR GIVING TALKS

We start with a simple dynamic programming example. The bioinformatics content is nonexistent but it is simple enough to illustrate the main ideas underlying the strategy of dynamic programming.

E.3.1 Problem Statement

You have been hired as a consultant to help Al Gore promote his campaign against global warming. He has already decided to tour various cities in the United States to lecture on this subject. The tour will take place over n consecutive days. For each of these days, he has prescheduled each meeting site and he knows exactly how many people will show up for each site. We will assume that the ith day has a meeting with $u(i)$ attendees. Although he would like to attend all these meetings, the demands of his campaign have weakened him to such an extent that he cannot attend meetings on two consecutive days and so some meetings must be skipped. (When he does not attend a meeting, there will be a free showing of the movie *An Inconvenient Truth*.)

Design an algorithm that will select the days when Gore will meet the attendees with the objective that, for the entire tour, he will meet the maximum number of these people. Rephrasing the meeting constraint: If he meets attendees on day $i - 1$, then he cannot meet attendees on day i. We will assume there is no meeting planned on the day before day 1.

E.3.2 Terminology: Configurations and Scores

We should always distinguish two aspects of optimization problems that are amenable to a dynamic programming solution. We ultimately want some type of *configuration* that provides an answer to the problem. In this problem, a configuration is simply some selection of days chosen from the available n days. The definition of a configuration will change from one problem to the next. The other aspect of the optimization problem is that we can assign a *score* to each configuration. In this problem, the score for a configuration will be the sum of all $u(i)$ values such that i ranges across the selected days that make up the configuration.

With this terminology in place, we can see that the solution of the optimization problem is the derivation of a configuration that provides the optimal score. For this problem, optimal means maximal. For other problems, we might want to find the configuration with the minimal score.

E.3.3 Analysis of Our Given Problem

As stated earlier, we find an optimal solution of our given problem by first finding an optimal solution for various subproblems. We consider the following definition of a subproblem: Find the configuration that gives the optimal score for a problem, which is limited to the first i days of the tour. We let this optimal score be represented by the function $S(i)$. Now we need an analysis of the situation:

For day i, Gore either meets attendees or he does not.

- If he does attend the meeting, then for day i the sum $S(i)$ includes $u(i)$, day $i - 1$ is skipped and moreover, the selection from the first $i - 2$ days must be an optimal selection. If it was not optimal, then we could replace this nonoptimal configuration of days with a better selection having a larger sum of attendees. This sum plus the value of $u(i)$ for day i would lead to a larger score for $S(i)$ in contradiction of the fact that $S(i)$ is already assumed to be optimal. So, Gore meeting attendees on day i implies $S(i) = u(i) + S(i - 2)$, where $S(i - 2)$ is the optimal score for the first $i - 2$ days.
- If Gore does not meet attendees on day i then, nothing has changed since the previous day. In other words: $u(i)$ is not part of $S(i)$, and $S(i) = S(i - 1)$.

Since we want to choose the scenario that yields the larger score $S(i)$, we simply insist that $S(i)$ is the maximum of these two possibilities, in other words:

$$S(i) = \max\{u(i) + S(i-2), S(i-1)\}$$

Pseudocode for the algorithm could be written as follows:

```
S[0] := 0; S[1] := u[1];
for i := 2 to n do
        if (S[i-1] < S[i-2] + u[i]) then
                S[i] := S[i-2] + u[i];
        else
                S[i] := S[i-1];
return S[n];
```

Note that the value of $S(n)$ will be the best score that is achievable for the entire n days. With a bit of thought you should also realize that computation of the $S(i)$ values must be completed for all $i = 1, 2, \ldots, n$ before we can attempt to derive the configuration that gives the solution of the problem. In fact, we must inspect both $S(n)$ and $S(n - 1)$ before discovering whether Gore will attend the meeting on day n. Once that is established, we can then determine whether the meeting on day $n - 1$ is skipped or attended, then we consider $S(n - 2)$, and so on. The pseudocode for this algorithm is as follows:

```
i : = n;
while i > 0
        if (S[i-1] <> S[i]) then
                write("Day" i "chosen");
                i : = i - 2;
        else
                i : = i - 1;
```

This completes the solution of the given problem. Now, let us extract from this solution some procedural steps that can be applied to all dynamic programming solutions.

The first step was the most critical because it identified the subproblems that were to be solved to get the solution of the full problem. This means that the optimal configuration for the given problem will be constructed from the optimal configurations of subproblems. These subproblems will have optimal configurations that are in some sense *smaller*. The challenge of designing a dynamic programming algorithm will be to understand how these smaller optimal configurations relate to larger optimal configurations and eventually to the largest optimal configuration that we ultimately need. However, if you look back at the analysis of the problem, you will see that this relation is formulated in terms of the scores derived from the optimal configurations. This produces the recursion that is at the heart of the solution.

In general, the score value in the recursion is dependent on various parameters that characterize the extent of the subproblems. In our given problem, S depended only on i. In more complex dynamic programming problems, there may be two (perhaps even three) parameters that act as independent variables for the score function. We will see that it is often advantageous to consider the values of the scoring function to be stored in an array that has as many dimensions as there are parameters.

Using this array, we can visualize the computation of the score function recurrence by stipulating where the initial values are stored and the order of computation when filling the cells of the array. The analysis of the problem will also specify the array cell containing the optimal score for the given problem. Finally, the analysis of the problem will dictate how the solution configuration is recovered. This final step is usually referred to as the traceback or solution recovery step.

Be sure to understand the significance of the word *back* in the traceback step. The optimal scores for the subproblems are typically generated by having one or more parameters of the scoring function progressing through values in a direction that we characterize as the *forward* direction. For example, in the Al Gore problem, we have the index i going from 1 to n when the score values are evaluated. After all these computations are completed, we recover the configuration for the optimal solution by reversing direction and letting the parameter (or parameters) progress through values in the opposite direction. In our given problem, i progresses from n down to 1 in the traceback part of the algorithm.

For students who are just learning about dynamic programming, it is crucial to stress that the formulation of a dynamic programming algorithm should deal first with the scoring function, and then later, with the derivation of the configuration. When faced with a new and complicated dynamic programming problem, inexperienced students will sometimes attempt to derive a final configuration much too soon—immediately trying to understand how the configuration is computed while treating the scoring function as if it is an evaluation to be done after configurations are somehow calculated. This often leads to confusion and failure.

E.4 RECIPE FOR DYNAMIC PROGRAMMING

To improve the chances of success when designing a dynamic programming algorithm, here is a summary of the steps that are usually taken:

1. Identify the subproblems and provide a supporting analysis that produces a recursion for the evaluation of the scoring function.

   We need to specify how the optimal score of a subproblem contributes to the optimal score of a larger subproblem.

2. Provide details that stipulate how the score values are evaluated.

Typically, we store the optimal scores of the subproblems in an array (possibly a big, multidimensional array). We need to

a. Specify the matrix dimensions.

b. Describe how the value in a cell of the array depends on the values of other cells in the array (thus specifying the order of computation).

c. Specify where the final optimal score for the full problem will be found in the array.

3. Set values for the base cases.

   There will be cells that require initial values because the recursion equation alone cannot define all the score values in the array.

4. Recover the solution configuration.

   a. Keep track of the configurations that provided the optimal values.

   b. Use a traceback strategy to determine the full configuration that gives the optimal value for the full problem.

E.5 LONGEST COMMON SUBSEQUENCE

Now that we have some dynamic programming methodology established, we apply it to a problem with some extra complexity. The longest common subsequence (LCS) problem usually arises from methods that need to compare strings of nucleotides. The solution of the problem serves to provide a measure of the amount of similarity between two nucleotide sequences. We start with some abstractions that define the problem.

Definition: A subsequence of a string X is a string that can be obtained by extracting some or all of the characters from X while still maintaining their sequence order.

For example, if X = "`supercalifragilisticexpialidocious`", then the characters T = "`cialis`" would be a subsequence. There are approximately 16 billion subsequences that can be extracted from X.

The reader should not confuse this with the notion of a *substring*, which is a more specialized extraction of characters: A substring is subsequence of *consecutive* characters that are extracted from X.

E.5.1 Problem Statement

We are given two strings: $X = x_1, x_2, \ldots, x_m$ and $Y = y_1, y_2, \ldots, y_n$. The problem is to find the longest string Z that is a subsequence of both X and Y.

For example, suppose

```
X = "supercalifragilisticexpialidocious" and
Y = "supercalafajalistickespeealadojus"
```

Then the LCS of X and Y is Z = "`supercalfalisticepaldous`".*

A small issue of concern is the presentation of strings X, Y, and Z so that a reader can easily verify that Z is indeed the LCS of X and Y. One could write Y below X and then draw lines between the matching characters that produce Z. This is somewhat cumbersome and is typically not done. The usual technique is to insert dash characters into both X and Y strings so that they have the same length and written out so that matching characters are in the same column. For example,

```
X:  supercalifragilistic-expi-alidocious
Y:  supercalaf-ajalistickespeealado—j us
Z:  supercal f a  listic e p  al do   us
```

Characters that produce the LCS have been rendered in a bold font.

The LCS is not necessarily unique. For example, let X = "`ATCTGAT`", Y = "`TGCATA`". Then both Z = "`TCTA`" and Z = "`TGAT`" are legitimate as an LCS. Naturally, both possibilities have the same optimal string length.

E.5.2 Prefixes

We compute the LCS using dynamic programming. Our strategy is to derive the LCS of the given strings X and Y, by first solving various subproblems. Each subproblem involves the computation of an LCS for some pair of *prefixes* of X and Y.

Definition: The ith prefix of a string X, denoted by $X(i)$, will be the substring of X made up from the first i characters of X.

* There is some history associated with this particular Y string. It is related to a legal action brought against The Walt Disney Company. See: http://en.wikipedia.org/wiki/Supercalifragi listicexpialidocious. Incidentally, the string Y is flagged by Microsoft Word as a misspelling, but X is accepted.

We use lower case subscripted letters to denote the characters in X. Formally, we denote the full string X as: $X = x_1x_2 \ldots x_m$. Then, the prefix $X(i) = x_1x_2 \ldots x_i$. Assuming X has m characters, we have $X = X(m)$.

Given input strings $X = x_1x_2 \ldots x_m$ and $Y = y_1y_2 \ldots y_n$, our subproblem will be to find the LCS of $X(i)$ and $Y(j)$. Note that the subproblem is parameterized using two indexes i and j with $1 \le i \le m$ and $1 \le j \le n$. The score value for such a subproblem, denoted by $S[i,j]$, is the length of the LCS of the prefixes. The configuration for this subproblem is the LCS itself.

Having defined this notation and the significance of the score function, we are about half way through the first step of our recipe for dynamic programming introduced in Section E.4. It is now necessary to provide the supporting analysis that leads to a recursion defining $S[i,j]$. Since prefixes are essentially defined by their last position, we will be interested in the following clever observations about the last character of an LCS.

E.5.3 Relations among Subproblems

Suppose we consider $X(i) = x_1x_2 \ldots x_i$ to be a prefix of X and $Y(j) = y_1y_2 \ldots y_j$ to be a prefix of Y. Let $Z(k) = z_1z_2 \ldots z_k$ be the LCS of $X(i)$ and $Y(j)$. There are two possibilities for the last characters of $X(i)$ and $Y(j)$. Either $x_i = y_j$ or $x_i \neq y_j$. Then the following observations can be made:

1. If $x_i = y_j$, then $z_k = x_i = y_j$ and $Z(k - 1)$ is an LCS of $X(i - 1)$ and $Y(j - 1)$.
2. If $x_i \neq y_j$, then we have the negation of ($z_k = x_i$ *and* $z_k = y_j$). This means that $z_k \neq x_i$ *or* $z_k \neq y_j$. This gives us two subcases:
   a. $z_k \neq x_i \Rightarrow Z(k)$ is an LCS of $X(i - 1)$ and $Y(j)$.
   b. $z_k \neq y_j \Rightarrow Z(k)$ is an LCS of $X(i)$ and $Y(j - 1)$.

Let us consider Case 1: If $z_k \neq x_i$, then we could simply append x_i to $Z(k)$ to obtain an LCS of $X(i)$ and $Y(j)$ with longer length than k, contradicting the assumption that $Z(k)$ was the *longest* common subsequence. So, $z_k = x_i = y_j$. What about $Z(k - 1)$, the characters of $Z(k)$ before z_k? We know that they are a common subsequence of $X(i - 1)$ and $Y(j - 1)$, but can we say that $Z(k - 1)$ is the *optimal* common subsequence of $X(i - 1)$ and $Y(j - 1)$? In other words, has our claim that $z_k = x_i = y_j$ somehow compromised $Z(k - 1)$ and it is not optimal? No. If there was an LCS W of $X(i - 1)$ and $Y(j - 1)$ that had k or more characters, then we would simply append x_i to it to get an LCS of $X(i)$

and $Y(j)$ longer than $Z(k)$ and this would be a contradiction. Consequently, we have the fact that $Z(k-1)$ is an LCS of $X(i-1)$ and $Y(j-1)$.

Now consider Case 2(a): If $z_k \neq x_i$, then $Z(k)$ must be a common subsequence of $X(i-1)$ and $Y(j)$. Using reasoning similar to that of Case 1, we see that $Z(k)$ must be the *longest* common subsequence of $X(i-1)$ and $Y(j)$. If there was a longer common subsequence of $X(i-1)$ and $Y(j)$, (say W with length greater than that of $Z(k)$) then, since W is also a subsequence of both $X(i)$ and $Y(j)$, we would have a common subsequence of $X(i)$ and $Y(j)$ longer than Z, and so derive a contradiction.

Reasoning for Case 2(b) is symmetric to that of Case 2(a).

E.5.4 Recurrence for the Longest Common Subsequence

The different cases just described have implications that define the score function $S[i, j]$. Recall that $S[i, j]$ is the length of the LCS of $X(i)$ and $Y(j)$, in other words, it is the length of $Z(k)$.

1. Case 1: The LCS of $X(i)$ and $Y(j)$ is the LCS of $X(i-1)$ and $Y(j-1)$ extended by one character. This implies that $S[i, j] = S[i-1, j-1] + 1$.
2. Case 2(a): The LCS of $X(i-1)$ and $Y(j)$ is the same as the LCS of $X(i)$ and $Y(j)$. This implies that $S[i, j] = S[i-1, j]$. A similar analysis for Case 2(b) gives us $S[i, j] = S[i, j-1]$.

We want to extract the longest common sequence and this means that for Case 2 our computation for $S[i, j]$ will be evaluated by selecting the maximum of the two options described. Consequently, the main recurrence is

$$S[i, j] = \begin{cases} S[i-1, j-1]+1 & \text{if } x_i = y_j \\ \max\{S[i-1, j], S[i, j-1]\} & \text{if } x_i \neq y_j \end{cases} \tag{E.1}$$

This completes Step 1 of our recipe for dynamic programming. For Step 2, we describe the data structure that retains the scores and the computational procedure to evaluate them. The $S[i, j]$ scores will be stored in an n by m array. Considering the recursion we see that score $S[i, j]$ held in cell $[i, j]$ will depend on score values stored in cells $[i-1, j-1]$, $[i-1, j]$, and $[i, j-1]$. In other words, $S[i, j]$ depends on score entries in a previous row or column. Consequently, it is reasonable to have an order of evaluation that fills in the matrix row by row, starting with the first row. The very last computation will be that of $S[m, n]$, which represents the length of the LCS for the full strings X and Y. This completes Step 2.

Step 3 handles the necessary initialization needed to begin the evaluation of the recursion. The recursion formula claims that entries in row $i = 1$ would require prior values to be set up in row $i = 0$. This can be done by supplying an extra row 0 and column 0 with all entries initialized to 0. These values are independent of the characters in the given sequences. We are really saying that the length of LCS is 0 when one or more of the strings is null (zero length).

Step 4 deals with recovery of the solution configuration. Filling the score matrix will only compute the length of the LCS. To get the actual LCS, we have to keep track of the cell that is responsible for producing the $S[i, j]$ value. For each $[i, j]$ cell we keep a back-pointer $D[i, j]$ to one of $[i - 1, j - 1]$, $[i - 1, j]$, or $[i, j - 1]$, depending on which cell was used to determine $S[i, j]$. We use D for *Direction* and give each direction a symbol that would be represented in the program using some type of encoding:

$$D[i,j]=\begin{cases}\text{up-left if } S[i,j]=S[i-1,j-1]+1\\ \text{up} \quad\quad \text{if } S[i,j]=S[i-1,j]\\ \text{left} \quad\quad \text{if } S[i,j]=S[i,j-1]\end{cases} \tag{E.2}$$

This completes all four steps of the recipe.

Here is a pseudocode version of the algorithm:

```
//Base cases
for i : = 0 to m do S[i,0] : = 0;
for j : = 0 to n do S[0,j] : = 0;

//Filling the S[] and D[] matrices
for i : = 1 to m do
 for j : = 1 to n do
  if (x[i] = = y[j]) then
 S[i,j] : = S[i-1, j-1] + 1;
 D[i,j] : = "up-left";
  else S[i,j] : = S[i-1,j];
           D[i,j] : = "up";
           if (S[i,j-1] > S[i-1,j]) then
                  S[i,j] : = S[i,j-1];
                  D[i,j] : = "left";

return S[m,n];
```

Figure E.1 presents an example of the score matrix with the direction matrix superimposed. The bottom rightmost cell gives the length of the LCS when the input strings are X = "LOGARITHM" and Y = "ALGORITHM".

To use the back-pointers, we simply start in the final location containing the highest score and then follow the pointers. This produces a path that defines the LCS. All other back pointers are ignored. Figure E.2 shows the path extraction. The main idea behind the path extraction is to go through the optimal configurations and while doing so keep track of the occurrences when an *up-left* pointer is used. The origin of the pointer will be a cell that is in a row and column such that x_i and y_j are the same. Considering our previous analysis, this is the situation that causes a character to be appended to the LCS. The code has been written so that the LCS is generated in reverse order. It is then reversed in the final step.

The pseudocode to do this traceback would be as follows:

```
row : = m; col : = n; lcs : = "";
while (row > 0 and col > 0) do
 if (D[row,col] = = "upleft") then //x[row] = y[col]
  lcs : = lcs.x[row];
  row : = row-1; col : = col-1;
 else if (D[row,col] = "up") then
  row : = row-1;
 else if (D[row,col] = "left") then
  col : = col-1;
reverse lcs;
return lcs;
```

This completes our introduction to dynamic programming.

		A	L	G	O	R	I	T	H	M
	0	0	0	0	0	0	0	0	0	0
L	0	0	1	1	1	1	1	1	1	1
O	0	0	1	1	2	2	2	2	2	2
G	0	0	1	2	2	2	2	2	2	2
A	0	1	1	2	2	2	2	2	2	2
R	0	1	1	2	2	3	3	3	3	3
I	0	1	1	2	2	3	4	4	4	4
T	0	1	1	2	2	3	4	5	5	5
H	0	1	1	2	2	3	4	5	6	6
M	0	1	1	2	2	3	4	5	6	7

FIGURE E.1 Score matrix with superimposed direction matrix.

		A	L	G	O	R	I	T	H	M
	0	0	0	0	0	0	0	0	0	0
L	0	0	1	1	1	1	1	1	1	1
O	0	0	1	1	2	2	2	2	2	2
G	0	0	1	2	2	2	2	2	2	2
A	0	1	1	2	2	2	2	2	2	2
R	0	1	1	2	2	3	3	3	3	3
I	0	1	1	2	2	3	4	4	4	4
T	0	1	1	2	2	3	4	5	5	5
H	0	1	1	2	2	3	4	5	6	6
M	0	1	1	2	2	3	4	5	6	7

FIGURE E.2 Producing the traceback path.

E.6 EXERCISES

1. Assume you have an n-by-n checkerboard. Each square of the checkerboard is labeled with a coordinate pair so that the square in the ith row and jth column is labeled with the pair (i, j). You must move a checker from the top-left corner square (1, 1) to the bottom-right corner square (n, n) using at most $2(n - 1)$ moves. Legal moves are as follows:

   a. Move of the checker down one square.

   b. Move of the checker right one square.

   c. Diagonal move of the checker down one square and to the right one square.

   In this problem, you will assume that on square (1, 1) you have a purse with zero dollars and this value may go up or down as you make moves on the checkerboard according to the following *payoff* scheme that dictates changes to the sum in your purse:

   a. A down move that leaves square (i, j) changes your sum by d_{ij} dollars.

   b. A right move that leaves square (i, j) changes your sum by r_{ij} dollars.

   c. A diagonal move that leaves square (i, j) changes your sum by g_{ij} dollars.

You are given all the d_{ij}, r_{ij}, and g_{ij} values as input data. These values may be negative, zero, or positive so it may be possible to have a negative sum in your purse at any time. You want to find a path from (1,1) to (n, n) using moves that get as much money as possible, or to end up with the least negative value in your purse, depending on the data.

Provide an efficient algorithm based on dynamic programming. Your answer should follow the *four steps* recipe discussed in this chapter.

2. Suppose you are given a sequence of real numbers $x_1, x_2, x_3, \ldots, x_n$. Use a dynamic programming strategy to find a sequence of **consecutive** numbers with the maximum sum.* In other words, we want to find i and j such that the following sum is maximized:

$$\sum_{k=i}^{j} x_k$$

If all the numbers are positive, we simply set $i = 1$ and $j = n$. To make the problem nontrivial we assume that the sequence of real numbers contains both positive and negative numbers. The algorithm should have execution time $\Theta(n)$. Your answer should follow the *four steps* recipe discussed in this chapter. You should supply a short explanation that justifies the formulation of the recursion.

3. Repeat Exercise 1 subject to the following conditions:

   a. As before, you are working with an n-by-n checkerboard and you are only allowed to make the legal moves described by Exercise 1.

   b. The input data is the same as that described by Exercise 1 and the rules for payoff are the same.

   c. You are allowed to start in any square with a purse of zero dollars.

   d. You may finish in any square as long as you follow the rules for legal moves.

   e. You should assume that at least one of the d_{ij}, r_{ij}, and g_{ij} values is positive.

   As before, you want to find a path that will give you the most benefit.

* In algorithm design courses, this problem is usually known as *Bentley's Problem*.

Index

A

Abstract graph, 347, 349
Accept button, 299
Acetylcholine binding site, 130
addArc function, 347
addDisk, 169
addNode function, 352
addPolygon function, 80, 85, 383
addVertex function, 347
Affine gap model, 212
Allosteric proteins, 23
Alpha carbon atoms, 10
Alpha helices, 10–12
Amino acids, 5–10
Amphiphilicity, 286
Angiogenin molecule, 26
angleToPlane(), 160, 173, 175
Antiparallel beta strands, 13
    atom numbering for, 83
    ribbon diagram for, 15
Apply button, callback function for, 309
Arbitrary movement of atom, 60
Arc class, 347
arccos function, 142
Arc object, syntax for, 352
Arginine, 264
    side-chain dihedral angles for, 265
aspect, 373
Atom, 61–62
    attributes related to display, 62
    coordinates, working with, 59–60
    identification of, 58–59
    objects, attributes for, 58–62
    properties, identification of, 58–59
    sets, in contact rearrangement interfaces, 284
atomDistThreshold, 225
atomNames() function, 57
atomPairEnergy(a,b), 255
atoms attribute, 63
atomsMap attribute, 57
Attributes in class, 385
Autotransporter adhesin protein, 84
Axes instantiation, 389
axes_L, 339
Axis function, 98
axisMinLength, 225

B

Bacterial mechanosensitive channels, 171
barChart function, 367
Bar charts, 366–367
Beta sheet, 14
    topologies, 34
Beta sheet-rich proteins, 115
Beta strands, 12–16
B-factor, 77
bfactor attribute, 59
Binding site, principal component of, 233
BLOSUM62 matrix, 204
Bond angle, 135–137
    in protein backbone, 166
Bond attributes, 63–64
Bonded atoms, 246
    dihedral angles, 248
    quadratic function, 247
bondedTo() function, 57
Bond objects, attributes for, 62–64
Bonds, 61–62
Borromean rings, 170
Bovine pancreatic trypsin inhibitor, 8

boxAfunc callback function, 305
Breakers, secondary structure, 25
Bulletin dialogs, 299
Button widget, 176

C

Callback functions, 314
Cancel button, 298
Canvas widget, 309–310
Carbon monoxide, heme affinity for, 35
Cardioid curve, 373
  parametric plot of, 374
center, 170
Changing perspective, Chimera advantages, 2
Charged residues, 8
checkBoxFrame parameter, 305
Checkbutton class, 305–307
Checkbutton instantiation statement, 305
Chemical structure, atom attributes related to, 60–61
Chimera, 2–4, 39, 40, 43, 62
  object model, 46
  PDB files by using, 67
  toolbar buttons for, 42
chimera.dihedral() function, 167
Chimera object hierarchy, 44–47
Chimera Programming Guide, 4
chiralityExaminer, 169
clearEmphasis() function, 68
C-loop movement, 228, 233
cmap variable, 369
Coil, with proline residue, 33
colorByName function, 355
Colored surfaces, data structure for, 382–385
colorName parameter, 343
Command line, user interact with Chimera, 4
Comparison algorithms, scores, 186
Complementary component, binding site, 233
Computation
  of dihedral angle, 139
  of normal, 139–142
Computational tools, Chimera, 3
computeRotationMatrix, 201
Condensation reaction, 5
Connected components, 358–359
connectsTo() method, 62
Consecutive hydrogen bonds, 13
Contact interface, 283
Contact maps, proteins
  distance map, 97–99
  distance matrix, 95
  helical structure, 97
  2FIF, chain A of, 96
Contact rearrangement network (CRN), 285, 290–292
  graph visualization of, 294
contains(an_atom) method, 63
coord() method, 59, 285
Coulombic force, 253
Coulomb's law, 249
CRN, *see* Contact rearrangement network
Cross product, 139
crossSections_L, 328
CurvePlotBundle class, 382
Cyclotides, 35
Cys–cys bridges, 9
CYS–Fe–CYS angles in 1BRF, 165
Cysteine bridge, in bucain, 9
Cytokines, 30

D

DALI, *see* Distance alignment
Data structure for colored surfaces, 382–385
D-configuration of amino acid, 9
Dead-end elimination (DEE), 275
Degradation, life cycle functionalities, 182
Dehydrons
  biological function, 129–132
  identifying dehydrons
    hBonds_L, 111
    import statement, 110
  motivation
    hydrogen bond formation, 108
    visual constructs for, 125–127
    wrapping deficiency, 110

Delaunay tessellation, 345
  function to produce, 356–357
Density plot, 368–370
Dialogs
  basic design guidelines, 298–299
  modal and modeless, 298
Dielectric function, 253, 254
`dihedralAngle` function, 167
Dihedral angles, 271
  computation of normal, 139–142
  defining, 137–139
  in disulfide bridge, 167
  mathematics for, 139
  phi, calculation, 142
  in script, calculation, 144
  in side chain, 266
  sign of, 142–143
Dimensionality of plot, 365
`diskScene.display()`, 170
`displaySurface` method, 326
Distal histidine, 21
Distance alignment (DALI), 219
Distance calculations
  applications, 90
  beta carbon atom, 115
  coloring H-bond plates, 127–129
  cysteine sulfur atoms, 88
  cysteine tetrahedrons, 112–113
  dehydrons
    biological function, 129–132
    identifying dehydrons, 110–111
    motivation, 108–110
    visual constructs for, 125–127
  distance geometry algorithms, generating test data for, 118–119
  disulfide bridges, 113–115
  helical hydrophobic moment, 122–124
  hydrogen bond plates and inertial axis tubes, 122, 123
  implementation strategy
    central cube, 93
    `Shell class`, 92
    3D bounding box, 91
  inertial axes
    `coordArray`, 105
    `distanceToAxis`, 107
    generate tube display, 120
    mathematical analysis, 100–104
    motivation, 99–100
    `projectionOnAxis`, 107
    recursive tubing, 121
    `tensorMatOneAtom`, 105
  oxygen, imidazole ring distances, 116–118
  `Point objects`, 89
  proteins, contact maps for
    distance map, 96–99
    distance matrix, 95
    helical structure, 97
    2FIF, chain A of, 96
  rubredoxin, 88
  `Shell object`, 91
  `Solids class`, 119–120
  testing, Shell class, 112
Distance maps, 98
`distanceToPlane()`, 160
`distThreshold`, 289
Disulfide bridges, 28
  dihedral angle in, 167
  1BQY, chain A of, 114
Dodecahedral disks, 170
Domain movement visualization, 240–243
Domains, protein, 31–32
D9X drawn with H-bond plates, 84
Dynamic programming, 395
  algorithm, 206, 215
  example, 406–407
  introduction, 405–406
  LCS, 410–416
  recipe for, 409–410

E

`Edge` object, 352
`Ellipsoid` class, 339
`endingNodeLabel`, 348
Endogenous cellular prion protein, 24
Energy calculations
  Chimera, Amber parameters in
    `base.Minimizer` call, 257
    `evaluateAmberEnergy`, 258
    molecular structures, energy minimization for, 256

Lennard–Jones formula, 255–256
linear repulsive energy terms, 254–255
Energy computations for rotamer, 276
Entry box widget, 311
Entry widget, 176
Enzymatic catalysis, 29–30
Equivalence, 188
Escherichia coli ribonuclease 1V74, 17
Euclidean plane, 372
evaluateAmberEnergy, 257, 258
Excess baggage, 23
excludedResidues, 258
Explicit solvent, 253
ExtrusionSurface class, 333

F

Favorites/Model Panel, 73
Fetch button, 273, 393
Fetch dialog for pie chart, 394
fetchSPB(), 390
Files, working with, 64–67
fillInUI function, 301
findAtom() method, 57, 64
findBond() method, 62
FindHBond class, 69
findResidue(2) expression, 52
Flexibility, 23
FrenetFrameSurface Class
GUI, 333
pie chart, 331
spine curve, 329
2D points, 328
Full dialog for pie chart, 394
Function dialogs, 299

G

g.addArc("alpha," "delta," Arc(4.0)), 350
g.addVertex("alpha," Vertex()), 350
Gap penalty, 206
Generate_hBond_plates, 127
getArcTupleList(), 348
getAtomsInShell, 91
getConnectedComponents function, 358
get() function, 306
getShortestPath method, 350
getShortestPathSubgraph function, 359
getSubgraph_EdgesRemoved function, 361
getSubgraph_NodeDegreeWithinRange function, 360
getSubgraph_NodeWeightsWithinRange method, 355
getSubgraph_VertexDegreeWithinRange function, 356
Global alignment, 203
algorithm, variations in, 212–213
Globin fold, 18
Glycine residue, Ramachandran plot for, 147
Goldstein's DEE
basic strategy, 276–277
more efficient strategy, 278
GraphBase class, 346
Graph class
introducing, 350–353
methods within, 353–362
Graphical user interface (GUI), 241, 297, 333, 376
implementation, 289
for multiple curve plots, 382
for Ramachandran plots, 177, 178
for side chain control panel, 179
for surface plot, 393
graph.py module, 356
graph.py script, 347, 350
Grid Manager, 302
GUI, *see* Graphical user interface

H

hBds.hBond_D dictionary, 70
hBondCategory_D dictionary, 70
HBondDicts class, 69–71, 80
H-bond plates
drawing, 81
D9X drawn with, 84
hBondSameChain_D dictionary, 70

HCI principles, *see* Human–computer interaction principles
Helical hydrophobic moment, 122–124
Helical wheel, 124
Helix, with hydrogen bonds, 12
Heme affinity for carbon monoxide, 35
Heme group
  cutaway view, 21
  and histidine residues, 20
  from 1MBO, 19
Heme rings, 175
Hemoglobin proteins, 18
  stickiness of, 26
Hemoglobin ring, polygonal plane within, 174
Heterodimer, 27
Hexamer structure of insulin, 29
`HiLiter` class, 67–68, 79, 80
Holo-state protein, C-loop, 237
Homotetramer, 27
Hooke's law, 247
Horseshoe tertiary structure, 26
Human–computer interaction (HCI) principles, 300
Hydrogen bond
  in alpha helix, 11, 12
  in antiparallel beta strands, 15
  in parallel beta strands, 13
Hydrophilicity, 184
Hydrophilic residues with zinc atoms, network of, 288
Hydrophobicity, 184
Hydrophobic residues, 7, 20
  network of, 288

I

`idatmType` attribute, 59
Immune system, 30
Immunoglobulins, 30
Implicit solvent, 253
Improper rotation, 196
Induced subgraph, 346
Inertial axes
  `coordArray`, 105
  `distanceToAxis`, 107
  generate tube display, 120
  mathematical analysis, 100–104
  motivation, 99–100
  projectionOnAxis, 107
  recursive tubing, 121
`InertialAxes` class, 104, 240
`InertialAxisAlignment` class, 224, 225
`InertialAxis` class, 173
`inertialAxisTolerance`, 225
Inertial tensor, 103
Informatics, 365
Instantiation of class, 388, 389
Insulin monomer, 28
Interaction, life cycle functionalities, 181
Interatomic distances, *see* Distance calculations
Intrinsic energy, 276
Inverse-Boltzmann formula, 90
`isHet` attribute, 54

K

Krait venom, 9
Kyte-Doolittle hydrophobicity value, 57

L

`LabelGroups` class, 341
`labelGroupsName`, 342
Labeling atoms, Chimera advantages, 3
Label widget, 178
Lagrange formulation, 157
`lambda` expression, 306
L-configuration of amino acid, 9
LCS, *see* Longest common subsequence
Least squares plane
  mathematical analysis, 154–159
  motivation, 154
  script for computing, 159–165
Lennard–Jones energy function vs. piecewise linear energy functions, 259–260
Lennard–Jones energy value, 272, 273
Lennard–Jones formula, 255–256
`LHBondDicts` class, 69–71
Life cycle functionalities, 181
Linear gap model, 212
Linear repulsive energy terms, 254–255
Linguini type of ribbon, 11

LJ_evaluation class, 255
Local sequence alignment, 214–215, 218
Longest common subsequence (LCS), 410
  prefixes, 401–402
  problem statement, 401
  recurrence for, 403–405
  subproblems, relations among, 402–403
Longitudinal axis for helix, 99, 100
Loops, 16–17

M

Macromolecular structures, 1
matchThreshold, 225
matplotlib module, 78, 366
matplotlib.pyplot, 371
matplotlibSurface function, 375
Matplotlib, with modeless dialog, 376–378
matrixA, 165
Maturity, Chimera, 3
Maximal overlap, 190–197
Menu invocations, user interact with Chimera, 4
Midnight zone, 214
minSize parameter, 348
miscFunctions module, 355
MMTK, *see* Molecular Modeling Toolkit
Modal dialogs, 298
  building, 314–316
Modeless dialogs, 298
  building, 313–314
  matplotlib with, 376–378
  widgets of, 299–301
    canvas, 309–310
    Checkbutton class, 305–307
    entry box, 311
    list box widget with scroll bar widget, 308–309
    normal buttons, 306
    preamble statements, 301
    radio buttons, group of, 307
    scale, 310–311
    text box, 312
    widgetBuffet dialog, testing, 313
    widget placement, 302–304
modelName, 389
Model Panel, 45, 85, 294, 352, 355, 357, 359, 360
Molecular ensemble, 251
Molecular Modeling Toolkit (MMTK), 257
Molecular recognition, 2
Molecular transport, 30
Molecule objects, attributes for, 48–49
Motivation, 1–2
Motor elements, 31
moveToOverlap method, 200, 201
Multiplatform, Chimera, 3
MULTIPLE selection mode, 309
multisample checkbox, 386
Myoglobin proteins, 18, 25

N

Needleman–Wunsch global alignment algorithm, 214
Neighboring residues, 345
  network of, 287
Nicotinic acetylcholine binding proteins, 229–232
Nicotinic acetylcholine receptor, 230
Nicotinic receptor, architecture of, 229
Node degree, subgraph extraction determined by, 361
Node object, syntax for, 352
Node weights, subgraph extraction determined by, 354–356
Nonbonded atoms, 246, 249–250
NP-hard [Ak97], 264
np.meshgrid function, 375
Nuclear magnetic resonance (NMR) experiments, 45
numTicks, 379

O

Object-oriented programming, 45
Omega (ω) angle, 138
1AG9, pre-proline residue in, 149
1A4Y, Ramachandran plot for, 146, 147
1BRF, CYS-Fe-CYS angles in, 165
1YJH, Ramachandran plot for, 151, 152
Optimal global alignment, 203
orderedPathLabels_L, 350
otherAtom(an_atom) bond method, 64
Overlapper class, 199, 225–227, 293

P

Pack Manager, 302
Parallel beta strands, 13
  ribbon diagram for, 13, 14
Parallel strands, atom numbering for, 82
`parametricPlot`, 373
Pauling, Linus, 11
`pcolor` function, 98
PDB, *see* Protein Data Bank
PDB ID, 44
Peptide backbone, angles in, 166
Peptide theory, 5
Percent identity, comparison, 213–214
Phi (f) angle, 137
Physeter catodon, 18
Piecewise linear energy functions vs. Lennard–Jones functions, 259–260
Place Manager, 302
`Planes` class, 173
Pleated sheet, 14
`plot.close()`, 367
`plot.ion()` function, 98
`plotMultipleSurfaces`, 391
`plot_surface`, 391
`plt.ion()`, 371
`Point objects`, 89
Polygonal plane, within hemoglobin ring, 174
Potential energy functions
  computational issues, 251–254
  empirical observations, 246
  energy calculations
    Chimera, Amber parameters in, 256–258
    Lennard–Jones formula, 255–256
    linear repulsive energy terms, 254–255
  interacting bodies, 245
  mathematical modeling
    bonded atoms, 247–248
    categories, 246
    nonbonded atoms, 249–250
    total potential energy, 250–251
  modeling forces, complexity in, 245
PPI, *see* Protein–protein interactions
Prefixes, 207
Pre-proline residue in 1AG9, 149
Pricing, Chimera, 3
`primaryBonds()` method, 61
`primaryNeighbors()` method, 61
Primary sequence, protein, 5–10
Principal component of binding site, 233
`printingClosestPairs`, 225
`printProgressiveQ`, 225
Process dialogs, 299
PROCHECK software, 146, 151
`projectionOnAxis`, 107
`projectionOnPlane()`, 160
Proline, 6
Proline residue
  coil with, 33
  Ramachandran plot for, 148
Property dialogs, 298
Prosthetic heme group, 18
Protein, 1
  domains, 31–32
  functionality, 29–31
  geometric relationships within, 186
  minimum energy of, 264
  1PYI and 1PYC, 222
  Ramachandran plot for, generating, 150–151
  stability of, 24
  3D overlap of, 197–198
Protein backbone, bond angles in, 166
Protein Data Bank (PDB), 87, 187, 257, 263, 283, 289
  code 6PTI, 8
  files, 43, 66
    with single model, writing, 76
    with single protein substrate, writing, 77
    by using Chimera, 67
Protein dimers, 130
Protein–protein interactions (PPI), 30
Protein sequence alignment
  computation
    global alignment, 211
    scoring alignments, 207–208
    subproblem specification, 207
    suitability of subproblem, 208–211
  gaps, 205–206
  partial matches, 203
  structural overlap, 202

Protein structure, overview of
 amino acids and primary sequence, 5–10
 domains, 31–32
 functionality, 29–31
 quaternary structure, 27–29
 secondary structure, 10
  alpha helices, 10–12
  beta strands, 12–16
  loops, 16–17
 tertiary structure, 17–18
  beyond binding pocket, 21–27
  of myoglobin, 18–21
Proximal histidine, 21
Pseudocode for algorithm, 397, 398
Pseudo-energy function, 254
Psi dihedral angle, 137–138
 calculation, 144
`pyplot` object, 371
Python
 programmer, 41
 skills, 40–41
Pythonista, 40
Python `numpy` module, 79
Python scripts, 3, 39, 41–44, 297, 382
 rotamers accessing using, 267–271
 snippets, 299
 user interact with Chimera, 4
Python Shell, 41–43, 49, 51, 68, 71, 297, 298, 313, 353, 367, 381
 interactions, 54

Q

Quality value (QV) function, 220
Quaternary structure, 27–29
QV function, *see* Quality value function

R

Radio buttons, group of, 307
`radius` attribute, 59
Ramachandran plots, 145–153
 background densities for, 151–152
 generating, scripts for, 149–151
 GUI for, 177, 178
 for residue, 168, 176
 3D, 152
Rearrangement factor, 284
Rearrangement networks, allostery and contact, 283–285
Regulation, protein function, 30
`resDistThreshold` parameter, 286
`residueAfter` method, 85
`residueBefore` method, 85
`ResidueCenter`, 285
`residueFletterickColor` function, 294
Residue maps for sequences, 52–53
Residue network, 281
 graphs for, 290
 for rearranged contacts, 292–295
 spheres and spindles for, 286–289
 3D visualization of, 282
Residue object, attributes for, 53–58
Residues, 32
 Chimera advantages, 3
`resMap` attriute, 52
`RevolutionSurface class`, 333, 335, 340
R group, 5
Rice cytochrome c, 34
Right-hand rule, 141
Root mean square deviation (RMSD), 198–199
Rotamericity, 265–267
Rotamers, 265, 271
 energy computations for, 276
 using python script, 267–271
`rotateAndMove` function, 334
`rotation3Dmats_L`, 336
Rotation matrix, code for computing, 202
Rotini extrusion, 334, 335
Rubredoxin, 88
Ruffle ruled surface, 338
`RuledSurface` class, 337
`runCommand` statements, 67

S

`savefig()`, 372
Scalability, Chimera, 3
Scale widget, 310–311
`scatter()`, 371
Scenographics
 categories, 317

Ellipsoid class, 339
LabelGroups class, 341
Lines class, 322
ParametricSurfaces class
base curve points, 323
Chimera model, 326
cross sections, 324
ExtrusionSurface class, 333
FrenetFrameSurface class, 327
mesh generation, 324
RevolutionSurface class, 335
RuledSurface class, 337
solids class
function headers, 320
import statements, 318
surfaces class, 321
Scores, similarities and optimizing, 186
Script C.3.
full graph produced by, 355
subgraph produced by, 356, 357
Script C.6., shortest path, 361
Script D_04 Plotting parametric curves, 372
Scripting capability, 3
Scrollbar widget, 309
Secondary structure, 10–17
alpha helices, 10–12
beta strands, 12–16
loops, 16–17
Secondary structure alignment program (SSAP), 219
Secondary structure elements (SSEs), 97, 220
structural alignment by overlapping pairs of, 220–226
selection.currentAtoms() function, 178
self.lstBx.curselection(), 308
Sequence alignment
script to do, 216–218
structural alignment, 220–226
SequenceLocalAlignment, 216
Sequence object, attributes for, 49–53
sequences() function, 49
identification of, 50
SequencesLocalAlignment class, 238
setCoord() method, 60
setVerbose(True) function, 68
Shell class, 285
Shell interaction, 63, 75
shellNeighbors, 91
shiftedOverlapCoordsP_L list, 201
shiftedOverlapCoordsQ_L list, 201
Shortest path, subgraph, 359–361
Sibling rotamers, 277
Sickle-cell anemia, 26
Side Chain Control Panel dialog, 271–273
Side-chain packing problem, 263, 275–278
computational issues, 264–265
Signal transmission, 30
Simple graph, 351
Single ligand molecule, writing MOL2 files with, 77
Single protein substrate, writing PDB files with, 77
Single side chain, 271–274
Singular value decomposition (SVD), 195
6PAX protein, 32
Smith–Waterman algorithm, 214
Solids class, 173, 240, 287, 351, 366
Solids graphics, 319
Sophisticated graphics, Chimera advantages, 2
Spheres representing residues, 285–286
Spine curve, 328
spinePts_L, 334, 335
sqlength() method, 64
Square ruled surfaces, 338
SSAP, *see* Secondary structure alignment program
SSEs, *see* Secondary structure elements
ssId attribute, 55
Stability, life cycle functionalities, 181
STAMP, *see* Structural alignment of multiple proteins
startingNodeLabel, 348
Stick structures, 14
Strip chart, 378–379
StructBio directory, 347
StructBio package, 67, 69, 286
StructBio.Plots.surfacePlotKit package, 387

Structural alignment of multiple proteins (STAMP), 219
Structural comparison
  problem specification, 184–185
  research in, 183
  techniques for, 185–186
Structural overlap, 189
  exercises, 226–242
  script to do, 216–218
Structural plasticity of family, 183
Structural superimposition, 199–202
Structural systems, 31
Structure comparison, 181
Structure overlap, of 1MBN and 1JEB, 197
Substitution matrices, 203
Substring, 400
Substructure, derivation of, 56–57
Sulfur atoms, 113
Superposition algorithms, 185
  characterization of, 188–199
  formal problem description, 189–190
  maximal overlap computations, 190–197
  overlap measurement, 198–199
  overview, 186–187
Surface Color Selection frame, 274
Surface plot bundle, 386
`SurfacePlotBundle` class, 382, 385
`SurfacePlotBundle` object, 384
Surface plots, Chimera to generate, 385–392
Surface plotting script, user interface for, 392–393
`Surfaces` class, 80, 85, 366
SVD, *see* Singular value decomposition

T

`tensorMatOneAtom`, 105
Tertiary structure, 17–18
  beyond binding pocket, 21–27
  of myoglobin, 18–21
Text box widget, 312
Three dimension (3D), 365
3D surface plots, 374–376
3ZY7 protein, 86
Tick list, 379–382
`tickList` function, 379
`TickLists`, 388
Tissue parsimony, 22
`tkFileDialog.askopenfilename()` method, 66
`tkFileDialog` module, 297
Tkinter, 302
Traceback path, 406
Transcription regulation, 30
Trimer of dimers, 27
`TubesForSSEs` class, 123, 173
Twilight zone, 214
Two dimension (2D), 365
Typical amino acid, 5

U

UCSF Chimera, 2, 4
Uncharged residues, 7
`undoMove` method, 200, 201
Uniqueness, life cycle functionalities, 181
`urllib2.Request()` method, 66
User interface
  dialogs, 298
  for pie chart script, 393–394
  for surface plotting script, 392–393
`useRotamer` function, 271
`userSelectsResidue` function, 270
`userSpecificationOfChains`, 291

V

Valine, 9
Vector alignment search tool (VAST), 220
Versatile selection abilities, Chimera advantages, 3
`Vertex` class, 347
Vestigial structural remnant, 23
Virtual atom, 285
Visual effects, Chimera advantages, 3
Visualization capabilities, Chimera, 2–3

W

`while` loop, 270, 271
Widget
  list box with scroll bar, 308–309
  placement, 302–304
  in `widgetBuffet` dialog, 304–313

WidgetBuffet class, 301
Widget buffet modeless dialog, 300
widgetBuffet.py script, 313
write command, 67

X

xform function, 74
xScale_L, 334, 336

Y

yScale_L, 334, 336

Z

Zero-based indexing, 52
zVals_A array, 369